ELECTRICAL SAFETY

A PRACTICAL GUIDE TO OSHA AND NFPA 70E®

2015 EDITION

atp AMERICAN TECHNICAL PUBLISHERS
ORLAND PARK, ILLINOIS 60467-5756

James R. White

Electrical Safety: A Practical Guide to OSHA and NFPA 70E® contains procedures commonly practiced in industry and the trade. Specific procedures vary with each task and must be performed by a qualified person. For maximum safety, always refer to specific manufacturer recommendations, insurance regulations, specific job site and plant procedures, applicable federal, state, and local regulations, and any authority having jurisdiction. The material contained is intended to be an educational resource for the user. American Technical Publishers, Inc. assumes no responsibility or liability in connection with this material or its use by any individual or organization. The material contained represents the author's understanding and technical opinions on OSHA and NFPA 70E requirements and should not be construed as an official interpretation of either OSHA, NFPA 70E, or any other documents referenced in the book. The user is encouraged to read the applicable documents for proper application.

American Technical Publishers, Inc., Editorial Staff

Editor in Chief:
    Jonathan F. Gosse
Vice President—Production:
    Peter A. Zurlis
Assistant Production Manager:
    Nicole D. Bigos
Digital Media Coordinator:
    Adam T. Schuldt
Art Supervisor:
    Sarah E. Kaducak
Technical Editor:
    Greg A. Gasior
Copy Editor:
    Jeana M. Platz

Cover Design:
    Nicole S. Polak
Illustration/Layout:
    Nicole S. Polak
    Nick W. Basham
    Robert M. McCarthy
    Bethany J. Fisher
Digital Resources:
    Robert E. Stickley

Kevlar and NOMEX are a registered trademarks of E. I. du Pont de Nemours and Company. National Electrical Code, National Fire Protection Association, NEC, NFPA, NFPA 70E, and Standard for Electrical Safety in the Workplace are registered trademarks of the National Fire Protection Association, Inc. National Electrical Safety Code and NESC are registered trademarks of the Institute of Electrical and Electronics Engineers, Inc. Velcro is a registered trademark of Velcro Industries. Quick Quiz and QuickLink are either registered trademarks or trademarks of American Technical Publishers, Inc.

© 2015 by American Technical Publishers, Inc.
All rights reserved

1 2 3 4 5 6 7 8 9 – 15 – 9 8 7 6 5 4 3 2

Printed in the United States of America

ISBN 978-0-8269-3589-2

 This book is printed on recycled paper.

ABOUT THE AUTHOR

James (Jim) R. White has been the Training Director of Shermco Industries Inc. in Irving, Texas, since 2001. Jim is certified by the National Fire Protection Association (NFPA) as a Certified Electrical Safety Compliance Professional (CESCP) and by the InterNational Electrical Testing Association (NETA) as one of approximately 140 Senior Certified Level IV technicians. He is the principal member for Shermco Industries on the NFPA Technical Committee for *NFPA 70B: Recommended Practice for Electrical Equipment Maintenance*. Jim represents the InterNational Electrical Testing Association (NETA) as an alternate member of the NFPA Technical Committee for *NFPA 70E®: Standard for Electrical Safety in the Workplace®*, is NETA's principle representative on the *National Electrical Code® (NEC®)* Code-Making Panel (CMP) 13, and represents NETA on ASTM International Technical Committee F18 on Electrical Protective Equipment for Workers. Jim is an IEEE Senior Member and in 2011 received the IEEE/PCIC Electrical Safety Excellence Award. In 2013 Jim received NETA's Outstanding Achievement Award. Jim was also the 2008 Chairman of the IEEE Electrical Safety Workshop and has participated in the development of the NFPA Certified Electrical Safety Worker (CESW) and Certified Electrical Safety Compliance Professional (CESCP) certification programs.

ACKNOWLEDGMENTS

The author and publisher are grateful to the following companies and organizations for providing technical information and assistance.

DuPont Company
Fluke Corporation
Ideal Industries, Inc.
Megger Group Limited
Milwaukee Electric Tool Corporation
OSHA
Salisbury
Shermco Industries

Technical Review provided by:
Dave McDonald
Automation Systems Estimator
Integrity Integration Resources

Hugh Hoagland
Partner
e-Hazard.com

CONTENTS

1 Electrical Hazards and Basic Electrical Safety Concepts — 1

OSHA and NFPA • NFPA 70E • NFPA Consensus Process • OSHA Electrical Regulations • Electrical Injuries • Recognized Hazards • Arc-Rated Clothing • Incident Energy

2 Multi-Employer Worksites and Electrical Safety Programs — 21

Multi-Employer Worksites • Employer Roles • Factors Relating to the Reasonable Care Standard • Electrical Safety Programs

3 Training of Qualified and Unqualified Workers — 39

Qualified Person • Requirements for a Qualified Electrical Worker • New Requirements and Informational Notes—NFPA 70E 130 • OSHA Table S-4 • Requirements for a Qualified Electrical Worker per OSHA 1910.332 • Arc Flash Hazards • Requirements for a Qualified Electrical Worker per OSHA 1910.333(c) • Types of Training

4 Approach Boundaries for Shock and Arc Flash Hazards — 57

Clearance Distances • Approach Boundaries • Arc Flash Boundary • Energized Electrical Work Permits • Exemptions to Energized Electrical Work Permits • Safe Work Zones

5 Performing a Risk Assessment — 83

Risk Assessment • Electrical Hazards • Assessing Hazards • General Duty Clause • Written Hazard Assessment • Determining the Hazard • Determining Risk • Methods for Performing a Risk Assessment

6 Establishing an Electrically Safe Work Condition — 111

Deenergized Parts • Placing Electrical Equipment in an Electrically Safe Work Condition • Electrical Lockout/Tagout • Verifying Conductors and Circuit Parts Are Deenergized • Establishing an Electrically Safe Work Condition per NFPA 70E • Temporary Protective Grounding Equipment

7 Working on Energized Conductors and Circuit Parts — 143

Determining Energized Work • Work Involving Electrical Hazards—NFPA 70E Article 130 • Energized Electrical Work Permits • Approach Boundaries to Energized Electrical Conductors or Circuit Parts for Shock Protection • Arc Flash Risk Assessment • Incident Energy Analysis • Equipment Labeling • Protective Equipment • Alerting Techniques • Work within the Limited Approach Boundary or Arc Flash Boundary of Uninsulated Overhead Lines

8 Portable Electric Tools and Flexible Cords — 197

Flexible Cords • Uses for Flexible Cords and Cables • Flexible Cord Types • Portable Electric Equipment • Visual Inspection of Portable Electric Tools and Flexible Cords • Grounding-Type Equipment • Receptacles, Cord Connectors, and Attachment Plugs

9 Choosing and Inspecting Personal Protective Equipment — 223

When to Use PPE • PPE Maintenance • Types of PPE • NFPA Table 130.7(C)(15)(A)(b), Arc-Flash Hazard PPE Categories for Alternating Current (ac) Systems • NFPA Table 130.7(C)(16), Personal Protective Equipment (PPE) • Clothing Material Characteristics • Clothing and Other Apparel Not Permitted • Care and Maintenance of Arc-Rated Clothing and Arc-Rated Flash Suits

10 Guidelines for Common Electrical Tasks — 265

Removing and Inserting Low- or Medium-Voltage Drawout-Type Circuit Breakers • Operating Medium-Voltage Air-Break Switches • Inserting and Removing Motor Control Center Buckets • Troubleshooting Circuits • Troubleshooting AC Drives • Testing for the Absence of Voltage • Operating Equipment Rated 240 V and Less and 240 V to 600 V • Operating NEMA E2 (Fused Contactor) Motor Starters • Removing Covers and Panels from Electrical Enclosures • Replacing Light Ballasts • Replacing Low- or Medium-Voltage Motors

DIGITAL LEARNER RESOURCES

Quick Quizzes®
Illustrated Glossary
Flash Cards

Media Library
ATPeResources.com

INTRODUCTION

Electrical Safety: A Practical Guide to OSHA and NFPA 70E® is a comprehensive overview of electrical safety in the workplace. Coverage of both OSHA regulations and NFPA 70E standards is organized to provide a clear overview of proper electrical safety procedures. When used with NFPA 70E, this textbook can be a valuable aid in preparing for the NFPA Certified Electrical Safety Worker (CESW) and Certified Electrical Safety Compliance Professional (CESCP) certification programs. Key topics include electrical hazards, electrical safety programs, approach boundaries, development of a risk assessment, and methods for choosing and inspecting personal protective equipment (PPE). The 2015 edition includes current OSHA regulations and NFPA 70E standards. Field notes supplement key topics and reinforce the importance of working safely on or near electrical equipment. Specific procedures that are used to reduce risks from the hazards involved with working on energized conductors and circuit parts are also discussed. Review questions at the end of each chapter help to assess comprehension of material.

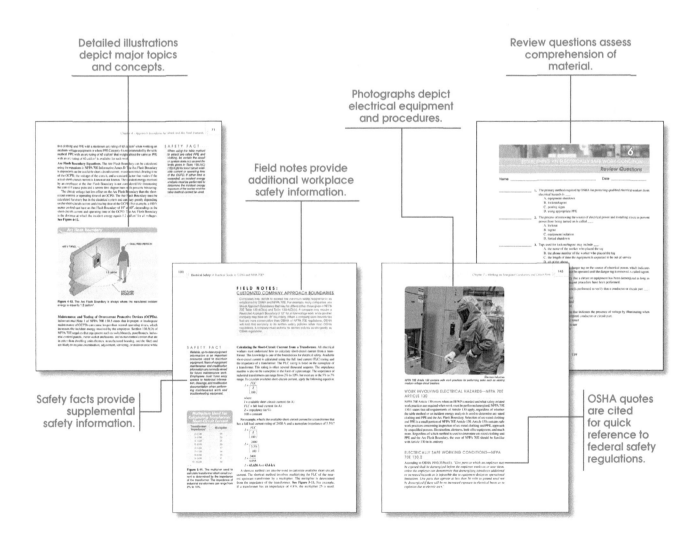

DIGITAL LEARNER RESOURCES

Electrical Safety: A Practical Guide to OSHA and NFPA 70E® also includes access to digital learner resources that reinforce textbook content and enhance learning. These online resources can be accessed using either of the following methods:

- Key ATPeResources.com/QuickLinks into a web browser and enter QuickLink™ code **705798**.
- Use a Quick Response (QR) reader app to scan the QR code with a mobile device.

Digital Learner Resources
ATPeResources.com/QuickLinks
Access Code: 705798

The Digital Learner Resources include the following:
- **Quick Quizzes®** that provide interactive questions for each chapter, with embedded links to textbook references and the Illustrated Glossary
- An **Illustrated Glossary** that serves as a helpful reference to commonly used terms, with selected terms linked to textbook illustrations
- **Flash Cards** that provide a self-study/review of common electrical safety terms and their definitions
- A **Media Library** that consists of videos and animations that reinforce textbook content
- **ATPeResources.com**, which provides access to additional online resources that support continued learning

To obtain information on other related training material including the eTextbook for this title, visit the American Technical Publishers website at www.atplearning.com.

The Publisher

PREFACE

The information contained in this book is based on years of field experience, my participation as an NFPA 70E committee member, teaching electrical safety classes, and my interaction with OSHA compliance personnel. The information is not meant as an official interpretation for OSHA or NFPA 70E and cannot be used as a defense in any court. In order to properly make use of this book, you will need to refer to the 2015 edition of NFPA 70E. This will help you become familiar with the entire document and also provide you with a greater understanding of the quoted material in the appropriate context. Examples provided in the Field Notes of this book are derived from real-life incidents and are included so that others may learn from common errors and attitudes and possibly prevent an electrical accident from occurring.

Having been trained by old-school technicians and engineers, it took me a while to appreciate the importance of electrical safety. The once common practice I resisted the most centered on the requirement of only working on deenergized equipment as was expressed frequently as, "Turn it off? Boy, we don't turn nothin' off!" With more and different personal experiences in the field, I began to reflect and question the motivation for being obstinate regarding safety and safe work practices. Working more safely wouldn't change my paycheck, and if caught violating one of the safe work practices, it could have resulted in being fired.

Following safe work practices added time to each project, but the company was willing to accept that extra cost. It was inconvenient, but lying in a hospital bed for several months was even more inconvenient. Each time I thought of an objection, a better reason for compliance with safe work practices would appear. However, it was when I became responsible for other workers that safety took its rightful place. It seemed that I was willing to accept the risks for myself, but I did not want to be responsible for the injury or death of someone else. Sadly, I had witnessed the catastrophic consequences more than once, and each time, the supervisor was greatly affected and remorseful, even if not directly involved. Safety had become personal.

Safe work practices are not something that can be written off as an elective group decision — responsibility and compliance is personal and not an option. OSHA regulations are federal law, and NFPA 70E works hand-in-hand with the OSHA regulations to enhance electrical safety for all workers. And this process continues to evolve, with NFPA 70E being improved with every edition and being recognized as the premier electrical safety document throughout the world.

My desire, more than anything else, is for workers to take responsibility for their own personal safety and not wait for someone else to do it. Yes, a company is the responsible party by law, and the company will be the one appearing in court. However, you are the one who will suffer the injury and your family will suffer with you. My friend Ken Mastrullo once told me that when he ran into supervisors stubbornly resisting compliance, he would ask them if they would let their children perform a task without the proper safety gear and work practices.

Throughout my career, several people have greatly influenced my personal and professional growth. My father, having little formal education, more than made up for it with his common sense and his truisms, such as "There's nothing common about common sense"; my wife, who has been an unfailing supporter for 39 years, and my father-in-law, Bill Ford, who has provided a great deal of guidance and stability.

I first became an NFPA 70E committee member in early 2001, just as the 2004 edition report on comments (ROC) was about to occur. During that meeting, and subsequent meetings since, there have been many people who have helped to shape my opinions and served as mentors to me, including Lanny Floyd, Danny Liggett, Mike Doherty, Vince Saporita, Tim Crnko, Palmer Hickman, Jim Dollard, David Wallis, Ken Mastrullo, Joe Sheehan, and Ray Crow — just to name a few. I am eternally grateful for these colleagues and friends who have consistently demonstrated a true commitment to electrical safety. There may be disagreement on the best way to achieve the goal of a safe workplace, but all of these dedicated professionals have that special drive to make each NFPA 70E edition better than the last.

Tony Demaria and Gary Donner of Tony Demaria Electric in Los Angeles, California, and their NETA crew over the years have helped me refine field applications of safety principles and practices. Both Tony and Gary promote innovative solutions in their commitment to protecting their employees from electrical hazards. Additionally, the company SUNOHIO (formerly located in Canton, Ohio), set the bar for safe work practices for me. Chuck Baker and Dale Bissonett made employee safety a top priority in the management of the company.

When I joined Shermco Industries in Dallas, I was introduced to an entirely new level of safety expectations. Ron Widup, President and CEO of Shermco, has also been a mentor to me and is one of the most safety-conscious managers I've even met. Ron has created an atmosphere where employees feel safe when refusing to perform unsafe work in the process of completing a job. Ron is the principal NETA representative on NFPA 70E and commands great respect from the other committee members. His leadership at Shermco Industries has resulted in the hardest-working, safest group of field-service technicians I have ever had the pleasure of working with.

James R. White

ELECTRICAL HAZARDS AND BASIC ELECTRICAL SAFETY CONCEPTS

Safety is fundamental for any electrical work. Electrical safety rules must be followed when installing and maintaining electrical equipment. Injuries, fires, and accidents can be prevented or minimized by using safe work practices and wearing appropriate personal protective equipment (PPE). Awareness of electrical hazards and knowledge of basic guidelines and requirements associated with accident prevention reduce risk of injury. Electrical safety has been advanced by the efforts of the U.S. Occupational Safety and Health Administration (OSHA), National Fire Protection Association (NFPA), and other safety organizations.

OBJECTIVES

- Identify the differences between OSHA regulations and NFPA 70E standards.
- Define Public Inputs (PIs) and discuss the meetings and actions involved in the NFPA consensus process.
- Explain the effects of electrical-related injuries.
- Describe the recognized hazards associated with the use of electricity.
- Explain the importance of arc-rated clothing.
- Define incident energy.

OSHA AND NFPA

The *Occupational Safety and Health Administration (OSHA)* is a federal government agency established under the Occupational Safety and Health (OSH) Act of 1970, which requires employers to provide a safe environment for their employees. Current OSHA regulations are included in Title 29 of the Code of Federal Regulations (CFR) Parts 1900–1999. OSHA 29 CFR Part 1910 Occupational Safety and Health Standards provides regulations that employers must follow to protect their employees in general industry. OSHA regulations are written in broad, regulatory, nonprescriptive, or performance-based language. The regulations are written this way because OSHA cannot foresee every possible problem or contingency that could arise during a workplace situation.

The *National Fire Protection Association® (NFPA®)* is a national organization that provides guidance in assessing the hazards of the products of combustion. The NFPA sponsors the development of the National Electric Code® (NEC®) and NFPA 70E®.

The *National Electrical Code®* is a standard on practices for the design and installation of electrical power systems, circuits, and components. It is a code designed to be adopted by local authorities having jurisdiction (AHJs). Once the NEC® is adopted, compliance can be enforced at the city, county, or state level. However, the NEC® is an installation code and does not address safe work practices. It is considered a safety code in that compliance ensures a safe installation and the chance of fires is greatly reduced.

NFPA 70E®, Standard for Electrical Safety in the Workplace® is a voluntary standard for electrical safety-related work practices. The standards in NFPA 70E are prescriptive and cover specific work practices required by electrical workers, both qualified and unqualified.

The International Electro-technical Commission (IEC) is a nonprofit organization that sets standards for electromechanical equipment in most of the world. In the U.S. there are other organizations, such as the Institute for Electrical and Electronic Engineers (IEEE), that set similar standards. There is an effort to merge these two sets of standards that may be concerning to U.S. technicians because IEC-compliant equipment can have substantial differences from the IEEE-compliant equipment.

NFPA 70E

In 1975, OSHA approached the NFPA about using its consensus process to develop a standard that could be used to create the OSHA electrical regulations. The NFPA 70E Committee was formed in 1976. In 1979, the NFPA published the first edition of *NFPA 70E, Standard for Electrical Safety Requirements for Employee Workplaces.*

In 1990, NFPA 70E was used as the basis for OSHA 29 CFR 1910 Subpart S—Electrical, which includes regulations 1910.331 through 1910.335, "Electrical Safety-Related Work Practices." Because the OSHA regulations were developed from NFPA 70E standards, the wording is identical in many instances.

Up until the 2009 edition, NFPA 70E contained a chapter titled "General Requirements for Electrical Installations," which included electrical safety installation requirements from the concurrent edition of the NEC®. Originally this was chapter 1 in the earlier editions of NFPA 70E. By the 2004 edition, it was moved to chapter 4 in order to allow more emphasis on the electrical safety-related work practices requirements, which the NFPA 70E Committee felt was the true focus of the standard. Finally it was deleted entirely in the 2009 edition of NFPA 70E. The NFPA 70E Committee did not feel it was a good idea to have a shortened version of the NEC® in the standard.

Tables in NFPA 70E 130.7 have been used to select personal protective equipment (PPE) since the 2000 edition of NFPA 70E was released in 1999. *Personal protective equipment (PPE)* is clothing and/or equipment worn by an employee to reduce the possibility of injury in the work area. For electrical workers, this can be arc-rated face shields, clothing, and other types of equipment, such as UV-rated safety glasses or goggles. Electrical workers may be required to use some types of PPE that are not arc-rated, such as respirators. Per NFPA 70E 130.7(C)(12) Exception 2, non-arc-rated PPE may be used if it is approved by the AHJ and shown to address the arc flash hazard by not contributing to injury.

FIELD NOTES:
THE IMPORTANCE OF NFPA 70E

Since the table method was introduced, there has only been one instance where an employee has suffered serious burns while wearing the recommended PPE. That employee would have had no injuries if he or she had not been wearing permanent-press clothing under the arc-rated PPE, which is in violation of NFPA 70E 130.7(C)(12) and OSHA 1910.269(l)(6)(iii). Both of these standards state that clothing made from meltable fibers cannot be worn. NFPA 70E is clearer, stating that meltable clothing cannot be worn either as outer layers or as underlayers. NFPA 70E 130.7(C)(9)(b) also states that the outermost layer must be arc-rated as well.

NFPA Consensus Process

The NFPA consensus process is very thorough, allowing full public input and comment on committee actions. Many people consider it to be the best methodology for standards publication being used by any organization. Public input is critical to the consensus process. Public Inputs (PIs), formerly called proposals, are solicited by NFPA using public announcements in news releases and in publications, such as print and online magazines and trade journals. Any person can submit a PI, either online at the NFPA website (www.nfpa.org) or using the form in the back of the appropriate standard or code book. All PIs are considered by the committee and either accepted or resolved. Resolved means they are no longer considered by the committee.

At the First Draft meeting, the committee has free reign to make any changes to the standard it believes are necessary. PIs are considered, but the committee is not bound by the content of the PI to make a change. Committee actions are First Revisions (FRs), formerly Report on Proposals, and are used to make the First Draft. The First Draft is circulated to committee members and posted on the NFPA website for public comments. It only takes a majority vote (51%) for an FR to be accepted by the committee during the meeting. However, it takes a 66% majority vote to approve an FR when it is formally balloted by the committee.

The Second Draft meeting is different in that the committee is bound by the First Draft and Public Comments (PCs) received. Any changes made at this meeting are Second Revisions (SRs). SRs are used to create the Second Draft. Any changes made to the First Draft that do not received PCs are considered resolved. No further action can be taken on them. People who made PCs and wish to address the committee concerning those PCs can do so if they schedule a meeting in advance with the committee chairperson. Once the Second Draft has been completed, the standard is balloted. It takes a 66% majority vote to approve an SR.

The Committee often breaks into task groups in order to consider all PIs that relate to a specific topic. For example, the Tables Task Group may consider all changes to NFPA 70E tables. The Word and Phrase Task Group is responsible for making the language in NFPA 70E consistent, easier to understand, and more usable. Sometimes the task groups will submit PIs in order to meet the scope of the task they were given by the chairperson. Each task group is required to submit a report to the NFPA 70E Committee, which considers the reports as it would any other PI.

The NFPA electrical oversight committee, the Correlating Committee, reviews the actions of the NFPA 70E Committee to ensure there are no major issues or inadvertent conflicts with earlier actions or other NFPA standards. When a Second Draft has been published and committee members or proposers still have disagreements, a Notice of Intent to Make a Motion (NITMAM) can be submitted to the NFPA. NITMAMs are published on the NFPA website for public review and are discussed at the annual NFPA World Safety Conference and Exhibition. Two NITMAMs were submitted during the 2015 revision cycle.

SAFETY FACT

One of the biggest issues the NFPA 70E Committee has experienced with PIs is when the required information on the form is incomplete. The Committee often has to reject a PI when the technical justification section has not been completed, which is a common occurrence.

Electrical Safety: A Practical Guide to OSHA and NFPA 70E

SAFETY FACT

During the First Draft meeting of the 2015 cycle, 448 Public Inputs were considered by committee members.

The final action that can be taken by the public is a Notice to Appeal. This does not occur often, as there is a very low probability that a Notice to Appeal will be considered once a NITMAM on the same committee action has already been considered. Once all NITMAMs and appeals are settled, the Correlating Committee approves the standard and it is edited by the NFPA staff and published. The NFPA staff must take all of the actions and implement them into the standard, which at times necessitates trying to interpret the intent of the Committee. The standard is then released for publication, usually in October of the year preceding its edition date. NFPA 70E becomes effective 15 days after it is approved by the NFPA Standards Council, which is usually in June.

FIELD NOTES: A CHANGING SAFETY CULTURE

In general, people don't seem to like change even if it's good change, unless they're getting pay raises or more vacation time. Often, it's the work culture that inhibits us from accepting new ideas or work practices. The work culture is also the hardest aspect to change.

When I was working in a power plant, we would work circuits and equipment energized, even when they could have been deenergized. Our supervisor would often say, "You're trained to do this hot, so do it." His other favorite saying was, "If you want to be here tomorrow, you'll get this done today."

Also, the concept of workers reporting every small injury or accident was completely foreign to the old-school worker. If someone cut themselves, short of losing a limb, they would wrap a rag around it and keep working. If someone was injured and went to the office to report it, unless it was really serious, the other workers would mock him and consider him to be a malingerer. Fortunately, the safety culture is changing and the things that were once acceptable on the job site could be grounds for dismissal today.

When some people perform a task incorrectly and are not killed or injured, the procedure they used becomes "their" way to do it from then on. Repeat performances of their incorrect and unsafe procedures only reinforce their belief that their way is the right way. The point they are missing is that tasks can be performed in an unsafe manner over and over until the right conditions occur or some small change happens, causing an accident. Once an accident begins, it is completely out of their control as to how it will end.

The DuPont Company has performed a study of 25 years of accident history. DuPont has thousands of workers in the U.S. and the study represents millions of work-hours. They found that for every 300,000 risky behaviors there were 30,000 near misses, 300 recordables, 30 lost-time accidents, and 1 fatality. See "General Work Population" triangle.

Risky behavior is defined as when a worker commits an unsafe act but nothing happens. For example, you might be working without safety glasses or goggles or tightening energized terminal screws with an uninsulated screwdriver and nothing happens to you. A near miss is when an accident does happen but no one is seriously injured. You may not be seriously injured when you melt your 6" Klein screwdriver into a 2" stub, but the potential for injury was there. A recordable means that there was an injury that required more than first-aid treatment. A worker performing electrical tasks is 30 times more likely to be killed than other workers. See "Electrical Workers" triangle. The bottom line is that what you do matters to you and everyone else on the job site.

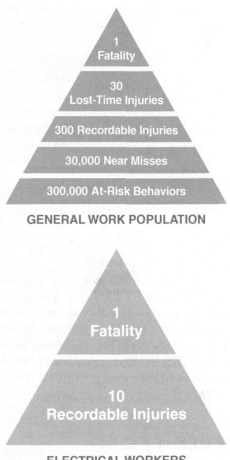

GENERAL WORK POPULATION
- 1 Fatality
- 30 Lost-Time Injuries
- 300 Recordable Injuries
- 30,000 Near Misses
- 300,000 At-Risk Behaviors

ELECTRICAL WORKERS
- 1 Fatality
- 10 Recordable Injuries

Source Data: "The Value of Electrical Incident Case Histories", Eastwood, K.; Hancharyk, B.; Pace, D. Petroleum and Chemical Industry Conference, 2003.

OSHA Electrical Regulations

According to the website for the U.S. Department of Labor (DOL), there were 14,000 job-related deaths annually before the OSH Act was passed. The work force in 1970 was about half the size it is today, meaning those numbers would effectively have to be doubled to put them into today's reference. In 2010 there were 4547 fatalities in the workplace. On a percentage basis, the number of work-related deaths today is significantly less than it was in 1970. Essentially, this means that the work force has doubled while the fatalities have been cut by about 80%. However, this also means that 4547 workers died just by going to their workplaces and trying to do their jobs. Most of these workers died needlessly through ignorance, lack of training, or willful neglect. In addition to fatalities, many more workers become disabled as a result of accidents.

OSHA concentrated on shock hazards when the regulations were first written. This is because arc flash and blast were recognized as hazards, but there were no methods to calculate or evaluate those hazards. Shock is still the biggest hazard electrical workers face in most industries, but electrical burns cause more severe injuries, resulting in more days away from work per injury. A table from "Trends in Electrical Injury, 1992–2002," Paper No. PCIC-2006-38, shows the median days away from work due to electric shock and electrical burn injuries. **See Figure 1-1.** This particular study did not differentiate between arc flash and electrical burns caused by current flow through the body because OSHA does not record arc flash injuries as a separate item.

SAFETY FACT

The Electrical Safety Foundation International (ESFI) reports that in 2010 there were 163 fatalities from electrical injuries. The number of fatalities has been steadily declining since 1995 when there were approximately 350 fatalities from electrical injuries.

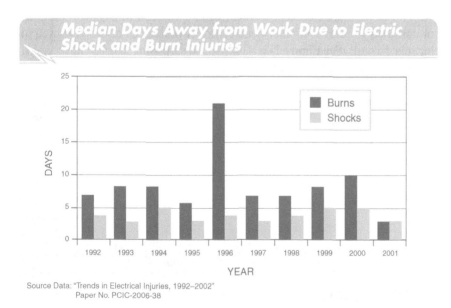

Source Data: "Trends in Electrical Injuries, 1992–2002" Paper No. PCIC-2006-38

Figure 1-1. Electrical burn injuries are the main cause of employees missing days from work.

Electrical Injuries

The Center for Disease Control (CDC) and the National Institute for Occupational Safety and Health (NIOSH) have presented several informative papers concerning accident statistics at the IEEE Industry Applications Society (IAS) Electrical Safety Workshop and the Petroleum and Chemical Industry Committee (PCIC) Conference. In 2006, James Cawley, PE (CDC), and Gerald T. Homce, PE (NIOSH), presented their paper, "Trends in Electrical Injuries, 1992–2002," Paper No. PCIC-2006-38, to the PCIC. Cawley currently works for the Electrical Safety Foundation International (ESFI) and continues to update the statistics on electrical injuries and fatalities. The most recent statistics from ESFI include the following:

- From 2003 to 2010, there were 42,882 fatalities from all causes in the workplace.
- From 2003 to 2010, there were 1738 fatalities from contact with electrical current.
- From 2003 to 2010, electrocution was the seventh leading cause of death in the workplace.
- From 1992 to 2010, there were 66,748 nonfatal injuries from contact with electrical current.
- From 1992 to 2010, the top three causes of electrocution in the workplace were (1) contact with overhead power lines (44%), (2) contact with wiring, transformers, or other electrical devices (27%), and (3) contact with machines, tools, and appliances or light fixtures (17%). *Note:* The fatality rate for these causes has been fairly stable, even though the total number of electrical contact fatalities has decreased.
- The top two causes of nonfatal electrical injuries were (1) contact with machines, tools, appliances, or light fixtures (37%) and (2) contact with wiring, transformers, or other electrical devices (35%).
- Construction workers had the highest fatality rate at 642, with 300 of those being electricians.
- In 2013, there were 139 fatalities from contact with electrical current.

It is expected that electrical workers will be exposed to electrical hazards, but unskilled workers and other nonelectrical workers can also be exposed to electrical hazards. **See Figure 1-2.** It may sound surprising, but groundskeepers, gardeners, truck drivers, and administrators are at a higher-than-average risk from electrocution. This is because these workers may encounter electrical lines and equipment during the course of their normal jobs and may not have any idea that they are in danger. For example, groundskeepers may encounter electrical lines when trimming trees or digging holes to plant new bushes or trees, truck drivers may raise their boxes or other apparatuses into overhead lines, and gardeners may hit buried lines for landscape lighting or outdoor lighting in trees and bushes.

CDC/NIOSH Study

At the 2004 IEEE IAS Electrical Safety Workshop, Kathleen Kowalski-Trakofler, PhD, presented a paper titled, "Non-Contact Electric Arc-Induced Injuries in the Mining Industry: A Multi-Disciplinary Approach." Even though the results shown were focused on the mining industry, the results of this CDC/NIOSH study could apply to most electrical workers. For example, 24% of electrical accidents involving electrical workers occurred during troubleshooting or voltage testing activities. **See Figure 1-3.**

SAFETY FACT

The Bureau of Labor Statistics (BLS) reports that in 2012 there were 156 fatalities from contact with electrical current.

Fatal Electrical Injuries to Nonelectrical Workers

Occupation	Fatal Injuries from 2003–2010
Construction Trade Workers	342
Installation, Maintenance, and Repair Workers	243
Grounds Maintenance Workers	113
Transportation and Materials Moving Workers	108
Workers in Other Management Occupations	76
Agricultural Workers	43
Subtotal	925

Source Data: "Trends in Electrical Injuries, 1992–2002"
Paper No. PCIC-2006-38

Figure 1-2. About one-half (56%) of fatal electrical injuries are to nonelectrical workers.

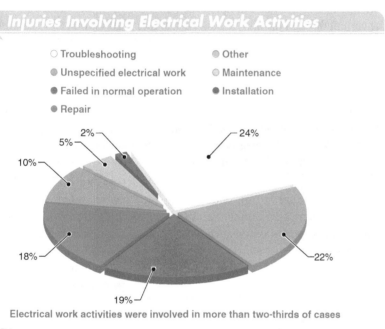

Injuries Involving Electrical Work Activities

- Troubleshooting
- Unspecified electrical work
- Failed in normal operation
- Repair
- Other
- Maintenance
- Installation

Electrical work activities were involved in more than two-thirds of cases

Source Data:
"Non-Contact Electric Arc-Induced Injuries in the Mining Industry: A Multi-Disciplinary Approach"
2004 IEEE IAS Electrical Safety Workshop

Figure 1-3. Electrical workers performing troubleshooting or voltage testing activities accounted for 24% of electrical accidents.

These results may be surprising because electrical workers often do not consider voltage testing a hazardous task. The truth of the matter is that voltage testing requires putting conductive objects (probes) and hands into a grounded metal enclosure. Inside the enclosure are components and wiring at 120 V to 480 V and only 1″ to 1¼″ clearance between energized conductors and ground. Measurements are often at the rear of the enclosure, which requires wires to be pushed out of the way. Much of the IEC-compliant equipment coming into the workplace is smaller and more compact than older equipment, which makes troubleshooting more difficult.

Effects of Electric Shock

IEEE 80, *Guide for Safety in AC Substation Grounding,* states that the average resistance of a 150 lb man is 1000 Ω. The average resistance for a woman is about 750 Ω, due to smaller physical size and somewhat different body composition. Most of the resistance is skin resistance at the contact point. Once past the skin, however, the human body is conductive. Muscles, nerves, and blood conduct current easily and the resistance drops to only about 100 Ω to 200 Ω per limb, which means current through the body increases greatly.

Typically when a person is shocked, it is from hand to foot. Even though the current passes through the core of the body, the resistance of the body is increased by wearing shoes and socks and standing on a surface that has resistance. Concrete has a fairly high resistance when it is dry, but a lower resistance when it is wet. These additional resistances decrease the current that will flow through the body.

Current through the human body is measured in milliamps (mA). A milliamp is one one-thousandth of an ampere. A current through the human body can be calculated using Ohm's law. *Ohm's law* is the relationship between voltage, current, and resistance. For example, if a man makes contact with a 120 V circuit, the current through his body would be 120 mA (120 V/1000 Ω = 120 mA). If a woman makes contact with a 120 V circuit, the current through her body would be 160 mA (120 V/750 Ω = 160 mA).

In 1961, Dr. Charles Dalziel of the University of California–Berkeley, published a paper titled "Deleterious Effects of Electrical Shock," which was based on his 1960 study. According to his research, the effects of electrical current on the body include the following:

- At 3 mA to 7 mA, most people realize they have been shocked and feel a painful sensation. **See Figure 1-4.**

- At 10 mA to 16 mA, most people reach the let-go threshold, where muscle contractions are strong enough that they cannot release from contact with an energized conductor. In actuality, both flexor muscles (those that contract) and extensor muscles (those that extend) are activated at the same time.

- At about 30 mA, the chest muscles contract and tighten against the lungs so hard that a person cannot breathe. This effect is known as respiratory paralysis. Once contact with the energized circuit is broken, breathing resumes normally.

- At 75 mA for 5 sec or 100 mA for 2.5 sec, the lower threshold limit for ventricular fibrillation is reached. Ventricular fibrillation occurs when the normal rhythm of the heart is interrupted and it begins beating rapidly without its two chambers in synchronization. During ventricular fibrillation, the heart cannot pump blood through the body. According to Dr. Dalziel's paper, at 75 mA there is a 1:200 chance of a person going into fibrillation. More current would increase the chances of fibrillation, whereas less current would decrease the chances.

The only practical way to correct this condition is by using an automated external defibrillator (AED), which many commercial facilities and all federal buildings have available on-site. An *automated external defibrillator (AED)* is a portable electronic device capable of interpreting a person's heart rhythm and automatically delivering a defibrillating shock to stop an

Effects of Electric Shock

Current (in mA)	Effects on Body
3 to 7	Painful sensation
10 to 16	Muscles contract
30	Respiratory paralysis
75 for 5 sec or 100 for 2.5 sec	Ventricular fibrillation

Figure 1-4. The severity of an electric shock depends on the amount of electric current going through the body and the length of time the body is exposed to the current.

irregular heart rhythm, thereby allowing the heart to reestablish a normal rhythm. **See Figure 1-5.** The AED passes an electrical current through the heart, causing the heart muscle to contract and briefly stop. When this current is stopped, the heart will begin to beat normally again unless it is too damaged to do so.

- At higher voltages (600 V and greater), the skin no longer presents appreciable resistance to electrical current flow. This is known as "skin-puncture voltage." Total body resistance decreases to the internal resistance of the body, between 200 Ω to 400 Ω. Reduced resistance equals greater current flow through the body, causing more damage. This is one reason why contact with higher voltages results in more fatalities.
- At about 4 A of current flow through the body, heart paralysis occurs.
- At 5 A or more, current flow through the body causes internal third and fourth-degree burning. In electrical burns, the tissue burns from the inside out. This is partly due to the high resistance of bone, which causes heating. Tissue with third-degree burns can no longer heal and skin grafts are necessary. Fourth-degree burns char bone.

SAFETY FACT

Exposure to high voltages will often cause electroporation. This is when cell membranes open due to the magnetic field around the current and what was inside the cells goes out and what was outside goes in.

Figure 1-5. An automated external defibrillator (AED) passes an electrical current through the heart, causing the heart muscle to contract. AEDs have been responsible for saving many lives that would have been lost through ventricular fibrillation.

DC and AC Power Systems

Men and women experience different effects of electric shock from DC and AC power systems. It requires a lot more DC to cause the same effect as 60 Hz of AC. **See Figure 1-6.** DC is more painful than AC, but AC will typically cause more damage to the body.

On average, it takes less electrical current to produce the same effect on women than it does on men. Again, this is due to the differences in body size and composition between men and women. It is imperative that all people, not just woment, be especially careful at all times.

Effects of Electric Shock on Men vs Women from DC and AC Power Systems

Effect	Milliamperes					
	Direct Current		Alternating Current			
			60 Hz		10,000 Hz	
	Men	Women	Men	Women	Men	Women
Slight sensation on hand	1	0.6	0.4	0.3	7	5
Perception threshold, median	5.2	3.5	1.1	0.7	12	8
Shock—not painful and muscular control not lost	9	6	1.8	1.2	17	11
Painful shock—muscular control lost by 0.5%	62	41	9	6	55	37
Painful shock—let-go threshold, median	76	51	16	10.5	75	50
Painful and severe shock—breathing difficult, muscular control lost by 99.5%	90	60	23	15	94	63
Possible ventricular fibrillation						
Three-second shocks	500	500	100	100		
Short shocks (T in sec)			$165/\sqrt{T}$	$165/\sqrt{T}$		
High voltage surges	50	50	13.6	13.6		

Source Data: Dr. Dalziel, "Deleterious Effects of Electric Shock"

Figure 1-6. It takes less electrical current to produce the same effect on women than it does on men because of body size and composition differences between men and women.

Electrical Burn Injuries

According to "Non-Contact Electric Arc-Induced Injuries in the Mining Industry: A Multi-Disciplinary Approach," burns of all types account for 59% of all lost workdays due to electrical injuries. Of that 59%, noncontact electric arc burns account for 35%, electric contact burns account for 12%, and other types of burns account for only 12%. **See Figure 1-7.** Noncontact electrical burns account for a much higher percentage than most people are aware. If such information were available from other industries, the seriousness of the arc flash hazard could be determined much more easily.

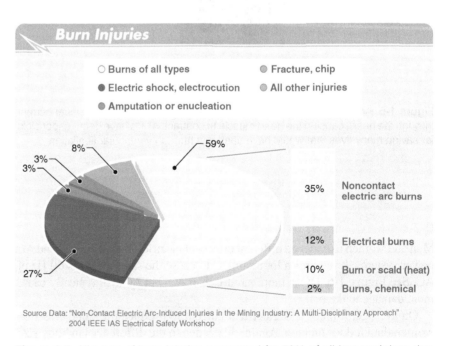

Figure 1-7. All types of burn injuries accounted for 59% of all lost workdays due to electrical injuries.

According to the same paper, 34% of all electrical accidents involving recordable injuries were caused by component failure and 19% were caused by equipment that failed in normal operation. **See Figure 1-8.** These statistics demonstrate that employees must always inspect equipment for anything that looks out of place prior to operating the equipment. It is good practice to always carefully inspect all equipment before using or operating it.

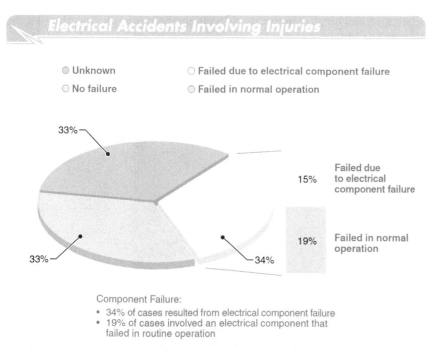

Figure 1-8. Equipment failing during normal operation and component failure together accounted for 34% of electrical injuries.

RECOGNIZED HAZARDS

The three recognized hazards associated with electricity are shock, arc flash, and arc blast. Any of these hazards can be fatal or can cause life-changing injuries. Electrical workers must use extreme caution when working on or near exposed energized conductors or circuit parts, as the hazards are present at all times. Some tasks, such as racking circuit breakers in and out of their cubicle, have a significantly higher risk than others.

Electric Shock

Electric shock occurs when a worker makes contact with an energized conductor or circuit part. Current flows through the worker's body, usually to ground, but possibly to another energized conductor. The higher the voltage is, the greater the amount of current that will flow through the body. Heightened awareness of the arc flash hazard is important, but workers must also be aware that electric shock kills and injures many more workers each year than arc flash. According to ESFI research on the Bureau of Labor Statistics (BLS) data, electrical arc flash fatalities appear to be rare. There were about one to four fatalities

SAFETY FACT

In 2012 there were 156 electrocutions. In 2013 that number dropped to 139 as reported by the Electrical Safety Foundation International (ESFI).

per year from 2003 to 2010. Injuries from electrical burns are more numerous. Data shows 39% of nonfatal electrical injuries are in the construction industry. Electrical burns are 57% of the total nonfatal electrical injuries in the construction industry. **See Figure 1-9.** What is not clear from the data is the number of electrical burns from contact (shock burns) versus the number of burns from noncontact (arc flash burns). An electrical burn can be caused by both shock and arc flash. The only way to protect workers from electrical hazards is to deenergize electrical equipment and systems before work is performed on them. While PPE provides only some protection from electrical hazards, placing equipment in an electrically safe work condition removes those hazards.

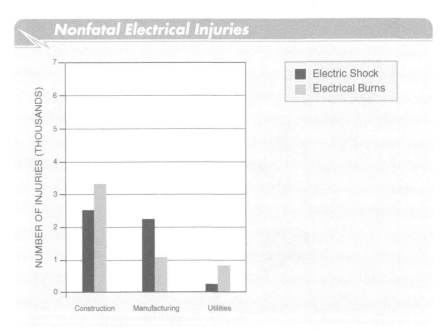

Figure 1-9. There were slightly more electrical burn injuries than electric shock injuries in the construction industry from 2003 to 2010.

Electrical Arc Flash

An *arc flash* is an extremely high temperature discharge produced by an electrical current that flows through air, usually to ground. Arc flashes typically occur when a conductive object makes contact between an energized conductor or circuit part and ground. About 95% of all arc flashes start out as phase-to-ground faults. The extreme heat of the arc ionizes the air creating an alternate path to ground. As long as there is sufficient voltage (about 100 V/in.) to push current across the arc, the arc will sustain itself.

Arc Blast

When an arc blast occurs, a pressure wave is created by the rapid heating of gases within the arc column, as well as vaporizing metal. The arc plasma ball (the superheated vaporized metal) ricochets off the sides and back of the enclosure, and what begins as a single-phase fault will usually end as a three-phase fault. This greatly magnifies the arc energy because the plasma source is almost continuous versus a single-phase arc. By some estimates, it is magnified up to three times what a single-phase open-air fault would generate. The arc plasma ball and ensuing heat is then focused to the front of the enclosure.

SAFETY FACT

Beware of companies that claim to have developed an arc blast protective suit. Since there currently are no industry-accepted equations to determine the intensity of the pressure wave created by an arc flash, these arc blast protective suits cannot be designed and rated for their intended use.

OSHA requires that workers be protected from recognized hazards. The term "recognized hazards" means that the industry, and therefore OSHA, recognizes them and not simply the company. Arc flash heat (incident energy) can be estimated using equations developed by the IEEE P1584 work group.

Even though it is not usually considered a recognized hazard, arc-flash-induced clothing fires are a serious issue. Most serious and disabling injuries from electrical arc flashes are from the ignition of clothing that workers are wearing. These workers may be untrained, undertrained, or careless. If an electrical arc flash contacts flammable clothing, that clothing will burn right next to the body at temperatures ranging from 800°F to 1400°F. This results in severe large-area burns. Often, these large-area second- and third-degree burns will become infected, further threatening the burn victim. The appropriate arc-rated clothing and PPE must always be worn when a worker is exposed to electrical hazards.

FIELD NOTES:
ELEVATED AWARENESS OF THE ARC FLASH HAZARD

In 2002, the NEC® added 110.16, which called for the labeling of all newly installed panelboards, switchboards, and motor control centers (MCCs) that may require servicing or maintenance while energized. This requirement followed the 1999 release of the 2000 edition of NFPA 70E. Prior to this, approximately between 1996 and 1997, improved methods of testing arc flash protective clothing and PPE began and older PPE was found to be ineffective. PPE manufacturers began making clothing rated in cal/cm^2 to comply with the requirements of the 2000 edition of NFPA 70E.

The 2011 edition of the NEC® added new Section 110.24, Available Fault Current, which corresponds to the NFPA 70E labeling requirements in Section 130.5(D) in the 2015 edition. NEC® 110.24 states that service equipment in other than dwelling units must be legibly marked with the maximum available fault current and the date the fault current calculations were performed. The label itself must be durable enough to withstand the environment. The exception for this section is for industrial installations where conditions of maintenance and supervision can ensure that only qualified persons service the equipment.

In the 2014 edition of the NEC®, an Informational Note was added to 110.24 that states that the available fault-current markings addressed in NEC® 110.24 are related to required short circuit current ratings of equipment. NFPA 70E provides help in determining the severity of potential exposure, planning safe work practices, and selecting personal protective equipment.

Together, these additions to the NEC® and NFPA 70E have elevated awareness of the arc flash hazard to an unprecedented level, and electrical workers and their companies continue to improve their electrical safety programs (ESPs). Electrical workers can now determine what arc-flash-rated PPE and other equipment are needed for specific types of installed equipment and the common tasks performed on them.

Arc-Rated Clothing

Arc-rated clothing is clothing that meets ASTM F1506, *Standard Performance Specification for Flame Resistant Textile Material for Wearing Apparel for Use by Electrical Workers Exposed to Momentary Electric Arc and Related Thermal Hazards*. *Arc rating* is the incident energy (in cal/cm^2) on clothing that has a 50% probability of resulting in a second-degree burn on bare skin underneath the clothing or material breakopen. *Material breakopen* is the incident energy

SAFETY FACT

According to ESFI, more than 2000 workers suffer traumatic electrically related burn injuries each year. "Traumatic" describes a person with second-degree burns on more than 50% of their body. Infection is the primary danger of large-area burns.

where clothing chars, crumbles, and falls apart, which allows flames and heat to penetrate it.

The intent of NFPA 70E is for electrical workers to wear clothing specifically designed and rated for its use (i.e., arc-rated), as opposed to firefighters, steel workers, or other workers who must also wear flame-resistant (FR) clothing. **See Figure 1-10.** This is because the heat created by an electrical arc is extreme but momentary, while the hazards faced by other workers wearing non-arc-rated FR clothing are typically of a lower temperature and longer duration.

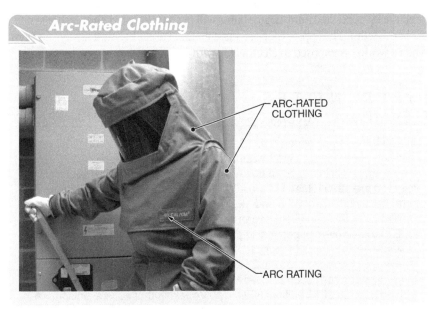

Figure 1-10. Arc-rated clothing is made specifically for electrical workers and is designed to protect from extreme heat for a brief amount of time.

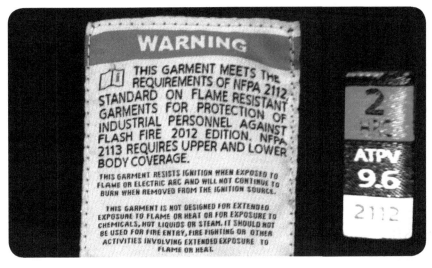

Beware of FR clothing that is not arc-rated but has a similar appearance. The presence of the arc thermal performance value (ATPV) and HRC-2 rating may cause workers to believe this clothing is arc-rated, but an examination of the label shows that it is only rated to NFPA 2112 and not ASTM F1506.

Some clothing is dual-rated, meaning it can be rated for flash fires by NFPA 2112, *Standard on Flame-Resistant Garments for Protection of Industrial Personnel Against Flash Fire*, and for arc flash by ASTM F1506. The difference is that refineries and some petrochemical facilities require all their workers to wear clothing rated for flash fires. This is due to the potential for petroleum-based flash fires. Electrical workers must wear clothing that is specifically designed and rated (arc-rated) to provide protection from electrical arcs. The problem arises when flash fire clothing is given markings similar to ASTM F1506 markings. Labels must be checked closely before purchasing.

FIELD NOTES: NETWORK PROTECTORS

When I was working in the power plant, one of my jobs was to calibrate the protective relays on network protectors. A network protector is basically a circuit breaker that is mounted directly to the secondary of a transformer. A network system typically uses three transformers that are tied to a common bus. This arrangement provides a very reliable, secure power system and is used by military bases, cities, municipalities, and other facilities and systems that require highly reliable equipment. If one of the transformers fails, the other two are sized to carry the full load of the system. Its major shortcoming is that if a fault develops behind one of the transformers, both of the other transformers will backfeed the fault, causing massive destruction.

The network protector has two relays: one is a voltage relay and the other is a reverse power relay. If the reverse power relay senses current flow going into a transformer instead of out of it, it opens the network protector, disconnecting it from the system.

At the time I was working in the power plant, it was common practice not to deenergize anything we worked on, even if we could. That was the old-school way. We would open the cover of the network protector, open the network protector using the long operating handle, activate a microswitch, unbolt the line-side fusible links using a 2½' T-bar wrapped in Scotch 33™, unbolt the load-side connections to the common bus, and slide the network protector out on its extension arms much like a drawout-type circuit breaker. The operating handle would only move about 1" or so, and it was easy to mistake its position. We would have to check the OPEN/CLOSE indicator to ensure it had really opened. We would perform maintenance and then I would calibrate the relays.

Putting the network protector back into service was similar to taking it out. Some of the last steps we would take before connecting it were to check the OPEN/CLOSE indicator to ensure it was open, install the fusible links, and then install the load-side connections. When I left the company, one of my old coworkers took over my tasks. He and another technician had just finished working on a network protector and were preparing to put it back into service. The second technician was cleaning up about 10' away when the network protector blew up.

NETWORK SYSTEM

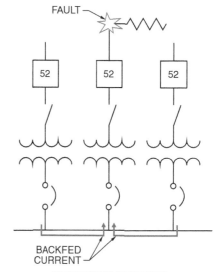

NETWORK SYSTEM

My friend suffered burns on 95% of his body. His nose, both his ears, and his eyelids were burned off. At the time we were issued permanent-press work uniforms, which melted to his body. He ran ¼ mile to an aid station and was transported to a burn unit. The second technician suffered burns on over 25% of his body, mostly on his back.

My friend lived nine months in a burn unit. One day his body temperature started to spike. The hospital staff placed him into an ice bath to try to save him, but he immediately expired. The cause of the original accident is unknown, but one theory is that the network protector was cycled a few times as part of the calibration procedure. It is probable that he did not recheck the OPEN/CLOSE indicator before installing the fusible links. Since the system was still energized, the network protector essentially blew up due to the massive current flow on contact with the fusible link. This is a sad ending to a sad story because my friend would have survived if they had deenergized the system before performing work on the network protector.

Incident Energy

Incident energy is a measurement of thermal energy at a working distance from an arc fault. Incident energy (in cal/cm^2) is used to compare the arc rating of PPE to the expected hazard. As long as the arc rating of the PPE is greater than the expected incident energy from the hazard, any injury should be minor or nonexistent.

Approximately 50% of the heat from an electrical arc is radiated heat and not convective heat (heat flow through air). When arc testing is performed, copper calorimeters are used to measure the total heat received by a surface (the face of a copper calorimeter). This value is expressed as cal/cm^2 for arc-rated clothing and PPE.

Incident energy decreases by the inverse square of the distance. As a person moves away from a potential arc source, the heat from an arc flash decreases rapidly. If there is an arc flash warning label on the equipment, it will have the working distance stated on it. The working distance for low-voltage electrical systems is the distance the incident energy was calculated to, usually 18″, for working at a comfortable arm's length with tools. For higher voltage (up to 15 kV) systems, the working distance is usually considered to be 36″. If incident energy decreases by the inverse square of the distance, the reverse is also true. As a person moves closer to the potential arc source, the incident energy increases by the square of the distance. Six to twelve inches can be the difference between a serious burn injury and almost no injury, so workers must be aware of their body position and the working distance as they perform energized work.

Each layer of clothing under arc-rated clothing decreases the heat received by the body by about 50%. Layering is one protective strategy given in NFPA 70E, but workers should keep in mind that layering flammable clothing under arc-rated clothing does not increase the effective arc rating of the clothing system. It is also important for workers to realize that incident energy is proportional to time. If the time of exposure is doubled, the incident energy received by a worker or any surface doubles.

Layering flammable clothing under arc-rated clothing does not increase the arc rating of the clothing system.

FIELD NOTES: OVERCURRENT PROTECTIVE DEVICES

It takes a short amount of time for an overcurrent protective device (OCPD) to operate. The table below provides estimates of operating times for certain OCPDs. The interrupting currents are typical to my observations of the thousands of circuit breakers I have worked on over the years. This table is for example only. The manufacturer's specifications for the specific device must be used.

For example, a thermal-magnetic molded-case circuit breaker has an operating time of 1.5 cycles to 2 cycles (0.025 sec to 0.33 sec). If this circuit breaker malfunctions due to lack of maintenance, which is a common problem, not only could it increase its operating time, it may not operate at all.

A related document to reference is NUREG/CR-5762 Wyle 60101, "Comprehensive Aging Assessment of Circuit Breakers and Relays." Conducted at the Davis-Besse Nuclear Power Station, this brief study was based on removing molded-case circuit breakers that had been in service for three to five years and performing tests to determine whether they still functioned to the manufacturer's specifications. Nearly 80% of the molded-case circuit breakers failed to operate the first time they were trip-tested for an overcurrent condition. My own, less scientific, estimate is that 50% of molded-case circuit breakers tend to malfunction after three to five years of service. This is based on my personal experience of testing thousands of circuit breakers since 1973.

Typical Interrupting Ratings and Operating Times for Overcurrent Protective Devices

Overcurrent Protective Device	Typical Interrupting Current*	Typical Clearing Time (Cycles)	Typical Clearing Time†
Thermal-magnetic branch circuit breaker	10,000	1.5 to 2	0.025 to 0.033
Thermal-magnetic molded-case circuit breaker	22,000 to 25,000	1.5 to 2	0.025 to 0.033
Thermal-magnetic insulated-case circuit breaker	85,000+	1.5 to 2	0.025 to 0.033
Low-voltage power circuit breaker	42,000 to 50,000	4	0.067
Low-voltage power circuit breaker with current limiting fuses	Up to 300,000	0.25 to 0.5	0.004 to 0.008
Medium-voltage air circuit breaker	—	6 to 8	0.1 to 0.13
Medium-voltage vacuum circuit breaker	—	3 to 5	0.05 to 0.08
Current limiting fuses	Up to 300,000	0.25 to 0.5	0.004 to 0.008

* in A
† in sec

Digital Learner Resources
ATPeResources.com/QuickLinks
Access Code: 705798

SUMMARY

- OSHA mandates that employers protect their employees from recognized hazards.
- The three recognized hazards associated with electricity are shock, arc flash, and arc blast. Any of these hazards can be fatal or can cause life-changing injuries.
- Electrical workers must use extreme caution when working on or near exposed energized conductors or circuit parts, as the hazards are present at all times. Some electrical hazards may also be present only during some tasks, such as racking circuit breakers in or out of their cubicles.
- About 95% of all arc flashes start out as phase-to-ground faults.
- Even though arc flash gets the majority of the attention, it is the shock hazard that injures and kills more people each year.
- Current flow through the body, even at 115 V, can be lethal.
- There is no such thing as a safe energized electrical task.
- Accidents are unexpected and unanticipated. Since there is no way to know when an accident is going to occur, it is important to be prepared at all times. This includes safe work practices and procedures and using appropriate PPE when performing tasks that could lead to injury.

ELECTRICAL HAZARDS AND BASIC ELECTRICAL SAFETY CONCEPTS

Review Questions

Name _____   Date _____

_____ 1. ___ is a voluntary standard for electrical safety-related work practices.
   A. OSHA
   B. NFPA 70®
   C. NFPA 70E®
   D. ASTM F855

T   F   2. Only NFPA committee members can draft a proposal and submit it to the NFPA.

T   F   3. The only way for workers to be protected from electrical hazards is for them to wear the proper PPE for the task.

_____ 4. When a person receives an electric shock, the current through the body is measured in ___.
   A. volts (V)
   B. ohms (Ω)
   C. milliamps (mA)
   D. kilovolt-amperes (kVA)

_____ 5. At ___, most people realize they have been shocked and feel a painful sensation.
   A. 3 mA to 7 mA
   B. 10 mA to 16 mA
   C. 30 mA
   D. 100 mA

T   F   6. Arc blast can be estimated using equations developed by the IEEE P1584 work group.

T   F   7. It takes less electrical current to produce the same effect on women than it does on men.

_____ 8. Which of the following is not considered a recognized hazard associated with electricity?
   A. shock
   B. arc flash
   C. arc blast
   D. electroporation

_____ 9. ___ occurs when a worker makes contact with an energized conductor or circuit part.
   A. Electric shock
   B. Epileptic seizure
   C. Electroporation
   D. Arc flash burn

_____ 10. A(n) ___ typically occurs when a conductive object makes contact between an energized conductor or circuit part and ground.
   A. arc flash
   B. backfed voltage
   C. induced voltage
   D. zone interlocking condition

_____ 11. A(n) ___ passes an electrical current through the heart, causing the heart muscle to contract and briefly stop.
   A. digital multimeter
   B. automated external defibrillator
   C. ohmmeter
   D. rescue hook

_____ 12. According to IEEE 80, *Guide for Safety in AC Substation Grounding,* the average resistance of a 150 lb man is ___.
   A. 500 Ω
   B. 750 Ω
   C. 1,000 Ω
   D. 1 MΩ

_____ 13. At ___, most people reach the let-go threshold, where muscle contractions are strong enough that they cannot release from contact with an energized conductor.
   A. 3 mA to 5 mA
   B. 6 mA to 8 mA
   C. 10 mA to 16 mA
   D. 30 mA to 40 mA

_____ 14. Incident energy ___ distance.
   A. decreases by the inverse square of the
   B. increases by the inverse square of the
   C. does not change with
   D. is proportional to the

_____ 15. Most of the heat from an electrical arc is ___ heat.
   A. convective
   B. conductive
   C. radiated
   D. potential

_____ 16. ___ is the biggest hazard electrical workers face in most industries.
   A. Arc flash
   B. Electric shock
   C. Arc blast
   D. Falling

MULTI-EMPLOYER WORKSITES AND ELECTRICAL SAFETY PROGRAMS

When OSHA regulations are violated on a multi-employer worksite, more than one employer may be cited. OSHA uses its Multi-Employer Worksite Policy to determine the responsibilities of an employer and whether the employer has met those responsibilities. Electrical safety programs (ESPs) are put in place to define the safety policies and procedures of a company. The contents of an ESP can vary depending on the tasks being performed, but there are common elements specified in NFPA 70E that must be included in all ESPs.

OBJECTIVES

- Explain the multi-employer worksite policy.
- List the four types of controlling employers.
- Explain the responsibilities of the host employer and the contract employer.
- Describe the purpose of an electrical safety program (ESP).
- Identify the items in an ESP.
- List the standards that address ESPs.

MULTI-EMPLOYER WORKSITES

Worksites where multiple employers have personnel can be especially hazardous. Often, if there is poor coordination between the employers, personnel will not know what others are doing, even though it can affect their tasks and safety. Even worse, some employers may contract out work they considered too hazardous for their employees without providing any oversight for the contractors doing the work. This lack of oversight allows some contractors to hire unqualified workers to perform hazardous tasks, which can lead to injury. OSHA addressed this situation in 1999 with CPL 2-0.124, *Multi-Employer Citation Policy,* section X, "Multi-Employer Worksite Policy."

There are two steps OSHA uses to enforce its Multi-Employer Worksite Policy. These steps include the following:
1. Determine the role the employer(s) may have had. The host employer is often cited as the controlling employer, as the equipment owner (host) has an obligation to ensure that the contractor only uses qualified persons to perform hazardous tasks. Employers may have multiple roles, which can lead to citations for each role.
2. Determine if the actions of each employer met their obligations. Controlling employers have fewer obligations toward contractor employees than they do for their own employees.

EMPLOYER ROLES

The Multi-Employer Worksite Policy from OSHA categorizes employers into four roles. These four roles include the creating employer, exposing employer, correcting employer, and controlling employer. **See Figure 2-1.** The definitions for these four roles are found in OSHA CPL 2-0.124.

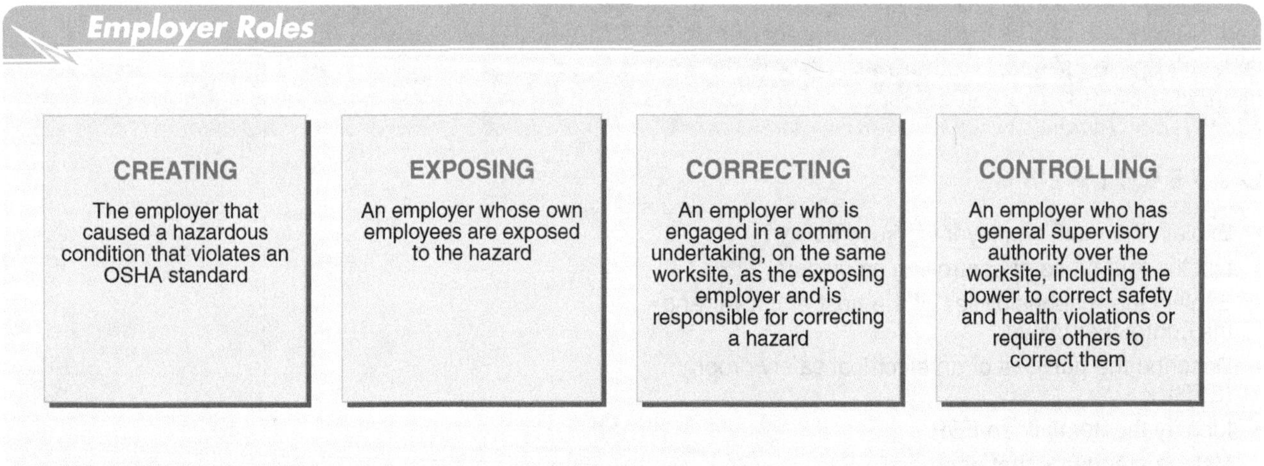

Employer Roles

CREATING	EXPOSING	CORRECTING	CONTROLLING
The employer that caused a hazardous condition that violates an OSHA standard	An employer whose own employees are exposed to the hazard	An employer who is engaged in a common undertaking, on the same worksite, as the exposing employer and is responsible for correcting a hazard	An employer who has general supervisory authority over the worksite, including the power to correct safety and health violations or require others to correct them

Figure 2-1. OSHA's Multi-Employer Worksite Policy categorizes employers into four roles: creating, exposing, correcting, and controlling.

The Creating Employer

"The employer that caused a hazardous condition that violates an OSHA standard. Employers must not create violative conditions. An employer that does so is citable even if the only employees exposed are those of other employers at the site."

The Exposing Employer

"An employer whose own employees are exposed to the hazard. See Chapter III, section (C)(1)(b) for a discussion of what constitutes exposure. If the exposing employer created the violation, it is citable for the violation as a creating employer. If the violation was created by another employer, the exposing employer is citable if it (1) knew of the hazardous condition or failed to exercise reasonable diligence to discover the condition, and (2) failed to take steps consistent with its authority to protect its employees. If the exposing employer has authority to correct the hazard, it must do so. If the exposing employer lacks the authority to correct the hazard, it is citable if it fails to do each of the following: (1) ask the creating and/or controlling employer to correct the hazard; (2) inform its employees of the hazard; and (3) take reasonable alternative protective measures. In extreme circumstances (e.g., imminent danger situations), the exposing employer is citable for failing to remove its employees from the job to avoid the hazard."

The Correcting Employer

"An employer who is engaged in a common undertaking, on the same worksite, as the exposing employer and is responsible for correcting a hazard. This usually occurs where an employer is given the responsibility of installing and/ or maintaining particular safety/health equipment or devices. The correcting employer must exercise reasonable care in preventing and discovering violations and meet its obligations of correcting the hazard."

The Controlling Employer

"An employer who has general supervisory authority over the worksite, including the power to correct safety and health violations itself or require others to correct them. Control can be established by contract or, in the absence of explicit contractual provisions, by the exercise of control in practice. Descriptions and examples of different kinds of controlling employers are given below. A controlling employer must exercise reasonable care to prevent and detect violations on the site. The extent of the measures that a controlling employer must implement to satisfy this duty of reasonable care is less than what is required of an employer with respect to protecting its own employees. This means that the controlling employer is not normally required to inspect for hazards as frequently or to have the same level of knowledge of the applicable standards or of trade expertise as the employer it has hired."

FIELD NOTES:
EMPLOYER OBLIGATIONS

When a host employer hires a contractor to perform work, the employer must ask for proof that the employees sent to the worksite are qualified. There have been cases when a company has submitted a project bid proposal listing their most-qualified employees, but when the project is started, the most-qualified employees are unavailable and the "B Team" shows up to do the work. It would be in the best interest of the host employer to ensure that the substitute workers are qualified before work begins.

OSHA is more aggressively pursuing criminal prosecution against companies it believes are negligent toward employee safety. OSHA has recently tried to involve parent groups of companies when a subsidiary is involved in fatalities. OSHA's actions should serve as a wake-up call to all employers that their actions, and the actions of subsidiary companies, can result in criminal prosecution of supervisors, managers, and others who would normally be considered "immune."

If OSHA finds that employers do not meet their obligations under the assigned roles, they will issue citations. However, OSHA states that only exposing employers can be cited for General Duty Clause violations. In some cases, the burden can be shifted to another party, such as when the host company has a general contractor oversee a construction project. In this case, the general contractor may be viewed as the "controlling employer" since the host company does not have anyone on the worksite and the general contractor will probably have the contractual obligation to ensure a safe worksite.

Factors Relating to the Reasonable Care Standard

"Factors that affect how frequently and closely a controlling employer must inspect to meet its standard of reasonable care include: a. The scale of the project; b. The nature and pace of the work, including the frequency with which the number or types of hazards change as the work progresses; c. How much the controlling employer knows both about the safety history and safety practices of the employer it controls and about that employer's level of expertise. d. More frequent inspections are normally needed if the controlling employer knows that the other employer has a history of non-compliance. Greater inspection frequency may also be needed, especially at the beginning of the project, if the controlling employer had never before worked with this other employer and does not know its compliance history. e. Less frequent inspections may be appropriate where the controlling employer sees strong indications that the other employer has implemented effective safety and health efforts. The most important indicator of an effective safety and health effort by the other employer is a consistently high level of compliance. Other indicators include the use of an effective, graduated system of enforcement for non-compliance with safety and health requirements coupled with regular jobsite safety meetings and safety training."

It is clear from this section of the Multi-Employer Worksite Policy that OSHA expects the host employer to be involved in the safety of the project. This ensures that the host employer cannot hire a contractor with the expectation that they will be shielded from responsibility simply because they do not do the work in-house. The five factors that OSHA provides affect how often and closely the host employer has to inspect the contractor. The most important factor is a consistently high level of compliance. Hiring contractors with poor safety records, no matter how reasonable their price may be, will result in a higher rate of incidents and probably greater fines and scrutiny from OSHA.

Evaluating Reasonable Care

"In evaluating whether a controlling employer has exercised reasonable care in preventing and discovering violations, consider questions such as whether the controlling employer: (a) Conducted periodic inspections of appropriate frequency (frequency should be based on the factors listed in E.3.); b. Implemented an effective system for promptly correcting hazards; c. Enforces the other employer's compliance with safety and health requirements with an effective, graduated system of enforcement and follow-up inspections."

In this section, OSHA states the method they use to determine reasonable care. The described method consists of inspecting for deficiencies, having a program to correct those deficiencies, and enforcing the contractor to follow appropriate safety and health guidelines. The action verbs "conducted," "implemented," and "enforces" illustrate OSHA's requirement that safety be a proactive process, not a reactive process. The host employer is expected to aggressively inspect, correct, and enforce safety on the job site.

SAFETY FACT

According to OSHA, a multi-employer worksite is when more than one employer is working at a single site.

Types of Controlling Employers

There are four types of controlling employers. Types of controlling employers include control established by contract, control established by a combination of other contract rights, architects and engineers, and control without explicit contractual authority.

Control Established by Contract. *"In this case, the Employer Has a Specific Contract Right to Control Safety: To be a controlling employer, the employer must itself be able to prevent or correct a violation or to require another employer to prevent or correct the violation. One source of this ability is explicit contract authority. This can take the form of a specific contract right to require another employer to adhere to safety and health requirements and to correct violations the controlling employer discovers."*

The controlling employer has the authority to correct safety violations and to ensure that tasks are performed in a safe manner.

Control Established by a Combination of Other Contract Rights. *"Where there is no explicit contract provision granting the right to control safety, or where the contract says the employer does not have such a right, an employer may still be a controlling employer. The ability of an employer to control safety in this circumstance can result from a combination of contractual rights that, together, give it broad responsibility at the site involving almost all aspects of the job. Its responsibility is broad enough so that its contractual authority necessarily involves safety. The authority to resolve disputes between subcontractors, set schedules and determine construction sequencing are particularly significant because they are likely to affect safety. (NOTE: citations should only be issued in this type of case after consulting with the Regional Solicitor's office)."*

Architects and Engineers. *"Architects, engineers, and other entities are controlling employers only if the breadth of their involvement in a construction project is sufficient to bring them within the parameters discussed above."*

Control without Explicit Contractual Authority. *"Even where an employer has no explicit contract rights with respect to safety, an employer can still be a controlling employer if, in actual practice, it exercises broad control over subcontractors at the site. NOTE: Citations should only be issued in this type of case after consulting with the Regional Solicitor's office."*

There are circumstances where the owner, who would usually be referred to as the host employer, is not legally responsible for worksite safety. The most common example is when a company hires a general contractor to oversee the day-to-day work and that contractor assumes responsibility for worksite safety by contract. The exception to this example would be if the owner continued to exert authority over the subcontractors and the general contractor did not have actual control.

Multiple Roles

"1. A creating, correcting or controlling employer will often also be an exposing employer. Consider whether the employer is an exposing employer before evaluating its status with respect to these other roles. 2. Exposing, creating and controlling employers can also be correcting employers if they are authorized to correct the hazard."

In this section, OSHA explains that an employer may be assigned more than one role, depending on the circumstances, specific structure of the worksite, and contractual agreements.

FIELD NOTES:
OSHA ACT, 29 USC 666 SECTION 17(E)

There is no regulation in OSHA 1910.301 through .308 or .331 through .335 that covers multi-employer worksites. However, the Occupational Safety and Health Act, 29 USC 666 Section 17(e), *Penalties*, states that criminal penalties can be assessed when all of the following conditions are met:
- There is a violation of a specific regulation.
- The violation is willful.
- An employee is killed.
- There is a causal relationship between the violation and the death.

Criminal penalties can be assessed against the person who is responsible for the death, which could include first-line supervisors, managers, or company owners. The maximum penalty provided for under Section 17 is a $10,000 fine, six months in prison, or both for a first time conviction. Besides any resulting lawsuits, further legal actions may be taken by other state and Federal authorities, depending on the specifics of a case.

On April 11, 2014, OSHA released a revision to OSHA 1910.269—Electric Power Generation, Transmission and Distribution. OSHA 1910.269(a)(3)—Information Transfer requires the host employer and the contract employer to provide certain types of information that are important for performing live-line and other work on or near utility-type equipment and lines. This part of the revision was effective upon release.

SAFETY FACT

All employers should be aware of OSHA's Multi-Employer Worksite Policy. Not having awareness and knowledge of this policy is financially dangerous, not only for the company but also for the supervisors and management.

Host and Contract Employers' Responsibilities

NFPA 70E 110.3 covers multi-employer worksites. The intent of this section is to have all parties communicate with each other. Good communication is critical for a safe work environment, especially when there are multiple companies or crews working in the same area. **See Figure 2-2.** For example, a serious danger to workers exists if the electrical services contractor decided to reenergize a circuit without planning and communicating this to the other affected parties.

The 2004 edition of NFPA 70E required that the pre-job safety meeting between the host and contractor be documented. In the 2009 edition of NFPA 70E, this requirement was removed. However, the 2012 edition of NFPA 70E reinstates that requirement, which is now 110.3 in the 2015 edition.

Host Employer Responsibilities. NFPA 70E 110.3(A) outlines the responsibilities of the host employer. A documented pre-job meeting is required when the host employer has knowledge of hazards that are related to the contract employer's work. This requirement is new to the 2015 edition of NFPA 70E. It ensures that small franchise-type shops and stores are not held responsible for information they may not have. For example, if the owner or manager of a small convenience store has no knowledge of electrical work or its hazards, he or she would not be required to have a documented meeting. However, for a large industrial or commercial site, engineers and maintenance staff would, or should, have knowledge and are required to have a documented meeting. The host employer is required to inform contracted personnel of known hazards related to the contractor's work and hazards that might not be recognized by the contractor or the contractor's employees. Furthermore, any information that might be needed by the contractor to properly assess the worksite for hazards must be communicated to the contractor by the host employer. This could include other work being performed in the same area by other contractors or the host employer's personnel. It may also include the location of classified areas, confined or enclosed spaces, and chemical or airborne hazards that might not be immediately recognized. If the work falls under OSHA 1910.269 or 1926 Subpart V, such as with overhead line work or construction of overhead lines, then NFPA 70E would not apply. The relevant OSHA regulations would be effective.

Figure 2-2. Good communication between multiple companies or crews working in the same area is critical for a safe work environment.

If the host employer observes any of the contractor's employees violating any of the safety policies in NFPA 70E, OSHA regulations, or the host company's policies, these violations are to be reported to the contractor responsible for those workers.

Contract Employer Responsibilities. According to NFPA 70E 110.3(B), any hazards that have been communicated to the contractor by the host employer must be communicated to the workers of the contractor as well. This is in addition to any safety training that is already required by NFPA 70E. The contractor is to ensure that their employees follow all safety-related work practices required by NFPA 70E and the host employer. The contractor is required to inform the host company of the following:
- unique hazards that may be created due to the work of the contractor
- unanticipated hazards that are discovered while the contractor is performing work

- measures that the contractor took to correct any safety issues or violations of safe work practices and the steps taken to prevent those violations from recurring

ELECTRICAL SAFETY PROGRAMS

An electrical safety program (ESP) establishes the safety basis for all electrical work performed at a job site. Per NFPA 70E, an electrical safety program is required to be implemented and documented. Although it should be part of the overall safety program, an ESP should also be a self-contained unit developed specifically for electrical work rather than an add-on or afterthought. Each worksite will have unique needs and circumstances that need to be addressed by an ESP.

Documenting an ESP

Numerous studies have shown that a written safety program is more effective than a verbal one. In addition, writing an ESP has two distinct advantages:
- It allows review by a qualified third party to catch mistakes or omissions.
- OSHA can review the hard copy. This will demonstrate good-faith effort on the part of an employer to create an ESP. The documentation of a safety program is proof that a safety program exists.

ITEMS IN AN ESP

Electrical hazards can vary depending on the type of work being performed. An ESP can vary as well, but there are some common elements that must be included in all ESPs. At minimum, an ESP should contain the following sections:
- scope
- philosophy
- responsibility
- safety team
- procedures
- controls
- principles
- standard of performance
- training
- auditing
- policies

Scope

The scope section identifies the employees to which the ESP applies, such as in-house personnel, contractors, and vendors. The scope also states whether the ESP is mandatory. Other elements included in the scope include boundaries or limits that are applied to a specific site, required documents such as single-line or ladder diagrams, and the individual responsible for implementing and maintaining the ESP.

Philosophy

The philosophy section identifies the basic beliefs or goals of a company. In this section, employees may be informed that working safely is a condition of employment or that unsafe acts are grounds for removal from the worksite. This section may also state that injuries and accidents are preventable or the goal is zero accidents for injuries at the worksite.

Responsibility

The responsibility section identifies the obligations of the employees. This section is where employees may be notified that they are responsible for their own personal safety. Other responsibilities of the employees may include reporting unsafe acts and knowing to whom to report the unsafe acts. In addition to the responsibilities of the employees, this section should clearly specify the responsibilities of managers and supervisors, personnel responsible for developing or administering the ESP, personnel who audit the program, personnel responsible for how the program will be funded, and unqualified workers.

Electrical safety programs may contain procedures for lockout/tagout.

Safety Team

The safety team section identifies the safety team members and their responsibilities. Also included is the reporting structure of the safety team. This includes the personnel to whom the safety team reports and the manner in which the reporting is executed.

Procedures

The procedures section identifies the written procedures that will be used for specific tasks. Procedures are used to document a series of operations in order to achieve a desired result. One of the main reasons procedures are documented is so they can be reviewed by another qualified person for accuracy. Written procedures must be site-specific; procedures written for one site may not be appropriate for another site, even if the same process is performed. Other elements that may impact procedures and cause them to be different include equipment changes, differences in equipment layout, and differences in software.

Employees are more likely to use procedures when they are clearly defined. Procedures should be designed to fit on a single piece of paper. Using a checklist format makes procedures easier for employees to use. For example, a checklist can make the procedures for work involving battery systems easier for employees. **See Figure 2-3.** Some of the more common procedures required for an ESP include the following:
- lockout/tagout
- placing equipment in an electrically safe work condition
- switching procedures
- emergency response procedures

- black-start procedures from a power outage
- hazard identification, risk assessment, and risk control implementation as required by NFPA 70E 110(1)(G)
- special operating procedures when electrical devices can produce dangerous incident energies
- application of personal protective grounds
- selection, use, and care of personal protective equipment (PPE)
- work involving battery systems or other systems that cannot be deenergized
- inspection and testing of portable electric tools and equipment
- use of mechanical equipment such as aerial platform lifts or bucket trucks near overhead lines
- requirements for excavations, including how to locate underground utilities and a method for coordinating with the utility company

Figure 2-3. A common procedure required for an ESP may include work involving battery systems or other systems that cannot be deenergized.

Additional procedures include tasks that can be performed on energized conductors and circuit parts, such as repair or maintenance. This should also mandate the use of an energized electrical work permit when it is needed and the steps required for completing it, as well as any exceptions. Other elements that must be included when working on energized circuits include the following:
- determining proper PPE
- determining safe approach distances for shock and arc flash
- determining when a safety backup is needed

- determining the required PPE the safety backup person is to wear and the required training he or she should have
- tools or equipment required for the tasks
- precautions to be taken for unqualified persons

SAFETY FACT

NFPA 70E 110.1 lists the common elements involved with an electrical safety program. Examples of these elements are located in NFPA 70E Informative Annex E.

Procedures include the necessary information for performing a task safely, such as whether insulated hand tools, live-line tools, voltage detectors, and other tools and equipment are needed. **See Figure 2-4.** Procedures typically contain the following information:

- purpose of the task being performed
- number of employees required for a task
- qualifications of the employees
- hazards associated with the task
- approach boundaries from the arc flash hazard analysis including the Limited Approach, Restricted Approach, and Arc Flash Boundaries
- required safe work practices such as implementing a test-before-touch (TBT) program, placing equipment in an electrically safe work condition, or using temporary insulating shielding
- required PPE appropriate for the hazards
- tools and equipment necessary to perform the task safely
- special precautionary techniques such as establishing a safe work zone
- single-line diagrams, schematics, or other drawings as required
- instruction books from the manufacturer, technical manuals, or other detailed information regarding the equipment being worked on
- information on any unique features or additions to the equipment being worked on, such as new components that may not have been installed by the manufacturer

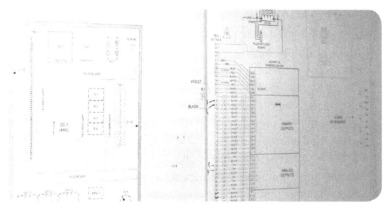

Information that may be included in an ESP includes single-line diagrams, schematics, or other drawings.

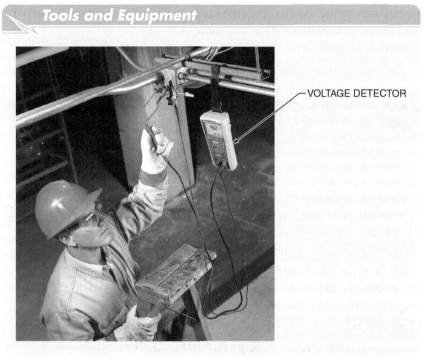

Fluke Corporation

Figure 2-4. The tools and equipment necessary for performing a task safely, such as insulated hand tools, live-line tools, and voltage detectors, are elements that must be included in the procedures for working on energized circuits.

Controls

The controls section identifies components that are used to reduce employee exposure to hazards. The 2015 edition of NFPA 70E added Informational Note No. 1 to section 110.1(A) stating that safety-related work practices such as verification of proper maintenance and installation, alerting techniques, auditing requirements, and training requirements are examples of administrative controls and are part of an overall ESP. Other controls may include the following:

- Substitution of less hazardous systems—For example, arc-resistant switchgear is used instead of standard NEMA-rated switchgear.
- Elimination of hazards—Redundant systems or equipment are installed in order to perform deenergized work. Elimination also means deenergizing equipment and systems.
- Engineering controls—An arc flash hazard analysis is performed, settings of overcurrent protective devices (OCPDs) are lowered to reduce the incident energy exposure to workers, and/or insulating rubber shields, barriers, or attendants are used to prohibit entry where hazards exist.
- Warnings on equipment—Arc flash hazard warning labels are applied to equipment upon completion of an arc flash hazard analysis. **See Figure 2-5.**
- Administrative controls—Audits are performed and records are reviewed to ensure that workers have the necessary training.

- PPE—When and how to use the required PPE is specified. The rating and the limits of PPE are identified. The maintenance, testing, and storage requirements for PPE are specified. Developing and implementing safe work practices, procedures, and training also are examples of administrative controls.

Warning Labels

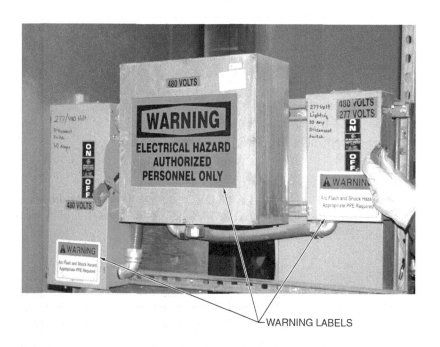

Figure 2-5. The controls section of an ESP may include placing warning labels on equipment.

SAFETY FACT

New to the 2015 edition of NFPA 70E is Section 110.1(B), Maintenance. This adds a requirement that the condition of maintenance of electrical equipment and systems must be considered in an ESP. This new section emphasizes the concerns the NFPA 70E Committee has about inadequate maintenance and its effects on employee safety.

PPE does not eliminate a hazard. It only reduces the effect that the hazard will have on the body. The most effective control is eliminating the hazard by deenergizing because it is the only way to remove the hazard completely so it can have no effect. This control is best implemented when the facility is in the design stage. Instead of employees trying to work around hazards, the electrical system is designed so hazards are either not present or can be eliminated by using redundancy.

Principles

The principles section identifies the safety values of the corporate culture of a company. Principles may vary depending on the company. A company may have the following principles:
- Plan every job.
- Anticipate unexpected events.
- Use the right tool for the job.
- Use procedures as tools.
- Isolate the equipment.
- Identify the hazard.
- Minimize the hazard.
- Protect the employee.
- Assess the abilities of employees.
- Audit the principles.
- Maintain the equipment.
- Document first-time procedures.
- Evaluate equipment before working on it.

Standard of Performance

If employees are to comply with regulatory requirements or standards such as OSHA 1910.331 through .335 or NFPA 70E, these standards should be listed in the standard of performance section.

Training

The training section identifies all required training, including training required for contractors and vendors. Task-specific training needs to be identified, including basic safety and skills training, any emergency rescue or extraction training such as pole-top rescue, and confined-space work that requires permits, as well as training for safety backups. The minimum number of hours, the frequency, and the scope of the training should be specified, if possible.

Training documentation should include a table of contents, training objectives, an outline of the course, the course curriculum, or a course syllabus. If using objectives, it is recommended that terminal and enabling objectives be developed. Terminal objectives state what the outcome should be. Enabling objectives state how the terminal objectives will be met.

Auditing

The auditing section provides feedback information to both the employer and employees. Field safety audits are critical for creating a safe work environment. Per NFPA 70E 110.1(I)(2), fieldwork by employees is to be audited annually. The ESP is to be audited every three years to verify the principles and procedures are in compliance with the current edition of NFPA 70E. The auditing section may include the following:
- Field safety audit forms giving the specific audit items should be prepared for each project.
- The person performing the audits and the person who receives the audit reports should be identified.

- The frequency of the audits should be stated.
- There should be periodic audits of each procedure, both for its use in the field and a review of the procedure itself to ensure it is still appropriate.
- Audits can be performed by in-house personnel or through a contractor. Day-to-day audits are best performed by in-house personnel, while overall audits are often best done by contracted personnel. The overall audit should cover the same inspection items an OSHA audit would cover and should be documented by video, photos, and an examination of documentation.

Policies

The policies section identifies the safety policies to be implemented at the worksite. Examples of safety policies include the following:
- The enforcement of the policies and the consequences of a failure to comply with them must be specified. If there is a zero tolerance policy on lockout/tagout or other safety procedures, it would be stated here. The consequences of violations, such as three strikes or immediate removal, should also be stated.
- The personnel with overall responsibility for site safety, procedures, and maintenance of the procedures must be identified.
- The devices or systems that are considered high risk or high hazard and require special operating procedures must be identified.
- The person that investigates incidents and near misses and the person who receives the reports must be identified.
- The person responsible for maintaining site-specific drawings must be identified and how those drawings are to be controlled and maintained must be specified, especially for electrical safety single-line diagrams. This also includes identifying the person that has access to the drawings and how they will be controlled.
- The person responsible for the maintenance and testing of electrical power system protective devices, the frequency of testing and maintenance, and the criteria used to evaluate the results must be specified.
- The person that maintains and secures standards and regulations and ensures that they are current must be identified.
- The person that needs to be notified prior to excavation must be identified. The excavation procedure to be followed must be specified. This includes phone numbers and contact names.

ESP Standards

There are many standards that address ESPs. Four standards were used to develop this section.
- NFPA 70E, *Standard for Electrical Safety in the Workplace,* Section 110.1 and Informative Annex E are both titled "Electrical Safety Program."
- ANSI/AIHA/ASSE Z10, *Occupational Health and Safety Management Systems,* provides guidelines for electrical safety programs.
- The old IEEE Color Book series (Gold, Buff, Red, etc.) has been broken up into smaller IEEE standards known as the "dot" standards. Previously, electrical safety was covered in IEEE 902 (the Yellow Book). Electrical

safety is now covered by IEEE 3007.1, *Recommended Practice for the Operation and Management of Industrial and Commercial Power Systems,* and IEEE 3007.3, *Recommended Practice for Electrical Safety in Industrial and Commercial Power Systems.* Chapter 5 of each standard covers ESPs.

- *Electrical Safety: A Program Development Guideline* is a safety guideline developed by Enform that provides information for developing an ESP.

There are several other publications, including *The Electrical Safety Program Book* from the NFPA, that provide detailed information about setting up an ESP. These publications tend to be easier to decipher than the standards.

Digital Learner Resources
ATPeResources.com/QuickLinks
Access Code: 705798

SUMMARY

- OSHA's Multi-Employer Worksite Policy categorizes employers as creating, exposing, correcting, or controlling.
- According to its Multi-Employer Worksite Policy, OSHA can hold more than one employer responsible when an accident occurs.
- When there are multiple companies or crews working in the same area, good communication is critical for a safe work environment.
- An electrical safety program (ESP) establishes the safety basis for all electrical work performed at a job site.
- Common elements that must be included in an ESP include scope, philosophy, responsibility, safety team, procedures, controls, principles, standard of performance, training, auditing, and policies.

MULTI-EMPLOYER WORKSITES AND ELECTRICAL SAFETY PROGRAMS

Review Questions

Name _____ Date _____

_____ 1. The Multi-Employer Worksite Policy from OSHA categorizes employers into ___ roles.
    A. 2
    B. 4
    C. 8
    D. 10

_____ 2. The responsibilities of the host employer are covered in NFPA 70E ___.
    A. 110.1(A)
    B. 110.1(B)
    C. 110.3(A)
    D. 110.3(B)

_____ 3. A(n) ___ establishes the safety basis for all electrical work performed at a job site.
    A. energized electrical work permit
    B. job briefing
    C. electrical safety program
    D. none of the above

T   F   4. The host employer is expected to aggressively inspect, correct, and enforce safety on the job site.

T   F   5. An employer cannot be assigned more than one role per worksite.

_____ 6. An electrical safety program should include a section on ___.
    A. responsibility
    B. procedures
    C. principles
    D. all of the above

_____ 7. The ___ section of an electrical safety program identifies the employees for which the electrical safety program applies, such as in-house personnel, contractors, and vendors.
    A. scope
    B. philosophy
    C. responsibility
    D. auditing

_____ 8. The responsibility section of an electrical safety program identifies the ___.
    A. safety team members
    B. obligations of the employees
    C. safety procedures
    D. goals of the company

T   F   **9.** A written, documented electrical safety program is much more effective than a verbal one.

T   F   **10.** The safety team section of an electrical safety program identifies the written procedures that will be used for specific tasks.

_____ **11.** The ___ section of an electrical safety program identifies components that are used to reduce exposure an employee may have to hazards.
      A. procedures
      B. controls
      C. training
      D. auditing

_____ **12.** NFPA 70E requires that the contract employer inform its employees of ___.
      A. the time they are to show up for work
      B. hazards the host employer has communicated to the contract employer
      C. information related to upcoming OSHA site audits
      D. any previous citations the host employer has been cited for by OSHA

_____ **13.** Electrical safety programs are addressed in ___.
      A. NFPA 70E
      B. ANSI/NETA MTS-2011
      C. IEEE C37.5
      D. NFPA 70

_____ **14.** In accordance with NFPA 70E Section 110.3(A), what is the host employer supposed to do if they observe a contracted employee violate safety policies?
      A. Write the contracted employee a ticket using the suggested form in NFPA 70E Informative Annex S.
      B. Remove the contracted employee from the worksite immediately.
      C. Call the local authorities and have the contracted employee arrested.
      D. Inform the contractor responsible for the employee of the violation.

T   F   **15.** One of the main reasons procedures are documented is so they can be reviewed by a qualified third party.

T   F   **16.** The training section of an electrical safety program identifies all training required for contractors and vendors.

TRAINING OF QUALIFIED AND UNQUALIFIED WORKERS

A qualified electrical worker must have the skills, knowledge, and experience for the technical aspects of his or her job as well as OSHA-required specific safety training in order to meet OSHA regulations. Even workers with 5, 10, or 15 years of experience who hold licenses, certifications, and degrees must be properly trained to meet OSHA requirements. Licensure, certification, and other types of training provide technical skills qualifications, but they may not provide the safety skills and knowledge needed to qualify a worker to perform tasks on or near exposed energized conductors or circuit parts.

OBJECTIVES

- List the requirements a qualified person must meet.
- List the requirements a qualified electrical worker must meet.
- Explain the new requirements and Informational Notes in NFPA 70E 130.
- Describe the types of training required by OSHA and NFPA 70E.

QUALIFIED PERSON

In August 2007, OSHA issued a revision to Subpart S—Electrical, standards 1910.302 to .308. At the same time, OSHA revised the definition of a qualified electrical worker in 1910.399, "Definitions Applicable to This Subpart." The previous definition for a qualified person stated, *"Qualified person. One familiar with the construction and operation of the equipment and the hazards involved."* The previous definition referred to two separate requirements: one requirement for technical skills and knowledge and another requirement for safety training and knowledge.

The current definition of a qualified electrical worker states, *"Qualified person. One who has received training in and has demonstrated skills and knowledge in the construction and operation of electric equipment and installations and the hazards involved."* This definition refers to the following requirements:

- technical skills and knowledge (*construction and operation of electric equipment*)
- training on the recognized hazards of electricity (*hazards involved*)
- training on construction and operation of involved equipment (*received training in*)
- proficiency (*demonstrated skills and knowledge*)
- function and operation of the electric power system (*and installations*)

Through their years of experience, most electrical workers should feel comfortable with and understand the technical skills and knowledge requirement.

Apprenticeships and other types of programs usually focus on the technical aspects of a job, while safety training is either given to employees on the job or through formal training provided by the employer after employees are hired.

Technical skills are critical for an employee to perform his or her duties safely. If the employee does not have this level of knowledge, mistakes can occur that cause serious or fatal injuries. An employee operating a drawout-type circuit breaker may not need to know the function of every component in the circuit breaker, but the employee must understand its basic construction and operation. The skills and knowledge required to be qualified to operate a drawout-type circuit breaker may include the following:

- An employee should know how to open and close the circuit breaker if it is electrically or manually operated.
- An employee should have the skills to operate the circuit breaker, including how to charge the closing springs, in an emergency.
- An employee should recognize specific items and situations to be aware of when operating a circuit breaker. These items and situations may be different from manufacturer to manufacturer, as well as between low- and medium-voltage circuit breakers and between air-magnetic and vacuum circuit breakers, even though they have very similar outward appearance and switchgear enclosures.
- An employee should know the proper steps for racking the circuit breaker in or out of its cubicle.

Different voltage classes of the same type of equipment may require additional training to ensure an employee understands the differences in construction and operation, as well as any additional hazards that may be present between the different voltage classes. An employee who has had training on the construction, operation, and hazards associated with low-voltage power circuit breakers would be considered qualified in regard to that type of circuit breaker from all manufacturers, provided the employee reviewed the literature provided by the manufacturer for the specific breaker they were about to operate. However, if the same employee were to work on a medium-voltage (2.3 kV to 15 kV) metal-clad air-magnetic circuit breaker, he or she would require additional training. **See Figure 3-1.** Even though the basic construction and operation of the two circuit breakers are similar, there are major differences in the potential hazards each pose.

FIELD NOTES:
QUALIFIED/UNQUALIFIED WORKER

Suppose a worker is qualified to work on low-voltage power circuit breakers and is tasked with racking a medium-voltage metal-clad breaker, which the worker has not done before. Although qualified to perform this task on low-voltage breakers, the worker may not be qualified to do so on medium-voltage circuit breakers. If the worker is not qualified, what options are available for qualification? In this specific case, the worker could be trained using on-the-job training (OJT) where a qualified electrical worker demonstrates how to perform the task and discusses what hazards the worker will be exposed to, what PPE is needed, approach distances for shock and arc flash, and other pertinent information necessary for performing the task safely. The other option for qualification is formal training that covers those same areas. Documentation is required if either training method is used.

Training for Equipment with Different Voltages

LOW-VOLTAGE POWER CIRCUIT BREAKER

MEDIUM-VOLTAGE METAL-CLAD AIR-MAGNETIC CIRCUIT BREAKER WITH INTERPHASE BARRIERS REMOVED

Figure 3-1. An employee who has had training on the construction, operation, and hazards associated with low-voltage power circuit breakers would need additional training in order to be qualified to work on medium-voltage (2.3 kV to 15 kV) metal-clad air-magnetic circuit breakers.

Requirements for a Qualified Electrical Worker

OSHA 1910.332, OSHA 1910.333, and NFPA 70E 110.2 provide guidance on what is required for an employee to be considered a qualified electrical worker.

According to OSHA 1910.332(a), *"The training requirements contained in this section apply to employees who face a risk of electric shock that is not reduced to a safe level by the electrical installation requirements of 1910.303 through 1910.308. Note: Employees in occupations listed in Table S-4 face such a risk and are required to be trained. Other employees who also may reasonably be expected to face comparable risk of injury due to electric shock or other electrical hazards must also be trained."*

SAFETY FACT

Retraining for qualified persons must be conducted at least every three years. This requirement was included in the 2012 edition of NFPA 70E so that training can coincide with its three-year publication cycle. Employers should train their employees using the most current edition of NFPA 70E. Every three years is the minimum training frequency. Training may be required at a greater frequency if OSHA standards change, new PPE or other equipment is being used, or site audits or supervision indicate that safe work practices are not being followed.

The electrical installation requirements listed in OSHA 1910.301 to .308 derive from the NEC®, specifically the safety-related installation requirements of the NEC®. For an installation to be NEC®-compliant, all covers must be installed, all fasteners secured, and contact with energized components or circuits cannot be possible. **See Figure 3-2.**

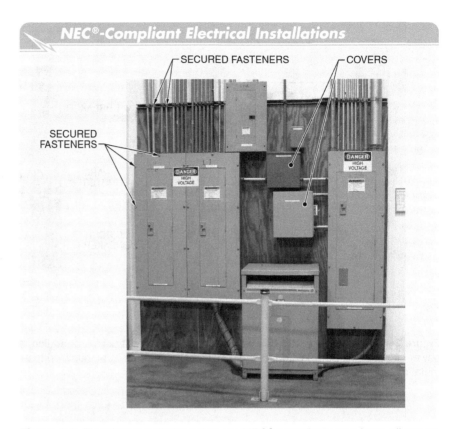

Figure 3-2. Electrical installations that are NEC®-compliant must have all covers installed, all fasteners secured, and contact with energized components or circuits cannot be possible.

A qualified person is always required when equipment is operated in a manner that could cause failure, such as inserting or removing (racking) a circuit breaker or installing or removing a motor control center (MCC) bucket. These tasks involve making and breaking electrical contacts with no arc suppression (arc chutes). If an arc is created for any reason, the switchgear would most likely not contain the arc flash or arc blast. The doors are likely to blow open and the bolted panels could rupture. Both of these tasks require the use of arc flash PPE and equipment. The qualified person may be an electrician or a person specifically trained in breaker operation and the safety procedures required for this operation.

When the covers to electrical equipment are opened or removed for maintenance, servicing, or troubleshooting, the equipment no longer is in compliance with OSHA 1910.301 to .308 and safety-related work practices, such as wearing proper PPE, must be used to protect the worker. **See Figure 3-3.**

SAFETY FACT

Safety-related installation requirements are in chapter 4 of the 2000 and 2004 editions of NFPA 70E, while safety-related installation requirements were in chapter 1 of previous editions of NFPA 70E. Chapter 4 was removed from the standard in the 2009 edition of NFPA 70E and does not appear in the 2009, 2012, and 2015 editions. See the most recent version of the NEC® for up-to-date installation requirements.

PPE for 480 V Low-Energy Equipment

Figure 3-3. Arc-rated PPE and equipment that would be appropriate for the hazard and the task must be worn to protect the worker.

New Requirements and Informational Notes— NFPA 70E 130

NFPA 70E 130.5, Informational Note No. 1, states that improper or inadequate maintenance can cause overcurrent protective devices (OCPDs) to operate slower, increasing incident energy. The 2015 edition of NFPA 70E added a sentence to this Informational Note that states that if equipment is not properly installed or maintained, arc-rated PPE selected using the table method or incident energy analysis method may not be adequate. This is because poorly maintained electrical equipment, especially OCPDs, will not function according to their manufacturer's specifications. Circuit breakers slow down due to lack of maintenance, which will increase the incident energy above what is anticipated on an electrical power system that is properly maintained.

NFPA 70E 130.5, Informational Note No. 2, cautions that smaller short circuit currents could cause higher incident energy due to the OCPD operating more slowly and increasing the operating time. NFPA 70E 130.5, Informational Note No. 3, refers the user to Annex O for equipment and installations that reduce risk to employees by designing electrical power systems to be safer.

NFPA 70E 130.7(A) contains several Informational Notes. Informational Note No. 1 states that the PPE required by NFPA 70E protects the employee from the heat of an arc flash but may not protect them from the pressure wave (arc blast).

Informational Note No. 2 states that electrical equipment rated 600 V or less that has been properly installed and maintained by qualified persons is not likely to expose workers to electrical hazards in normal operation. Normal operation has a dual meaning in Informational Note No. 2. If equipment is normally operating, arc-rated PPE and clothing is not necessary when working in close proximity to it, provided all covers are on and properly latched and secured. The second meaning is that arc-rated PPE and clothing is not needed if the equipment is operated in accordance with the manufacturer's instructions. If, however, the manufacturer states that a pushbutton-type start-stop station should be used to operate electrical equipment and a worker uses a circuit breaker, then that would not be normal operation per the manufacturer's instructions and arc-rated PPE is required.

It is important to note that Informational Note No. 2 states that the equipment "is not likely" to expose the worker to an electrical hazard. "Not likely" does not mean "never" or "impossible." It is critical that the person about to perform a task evaluate the conditions and risks that may be present at the time that task is about to be performed. If a worker feels it is necessary to wear arc-rated PPE and clothing to operate a circuit breaker, the NFPA 70E Committee does not discourage that practice.

One other important point is that if a worker is troubleshooting electrical equipment, it is no longer considered normally operating. That equipment is in distress and full arc-rated PPE and clothing should always be worn. Racking circuit breakers (inserting or removing them from their enclosures) and inserting and removing MCC buckets, bus duct plugs, and fuses is also not considered normal operation. These tasks involve making and breaking the connection to an energized bus with no arc extinguishers. If an arc starts, it will cause a large and possibly extended arc flash.

NFPA 70E 130.2(A)(4) was added to the 2015 edition of NFPA 70E and states that normal operation is permitted when the equipment has been properly installed and maintained, doors and covers are latched and secured, and there is no evidence of impending failure. By being in the body of the text and not in an Informational Note, this section is part of the standard's requirements. NFPA 70E 130.7(C)(15)(B), Informational Note No. 2, states that closed doors on electrical equipment will probably open during an arc flash and cannot be relied on to protect a worker.

OSHA Table S-4

OSHA requires that an employee receive training if he or she will perform any task on or near exposed energized circuits or equipment 50 V or greater to ground. This includes HVAC technicians, instrument and control technicians, electricians, and other employees who are considered qualified electrical workers by OSHA. It also includes the supervisors of these employees, even if the supervisor does not perform the task. OSHA wants to verify that employees are not pressured by their supervisors to perform work on energized equipment just to get work done in a timely manner.

OSHA 1910.332, Table S-4–Typical Occupational Categories of Employees Facing a Higher Than Normal Risk of Electrical Accident contains a list of occupations involving some type of energized work. These occupations include the following:
- blue collar supervisors (1)
- electrical and electronic engineers (1)
- electrical and electronic equipment assemblers (1)
- electrical and electronic technicians (1)

- electricians
- industrial machine operators (1)
- material handling equipment operators (1)
- mechanics and repairers (1)
- painters (1)
- riggers and roustabouts (1)
- stationary engineers (1)
- welders

Most of these occupations are followed by footnote (1). Footnote (1) states, *"Workers in these groups do not need to be trained if their work or the work of those they supervise does not bring them or the employees they supervise close enough to exposed parts of electric circuits operating at 50 volts or more to ground for a hazard to exist."* However, all electricians and welders require qualified personal training.

Requirements for a Qualified Electrical Worker Per OSHA 1910.332

In order for an electrical worker to be considered qualified, he or she must meet the requirements in OSHA 1910.332. According to OSHA 1910.332(b)(3), *"Qualified persons (i.e. those permitted to work on or near exposed energized parts) shall, at a minimum, be trained in and familiar with the following: (i) The skills and techniques necessary to distinguish exposed live parts from other parts of electric equipment, (ii) The skills and techniques necessary to determine the nominal voltage of exposed live parts, and (iii) The clearance distances specified in 1910.333(c) and the corresponding voltages to which the qualified person will be exposed. Note 1: For the purposes of 1910.331 through 1910.335, a person must have the training required by paragraph (b)(3) of this section in order to be considered a qualified person."*

The intent and meaning of the three requirements in this regulation can be clarified into the following:

- *"...(i) The skills and techniques necessary to distinguish exposed live parts from other parts of electric equipment..."* Distinguishing exposed live parts from other parts of electric equipment is referred to as absence-of-voltage testing in NFPA 70E. This phrase is used because the exposed energized circuit or part is supposedly deenergized when it is tested. Absence-of-voltage testing is used to verify the circuit is deenergized. Circuits have to be considered energized until tested and proven deenergized. According to OSHA 1910.333(b)(1), *"Conductors and parts of electric equipment that have been deenergized but have not been locked out or tagged in accordance with paragraph (b) of this section shall be treated as energized parts, and paragraph (c) of this section applies to work on or near them."*
- *"...(ii) The skills and techniques necessary to determine the nominal voltage of exposed live parts..." Nominal voltage* is the normal electrical system design voltage. Examples of nominal voltages are 120/208 V, 277/480 V, and 2.3/4.16 kV. If OSHA had used the word "actual" instead of "nominal," an employee would be required to test a circuit before they knew the actual voltage. This could lead to 600 V testers being connected to 4.16 kV circuits.

OSHA uses the term "nominal voltage" to direct employees to nameplates, single-line diagrams, and schematics to find the nominal voltage of the circuit or equipment being worked on or near. For example, electric motors have nameplates with the nominal voltage listed on them. **See Figure 3-4.** By determining the nominal voltage, the worker can choose the proper rubber insulating gloves and voltage detector without having to make contact with the circuit first.

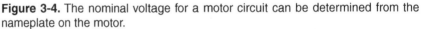

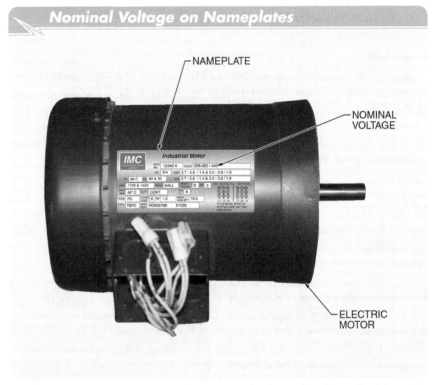

Figure 3-4. The nominal voltage for a motor circuit can be determined from the nameplate on the motor.

- *"…(iii) The clearance distances specified in 1910.333(c) and the corresponding voltages to which the qualified person will be exposed."* OSHA uses the term "clearance distances" in this regulation to refer to the safe approach distance for qualified electrical workers. This is the distance a qualified electrical worker can approach exposed energized electrical circuits or parts without using rubber insulating gloves, having the circuit parts or conductors insulated to the circuit voltage, or being in a bucket truck insulated for the voltage. **See Figure 3-5.**

The minimum approach distances (MADs), as well as the voltage ranges specified by OSHA 1910.269 and 1926 Subpart V, have changed. For example, in the previous Table R-6 the nominal phase-to-phase voltages were 50 V to 1000 V and 1100 V to 15,000 V. The new Table R-6 has voltage ranges of 50 V to 300 V, 301 V to 750 V, 751 V to 5000 V, and 5100 V to 15,000 V. This allows more minimum approach distances to better meet working conditions. Table S-5 is unchanged in OSHA 1910.333. However, NFPA 70E Table 130.4(C)(a) has also been modified in the 2015 edition and is now Table 130.4(D)(a).

The phase-to-phase voltage range 50 V to 300 V has been changed to 50 V to 150 V to allow for circuits and devices that operate at 120 V to ground. The voltage range 301 V to 751 V has been changed to 151 V to 751 V. The rest of the voltage ranges remain the same as in the 2012 edition of NFPA 70E. Most commercial and industrial facilities use NFPA 70E Table 130.4(D)(a) to determine minimum approach distances (Restricted Approach Boundary in NFPA 70E). Utilities and large industrial facilities that have distribution systems should refer to Table R-6 in 1910.269 or use the tables in the National Electrical Safety Code® (NESC®), C2.

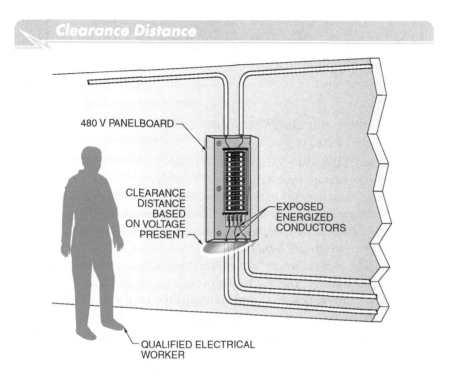

Figure 3-5. Clearance distances are listed in OSHA Table S-5.

Arc Flash Hazards

The requirement for knowing the safe approach distances for electric shock hazards, but not arc flash hazards, is contained in the OSHA regulations. Subpart S (regulations 1910.331 through .335) became federal law in 1990. At that time, although the arc flash hazard was recognized, there were no equations for determining minimum approach distances for arc flash. It is a similar situation to arc blast today. Arc blast is recognized as a hazard, but the effects cannot be calculated and the minimum approach distances cannot be determined. Based on historical data, few fatalities can be linked to arc blast, but this hazard must be considered. High fault current and enclosure characteristics are the main considerations for an arc blast pressure wave.

The minimum distances for the arc flash hazard can be calculated by using NFPA 70E and IEEE 1584, *Guide for Performing Arc Flash Hazard Calculations*. These documents allow workers to choose the proper PPE and equipment for arc flash based on incident energy, which is the heat generated by an arc flash. NFPA 70E also provides guidance for choosing arc-rated PPE and equipment in Tables 130.7(C)(15)(A)(b) (used for AC power systems) and 130.7(C)(15)(B) (used for DC power systems). Because it is possible to determine the heat generated by an electrical arc, and therefore the level of PPE that will provide protection, OSHA requires employees be provided with the proper protection for the hazard they are exposed to. Once arc blast equations are developed, OSHA will require employers to protect their employees from this hazard as well, even though it is not specifically listed in the current regulation.

SAFETY FACT

NFPA 70E has two tables for approach distances. Table 130.4(D)(a) is for approach distances for AC power systems. Table 130.4(D)(b) is for approach distances for DC power systems.

Requirements for a Qualified Electrical Worker Per OSHA 1910.333(c)

According to OSHA 1910.333(c)(2), *"Only qualified persons may work on electric circuit parts or equipment that have not been deenergized under the procedures of paragraph (b) of this section. Such persons shall be capable of working safely on energized circuits and shall be familiar with the proper use of special precautionary techniques, personal protective equipment, insulating and shielding materials and insulated tools."*

When OSHA uses the words "use of," they do not mean the technical aspect of using a tool or device. The words "use of" refer to the following:

- Setting up a safe work zone to exclude unqualified or unaware workers from coming into danger. This includes signage and barricades used to limit access, using an attendant, and knowing when a safety backup is needed. OSHA 1910.335 gives further guidance on how to set up a safe work zone.
- Choosing the right tool for the job or the proper PPE and equipment for the hazards an employee is being exposed to.
- Inspecting tools, equipment, or PPE to make sure they are safe to use.
- Properly using tools or equipment needed to do a job safely. For example, if using a voltage detector, an employee must understand how to interpret the readings. Also, when wearing PPE, the employee must understand the proper way to wear the PPE.
- Knowing and understanding the limitations of the tools, equipment, or PPE. All tools, equipment, and PPE have limitations. By understanding these limitations, an employee may prevent misuse that could result in injury or death.

TYPES OF TRAINING

The type of training required by both OSHA and NFPA 70E is either instructor-led classroom training or on-the-job training (OJT). The best training is a combination of the two. **See Figure 3-6.**

Welders are considered qualified persons according to OSHA 1910.332, Table S-4.

Types of Training

INSTRUCTOR-LED CLASSROOM TRAINING

ON-THE-JOB (OJT) TRAINING

Figure 3-6. OSHA recommends a combination of both instructor-led classroom training and on-the-job training (OJT).

FIELD NOTES: TRAINING

OSHA's letters of interpretation provide information or additional requirements that may not be clearly stated in the regulations. The following example is a letter of interpretation from OSHA.

November 22, 1994 [Corrected 6/2/2005]
Dear Employer:

Thank you for your letter of September 2, forwarded to the Occupational Safety and Health Administration's (OSHA's) Directorate of Compliance Programs from our Area Office in Baton Rouge, Louisiana, concerning the use of computer-based training to satisfy OSHA training requirements. In your letter, you ask a series of questions requesting clarification on whether the use of computer-based training is sufficient to comply with the minimum training requirements for initial employee training and retraining, in particular with regard to the number of hours of training required. We assume that your primary interest is in the training requirements of OSHA's Hazardous Waste Operations and Emergency Response Standard (HAZWOPER, 29 CFR 1910.120 and 1926.65), although your questions are also relevant to the training requirements of other OSHA standards. Each of the questions in your letter are answered in turn below.

Question 1. What is OSHA's position on computer-based training programs for cognitive training?

Answer: In OSHA's view, self-paced, interactive computer-based training can serve as a valuable training tool in the context of an overall training program. However, use of computer-based training by itself would not be sufficient to meet the intent of most of OSHA's training requirements, in particular those of HAZWOPER. Our position on this matter is essentially the same as our policy on the use of training videos, since the two approaches have similar shortcomings. OSHA urges employers to be wary of relying solely on generic, "packaged" training programs in meeting their training requirements. For example, training under HAZWOPER includes site-specific elements and should also, to some degree, be tailored to workers' assigned duties.

Safety and health training involves the presentation of technical material to audiences that typically have not had formal education in technical or scientific disciplines, such as in areas of chemistry or physiology. In an effective training program, it is critical that trainees have

FIELD NOTES: TRAINING (CONTINUED)

the opportunity to ask questions where material is unfamiliar to them. In a computer-based program, this requirement may be providing a telephone hotline so that trainees will have direct access to a qualified trainer.

Equally important is the use of hands-on training and exercises to provide trainees with an opportunity to become familiar with equipment and safe practices in a non-hazardous setting. Many industrial operations, and in particular hazardous waste operations, can involve many complex and hazardous tasks. It is imperative that employees be able to perform such tasks safely. Traditional, hands-on training is the preferred method to ensure that workers are prepared to safely perform these tasks. The purpose of hands-on training, for example in the donning and doffing of personal protective equipment, is two-fold: First, to ensure that workers have an opportunity to learn by experience, and second, to assess whether workers have mastered the necessary skills. It is unlikely that sole reliance on a computer-based training program is likely to achieve these objectives.

Thus, OSHA believes that computer-based training programs can be used as part of an effective safety and health training program to satisfy OSHA training requirements, provided that the program is supplemented by the opportunity for trainees to ask questions of a qualified trainer, and provides trainees with sufficient hands-on experience.

Question 2. How will computer-based training be compared to required hour training as set forth in 1910.120?

Answer: Where OSHA has specified a required duration represents, in OSHA's view, the minimum amount of training that will be needed for most trainees to acquire the necessary basic skills. For the reasons stated above, OSHA does not believe that the use of computer-based training will, in the majority of cases, enable trainees to achieve competency in substantially less time than the required minimum duration for training. Therefore, the use of computer-based training will not relieve employers of their obligation to ensure that employees receive the minimum require amount of training specified under HAZWOPER and other OSHA standards.

Question 3. Will a computer-based program's outline and development material suffice for conventional training material documentation?

Answer: OSHA standards, and HAZWOPER in particular, do not specify the kinds of materials that must be developed and maintained to document that a course meets the minimum requirements for course content. Employers may use whatever documentation is necessary to document the content of a training course.

Question 4. Will computer-based tracking of training competence levels be documentation enough for the training or will a hard copy, signed document be required?

Answer: OSHA standards that require training generally contain a requirement for the employer to maintain records of employee training; these records may be kept in any form deemed appropriate by the employer, so long as the records are readily accessible to the employer, employees and their representatives, and to OSHA. However, note that the HAZWOPER standard contains a unique requirement in that employees must be provided a certificate upon the successful completion of initial training; this is best accomplished by the use of hard copy.

We hope that this information is helpful. If you have any further questions, please feel free to contact us at the Office of Health Enforcement.

Sincerely,
Director, Office of Health Enforcement

This letter of interpretation demonstrates that instructor-led, hands-on training is preferred by OSHA when work involves exposure to hazards. Computer-based training (CBT), when used, must have a mechanism to answer attendee's questions. Some types of training, such as the initial training for a qualified person would not be acceptable for CBT, regardless of the support given to it.

Training for Unqualified Persons

An *unqualified person* is a person who is not qualified to perform a specific task. Anyone not meeting the definition in OSHA 1910.399 and the requirements of 1910.332 and 1910.333 would be unqualified to work on or near exposed energized conductors or circuit parts. Occupations such as painters, carpenters, plumbers, apprentices, helpers, and laborers involve tasks that may bring employees near energized conductors or circuit parts. These tasks may involve cleaning, painting, or other tasks that could expose the employees to electrical hazards. If this is the case, they would require training. That training would include understanding the hazards of electricity, knowing how to recognize those hazards, and knowing how to avoid the hazards.

FIELD NOTES:
RECOGNIZING ELECTRICAL HAZARDS

> While in Petersburg, VA, to do safety training for one of our customers, I was reading an article on the front page of a local newspaper. The article was about a commercial painter who had constructed a three-story scaffold to paint a building. He was using an aluminum extension handle for his paint roller when it came into contact with the service entrance conductors, electrocuting him. When he did not return home from work, his wife called the company he worked for and they found his body on top of the scaffold. Most painters or other tradesmen have no desire to work on or near electrical circuits. They recognize that they are not trained or equipped to do electrical work and avoid it. However, this incident highlights the fact that unqualified workers often do not recognize when they are in danger. The commercial painter might still be alive today if he had training on the hazards and how to avoid them.

Additional Requirements for Unqualified Persons

According to OSHA 1910.332(b), *"Employees who are covered by paragraph (a) of this section but who are not qualified persons shall also be trained in and familiar with any electrically related safety practices not specifically addressed by 1910.331 through 1910.335 but which are necessary for their safety."*

Unqualified persons must have some level of awareness training if they work around energized electrical equipment. This includes workers who use portable electric equipment, such as tools and extension cords, or workers who perform tasks such as cleaning or painting on or near energized electrical equipment. The type and level of training required depends on their exposure to electrical hazards.

For example, if a helper or laborer is working in a room that has operating electrical equipment, training would be required that includes how to recognize hazards (the effects of electric shock, arc flash, and arc blast, as well as how these may occur), tasks that may not be performed (such as opening doors to electrical equipment, inserting objects into any openings of operating electrical equipment, or cleaning bushings or other components that may be energized), who to contact in an emergency, and the importance and rules of electrical lockout/tagout. Training must also include hazards that may be encountered by an unqualified person using portable electric equipment, such as portable electric tools and extension cords.

Training Requirements Per NFPA 70E

The 2015 edition of NFPA 70E generally repeats the OSHA regulations concerning training requirements. The 2015 edition of NFPA 70E made numerous revisions to the training requirements in Section 110.2. Automatic external defibrillator (AED) training is required for facilities that include AED use in their emergency response plan. Employers have to verify that employees have the proper training annually. The 2012 edition stated the training had to be "certified," which could place an undue burden on the employer. The employer has to document that the training occurred.

NFPA 70E 110.2(D)(1)(b)(4) has been changed to include that a qualified person must be trained in the decision-making process in order to plan the job, identify the electrical hazards that may be present, assess the associated risk involved, and select appropriate risk controls. The changes made to the 2015 edition of NFPA 70E clarify the intent of NFPA 70E and assist employers in meeting OSHA and NFPA 70E requirements. Section 110.2 provides more details by compiling the OSHA regulations and adding some additional requirements. The following are NFPA 70E requirements for qualified persons and any OSHA regulations related to them:

- Only qualified persons can work on energized electrical equipment or circuits—OSHA 1910.333(c).
- The qualified person must be capable of working safely—OSHA 1910.333(c).
- The qualified person must be familiar with special precautionary techniques, personal protective equipment, insulating shielding materials, and insulated tools—OSHA 1910.333(c).
- The qualified person must be able to recognize and avoid specific hazards associated with electric energy—OSHA 1910.333(c) and 1910.399 (definition of qualified person).
- The qualified person must be trained in safety-related work practices and procedural requirements—OSHA 1910.333(a).
- The qualified person must be trained in emergency procedures, such as how to release victims of electric shock—OSHA 1910.269(a).
- Training is required annually for CPR and first aid—OSHA 1910.269(b) and 1910.269(a) for annual requirement. In a letter of interpretation dated April 15, 1999, OSHA recommends annual refresher training for CPR and first aid. The 2012 edition of NFPA 70E added the requirement for annual AED training in Section 110.2(C). The 2015 edition clarifies that AED training is only required when the employer's response plan includes the use of AEDs. Previously, the requirement applied to employers who did not have AEDs. This article also requires training in emergency procedures for anyone who is exposed to electric shock hazards and for employees who are responsible for taking action during an emergency.
- The qualified person must be trained in the construction and operation of the equipment—OSHA 1910.339.
- The qualified person must be trained in the skills and techniques to determine exposed live parts from others—OSHA 1910.332.
- The qualified person must be trained in the skills and techniques to determine nominal voltages—OSHA 1910.332.
- The qualified person must be trained in the approach distances in Table 130.4(D)(a) or (b)—OSHA 1910.332 and 1910.333.

- The qualified person must be trained in the decision-making process—NFPA 70E only.
- The qualified person must be trained to select an appropriate voltage detector. He or she must understand the voltage detector's indications and must demonstrate its proper use for determining the absence of voltage. The qualified person also must understand all the limitations of that voltage detector. If several voltage detectors are used, training must include all types used—specifically in NFPA 70E and implied by OSHA 1910.332.

Even though OSHA does not specifically mention the decision-making process in its regulations, it is critical for worker safety. NFPA 70E requires that a worker be able to perform a risk assessment, choose the proper PPE and equipment, and plan the job. If a worker cannot perform these decision-making processes, he or she cannot perform the job safely and would not be considered qualified by either OSHA or NFPA 70E. Some employers have an additional requirement to demonstrate that their electrical workers have the necessary maturity to perform their job assignments safely. For example, this requirement is designed to eliminate those individuals who engage in horseplay or practical jokes on the job site.

SAFETY FACT

Being capable of working safely is not only about technical skills, but also physical and mental conditions. If a worker is excessively fatigued, ill, emotionally distraught, or otherwise impaired, he or she may not be capable of concentrating on the task at hand and an accident is much more likely to occur.

SUMMARY

- NFPA 70E states the same basic training requirements for qualified and unqualified persons as OSHA 1910.332.
- Qualified and unqualified persons are required to have sufficient training so they can perform their job tasks safely.
- When the covers to electrical equipment are opened or removed for maintenance, servicing, or troubleshooting, safety-related work practices, such as wearing proper PPE, must be used to protect the worker.
- The type of training required by both OSHA and NFPA 70E is either instructor-led classroom or on-the-job training (OJT). Both types of training must be documented.
- The employer must determine what training is necessary and provide it.
- The preferred method of training for workers exposed to hazards per OSHA is instructor-led, hands-on training.
- A qualified person must be trained annually in CPR, the use of an AED, and first aid.

Digital Learner Resources
ATPeResources.com/QuickLinks
Access Code: 705798

TRAINING OF QUALIFIED AND UNQUALIFIED WORKERS

Review Questions

Name _____ Date _____

_____ 1. NFPA 70E recommends ___ training for electrical workers.
   A. instructor-led classroom
   B. on-the-job
   C. computer-based
   D. both A and B

_____ 2. Employers that include AED use in their emergency response plan must verify that employees have the proper AED training every ___.
   A. year
   B. 2 years
   C. 3 years
   D. 5 years

_____ 3. Minimum approach distances for qualified and unqualified persons for AC electrical power systems are found in NFPA 70E Table ___.
   A. 130.4(D)(a)
   B. 130.7(C)(7)(c)
   C. 130.7(C)(15)(a)
   D. 130.7(C)(15)(b)

_____ 4. Unqualified workers must be trained in ___.
   A. the skills and techniques necessary to determine energized parts from deenergized parts
   B. how to identify hazards and how to avoid those hazards
   C. the use of special precautionary techniques, temporary insulating and shielding materials, and insulated hand tools
   D. none of the above

T   F   5. A qualified person must be trained in the skills and techniques to determine nominal voltage.

T   F   6. OSHA uses phase-to-phase voltages for unqualified approach distances to overhead power lines.

_____ 7. Occupations that expose employees to electrical hazards must have training that includes all of the following except ___.
   A. understanding the hazards of electricity
   B. knowing how to recognize hazards
   C. knowing how to avoid hazards
   D. knowing how to organize materials

_____   8. The type and level of training required for an unqualified person working around energized electrical equipment depends on the ___.
   A. exposure to electrical hazards
   B. experience of the worker
   C. level of PPE worn
   D. type of equipment

_____   9. The requirements for knowing the minimum approach distances for a(n) ___ hazard are contained in OSHA regulations.
   A. arc blast
   B. arc flash
   C. electric shock
   D. fire

_____ 10. Training of qualified and unqualified workers is covered in NFPA 70E ___.
   A. 110.4
   B. 110.2
   C. 120.5
   D. 130.7

T   F   11. An employee who has had training on the construction, operation, and associated hazards of low-voltage power circuit breakers would be considered qualified to work on medium-voltage metal-clad air-magnetic circuit breakers.

_____ 12. ___ voltage refers to the design voltage of an electrical system.
   A. Nominal
   B. Actual
   C. Ghost voltage
   D. Simulated

_____ 13. In order for an electrical worker to be considered qualified, he or she must meet the requirements in OSHA ___.
   A. 1910.335
   B. 1910.334
   C. 1910.333
   D. 1910.332

_____ 14. OSHA requires that an employee receive training if he or she will perform any task on or near exposed energized circuits or equipment ___ V or greater to ground.
   A. 50
   B. 120
   C. 277
   D. 480

_____ 15. OSHA defines a qualified person (electrical worker) as one who ___.
   A. has received training in and has demonstrated skills and knowledge in the construction and operation of electric equipment and installations and the hazards involved
   B. has the skills and knowledge to perform a task safely
   C. is licensed by their state of residence to perform electrical work, such as a Master or Journeyman electrician, or who has a minimum of a Bachelor of Science degree in electrical technologies or engineering
   D. is certified by the U.S. Department of Labor to perform electrical tasks

APPROACH BOUNDARIES FOR SHOCK AND ARC FLASH HAZARDS

In order to be considered a qualified person per NFPA 70E and OSHA regulations, that person must receive training on and demonstrate the skills and knowledge needed to determine the safe approach distances for electric shock, among other requirements. OSHA 1910.333, Table S-5, has minimum approach distances for utilization equipment, which would include most commercial and industrial facilities. These minimum approach distances correspond to the Restricted Approach Boundary in NFPA 70E. In addition to the Restricted Approach Boundary, NFPA 70E also has the Limited Approach Boundary and the Arc Flash Boundary. In the 2015 edition of NFPA 70E, the Prohibited Approach Boundary was eliminated from Table 130.4(C)(a), which is now 130.4(D)(a), because the committee did not believe it served a useful purpose any longer.

OBJECTIVES

- Explain OSHA clearance distances.
- List the approach boundaries for shock hazards per NFPA 70E.
- Define Arc Flash Boundary and explain how to calculate boundaries for arc flashes.
- Explain the requirements of an energized electrical work permit (EEWP).
- Describe the purpose of a safe work zone.

CLEARANCE DISTANCES

OSHA 1910.332(b)(3)(iii) states that a qualified person must be trained in and familiar with the clearance distances specified in OSHA 1910.333(c) and the corresponding nominal voltages to which the qualified person will be exposed. OSHA 1910.333 Table S-5 – Approach Distances for Qualified Employees – Alternating Current lists the clearance (safe approach) distances to energized conductors and circuit parts. **See Figure 4-1.** Although this table is found in a section that concerns energized overhead power lines, OSHA 1910.332 emphasizes that these distances also apply whenever qualified personnel are working on or near exposed energized conductors or circuit parts.

Minimum Approach Distances To Energized Conductors — OSHA 1910.269

OSHA 1910.269 provides provisions that apply to the operation and maintenance of power generation, transmission, and distribution installations, as well as any related equipment used for communicating or metering. This includes installations and equipment of electrical utilities and utility-like installations at industrial establishments, which should only be accessible to qualified employees.

Approach Distances for Qualified Employees— Alternating Current (OSHA 1910.333 Table S-5)

Voltage Range (Phase-to-Phase)	Minimum Approach Distance
0 V – 300 V	Avoid contact
Over 300 V – 750 V	1'-0"
Over 750 V – 2 kV	1'-6"
Over 2 kV – 15 kV	2'-0"
Over 15 kV – 37 kV	3'-0"
Over 37 kV – 87.5 kV	3'-6"
Over 87.5 kV – 121 kV	4'-0"
Over 121 kV – 140 kV	4'-6"

Figure 4-1. OSHA 1910.333 Table S-5 is used to determine the clearance (safe approach) distances for a qualified employee working on or near exposed energized parts of electrical equipment.

Many commercial and industrial companies are required to follow OSHA 1910.269 for parts of their installations. For example, large industrial facilities that have high- and medium-voltage distribution systems or substations are placed under this regulation. These facilities fall under the note in OSHA 1910.269(a)(l)(i)(A) which states, *"The types of installations covered by this paragraph include the generation, transmission, and distribution installations of electric utilities, as well as equivalent installations of industrial establishments."* OSHA 1910.269 contains similar language concerning approaching energized conductors as OSHA 1910.333.

According to OSHA 1910.269(l)(6)(ii), *"The employer shall train each employee who is exposed to the hazards of flames or electric arcs in the hazards involved."*

This OSHA regulation states that employees must be trained to protect themselves against arc flash hazards. Although this regulation focuses on power generation, transmission, and distribution, OSHA expects all electrical workers to be protected and trained for all recognized hazards. Protection against arc flash hazards includes knowing the proper safe approach distance, called the Arc Flash Boundary. Training should include how to determine the Arc Flash Boundary for specific tasks and circumstances.

According to OSHA 1910.269(l)(3)(iii), *"The employer shall ensure that no employee approaches or takes any conductive object closer to exposed energized parts than the employer's established minimum approach distance, unless: (A) the employee is insulated from the energized part (rubber insulating gloves or rubber insulating gloves and sleeves worn in accordance with paragraph (l)(4) of this section constitutes insulation of the employee from the energized part upon which the employee is working provided that the employee has control of the part in a manner sufficient to prevent exposure to uninsulated portions of the employee's body), or (B) the energized part is insulated from the employee and from any other conductive object at a different potential, or (C) the employee is insulated from any other exposed conductive object in accordance with the requirements for live-line barehand work in paragraph (q)(3) of this section."*

The distances in OSHA 1910.269 Table R-6 – Alternative Minimum Approach Distances For Voltages of 72.5 kV and Less are similar to the distances found in NFPA 70E Table 130.4(D)(a) for AC voltages. OSHA recently revised parts of 1910.269. The minimum approach distance (MAD) tables were part of that revision. The nominal phase-to-phase voltage ranges changed, as well as the approach distances. **See Figure 4-2.** The low- and medium-voltage ranges are now 50 V to 300 V, 301 V to 750 V, 751 V to 5000 V, and 5100 V to 15,000 V. Most of the other phase-to-phase voltage ranges are the same as before.

Chapter 4 – Approach Boundaries for Shock and Arc Flash Hazards

OSHA Table R-6 and R-7 – Minimum Approach Distances

Table R-6 – Alternative Minimum Approach Distances for Voltages of 72.5 kV and Less[1]
(in meters or feet and inches)

Nominal Voltage (kV) Phase-to-Phase	Distance			
	Phase-to-Ground Exposure		Phase-to-Phase Exposure	
	m	ft	m	ft
0.50 to 0.300[2]	Avoid Contact		Avoid Contact	
0.301 to 0.750[2]	0.33	1.09	0.33	1.09
0.751 to 5.0	0.63	2.07	0.63	2.07
5.1 to 15.0	0.65	2.14	0.68	2.24
15.1 to 36.0	0.77	2.53	0.89	2.92
36.1 to 46.0	0.84	2.76	0.98	2.22
46.1 to 72.5	1.00	3.29	1.20	3.94

[1] Employers may use the minimum approach distances in this table provided the worksite is at an elevation of 900 meters (3,000 feet) or less. If employees will be working at elevations greater than 900 meters (3,000 feet) above mean sea level, the employer shall determine minimum approach distances by multiplying the distances in this table by the correction factor in Table R-5 corresponding to the altitude of the work.
[2] For single-phase systems, use voltage-to-ground.

Table R-7 – Alternative Minimum Approach Distances for Voltages of More Than 72.5 kV[1 2 3]
(in meters or feet and inches)

Voltage Range Phase-to-Phase (kV)	Phase-to-Ground Exposure		Phase-to-Phase Exposure	
	m	ft	m	ft
7.26 to 121.0	1.13	3.71	1.42	4.66
121.1 to 145.0	1.30	4.27	1.64	5.38
145.1 to 169.0	1.46	4.79	1.94	6.36
169.1 to 242.0	2.01	6.59	3.08	10.10
242.1 to 362.0	3.41	11.19	5.52	18.11
362.1 to 420.0	4.25	13.94	6.81	22.34
420.1 to 550.0	5.07	16.63	8.24	27.03
550.1 to 800.0	6.88	22.57	11.38	37.34

[1] Employers may use the minimum approach distances in this table provided the worksite is at an elevation of 900 meters (3,000 feet) or less. If employees will be working at elevations greater than 900 meters (3,000 feet) above mean sea level, the employer shall determine minimum approach distances by multiplying the distances in this table by the correction factor in Table R-5 corresponding to the altitude of the work.
[2] Employers may use the phase-to-phase minimum approach distances in this table provided that no insulated tool spans the gap and no large conductive object is in the gap.
[3] The clear live-line tool distance shall equal or exceed the values for the indicated voltage ranges.

OSHA

Figure 4-2. Table R-6 and R-7 from OSHA 1910.269 are used to determine minimum approach distances for voltages 72.5 kV and less and more than 72.5 kV.

Table S-5 is used for utilization equipment and lists approach distances that are slightly different from Table R-6. A comparison between these voltages and approach distances on each table may seem contradictory, but they are estimated distances and not measured distances. When OSHA Table R-6, OSHA Table S-5, and NFPA 70E Table 130.4(D)(a) are compared, from 50 V through 15,000 V the voltage ranges are somewhat different and the approach distances are not exactly the same. However, considering these distances are estimated, the differences are not significant. **See Figure 4-3.**

NFPA 70E Table 130.4(D)(b) is used for determining safe approach distances for DC voltages. The safe approach distances for DC voltages are somewhat different from those for AC voltages and are based on the nominal potential difference between conductors.

SAFETY FACT

When using OSHA 1910.269 to determine minimum approach distances 72.5 kV and above, OSHA requires that the maximum per unit transient overvoltage factor be calculated. Since these voltages typically involve overhead power lines and associated equipment, it is outside of the scope of this text. OSHA 1910.269, Tables R-6 and R-7, have the maximum per unit transient overvoltage factor included.

Phase-to-Phase Voltage	OSHA Table S-5	OSHA Table R-6	NFPA 70E Table 130.4(D)(a)
50 V to 300 V		Avoid contact	
50 V to 150 V			Not specified
0 V to 300 V	Avoid contact		
301 V to 750 V		1.09 ft	
151 V to 300 V			Avoid contact
301 V to 750 V			1 ft
300 V to 750 V	1 ft		
751 V to 5 kV		2.07 ft	
5.1 kV to 15 kV		2.24 ft	
751 V to 15 kV			2 ft 2 in.
Over 750 V to 2 kV	1 ft 6 in.		
Over 2 kV to 15 kV	2 ft		

Figure 4-3. When comparing OSHA Table R-6, OSHA Table S-5, and NFPA 70E Table 130.4(D)(a), the voltage ranges are not the same, but they are reasonably close.

Approach Distances for Unqualified Persons to Overhead Power Lines

Approach distances for unqualified persons to exposed, overhead power lines are shown in NFPA 70E Table 130.4(D)(a), column 2. These approach distances are referred to as the Limited Approach Boundary. The voltage values in column 1 are listed as phase-to-phase voltages.

OSHA 1910.333(c)(3)(i)(A)(1) states that a distance of 10′ is required for voltages ranging from 50 V to ground to 50 kV to ground. According to OSHA 1910.333(c)(3)(i)(A)(2) an additional clearance of 4″ for every 10 kV above 50 kV to ground. The approach distances listed in OSHA 1910.333(c)(3)(i)(A)(1) and (2) are phase-to-ground voltages. Phase-to-ground voltages can be converted to phase-to-phase voltages, as given in Table 130.4(D)(a), by multiplying by 1.732.

These regulations also apply to mechanical equipment that may come into contact with overhead power lines, such as line trucks, bucket trucks, and cranes. If the boom is lowered on mechanical equipment so it cannot come into contact with an overhead power line while the mechanical equipment is in transit, those distances can be reduced to 4′ for 50 kV to ground. However, the minimum safe clearance distance must be increased by 4″ for every additional 10 kV above 50 kV during transit with a lowered boom. These distances are used to protect against shock, but the distance from the equipment that is sufficient to ensure exposure of less than 2.0 cal/cm^2 may be greater than these distances. NFPA 70E calls this distance the Arc Flash Boundary. While the term Arc Flash Boundary is not used by OSHA, the concept is enforced.

SAFETY FACT

Overhead power line conductors rated 15 kV or higher are often uninsulated, even though they may have a protective jacket. This jacket is to protect the conductor from physical damage but does not provide any protection from electric shock.

Trends in Electrical Injury. James Cawley, PE, formerly with the National Institute for Occupational Safety and Health (NIOSH), now works with Electrical Safety Foundation International (ESFI). Cawley previously presented papers through the IEEE concerning injury and fatality rates for the U.S. workforce while employed with NIOSH and has continued that work with ESFI. Data has been compiled that shows electrical fatality rates from specified events dating from 1992 through 2010. **See Figure 4-4.** The specific events include contact with overhead power lines, contact with wiring, transformers, or other electrical components, and contact with the electric currents of machines, tools, appliances, or light fixtures. Although there have been some deviations, the trends for these events have remained fairly stable over the years despite the fact that the overall fatality rate and the electrical fatality rate have both decreased.

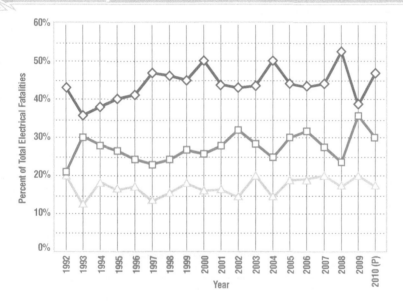
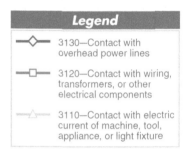

Figure 4-4. Electrical fatality rates from the three largest fatality categories have remained virtually the same since 1992.

Contact with overhead power lines was the largest cause of electrocution in the workplace with 44%. Approximately 50% of those fatalities were from mechanical equipment (cranes and man lifts) coming into contact with overhead power lines. Contact with wiring, transformers, or other electrical components had the next highest number of fatalities with 27%. Contact with electric currents of machines, tools, appliances, or light fixtures was the third highest cause of electrocution with 17%. These are not electrical injuries, they are fatalities.

Noncontact Electric Arc-Induced Injuries. At the 2004 IEEE IAS Electrical Safety Workshop, Kathleen Kowalski-Trakofler, PhD, presented a paper titled "Non-Contact Electric Arc-Induced Injuries in the Mining Industry: A Multi-Disciplinary Approach." Even though the mining industry is fairly specialized in many areas, the hazards of electric shock and arc flash apply to all industries performing the same type of maintenance work.

The paper contained a chart that showed that 24% of electrical accidents were caused by troubleshooting activities such as voltage testing and diagnostics. **See Figure 4-5.** Equipment failure during normal operation accounted for 19% of accidents. Repairs or repair-related activities accounted for 18% of accidents. These three categories accounted for the majority of the accidents because qualified electrical workers troubleshoot, operate, and repair electrical equipment every day. However, only 5% of recorded accidents were caused by maintenance-related activities. This is because maintenance is typically performed after the electrical equipment or power system is placed into an electrically safe work condition. Most of the maintenance-related incidents occurred because the employee mistakenly assumed that the electrical power was deenergized when it was not.

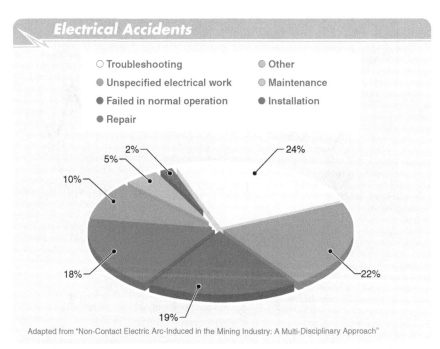

Figure 4-5. The most electrical accidents occur during troubleshooting activities.

Shermco Industries
There is always danger involved when working on energized equipment. This is especially true when troubleshooting inside small electrical enclosures.

The paper also illustrated that there is no such thing as completely safe energized work, even for something as simple as voltage testing or troubleshooting. Many times voltage or current measurements are taken from components in the back of an enclosure where there may be only 1″ or 1¼″ between energized components and ground. IEC-style equipment is even smaller and more compact, making troubleshooting even more of a hazard. Employees most often place their hands, tools, or probes in grounded metal enclosures that have components rated at 480 V.

APPROACH BOUNDARIES

NFPA 70E identifies approach boundaries to protect against electric shock. These boundaries are the Limited Approach Boundary and the Restricted Approach Boundary. Each boundary is viewed as a sphere, extending 360° around an exposed energized conductor or circuit part. The size of each boundary is based on the phase-to-phase nominal voltage of the energized conductor or circuit part. NFPA 70E Table 130.4(D)(a) or 130.4(D)(b) can be used to determine approach boundaries. For example, per Table 130.4(D)(a) an energized conductor with a nominal voltage of 277/480 VAC will have a Limited Approach Boundary of 42″ and a Restricted Approach Boundary of 12″. **See Figure 4-6.** Table 130.4(D)(b) is used for DC voltages and uses the same Limited and Restricted Approach Boundaries for voltages from 301 V to 1000 V, but it uses the nominal potential difference between conductors instead of nominal phase-to-phase voltages.

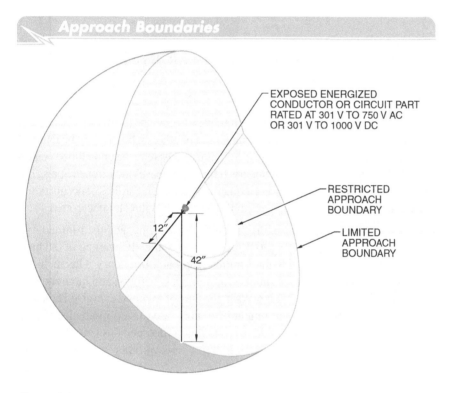

Figure 4-6. Approach boundaries are viewed as a sphere, extending 360° around an exposed energized conductor or circuit part. The size of each boundary is based on the nominal voltage of the energized conductor or circuit part.

Limited Approach Boundary

The *Limited Approach Boundary* is the distance from an exposed energized conductor or circuit part at which a shock hazard exists and is the closest distance an unqualified person can approach. The *Limited Space* is the area between the Limited Approach Boundary and the Restricted Approach Boundary. As an exception to the basic rule, an unqualified person may enter the Limited Space only if they are advised of the specific hazards and are continuously escorted by a qualified person. **See Figure 4-7.** Both the escort and the unqualified person are required to wear the appropriate personal protective equipment (PPE).

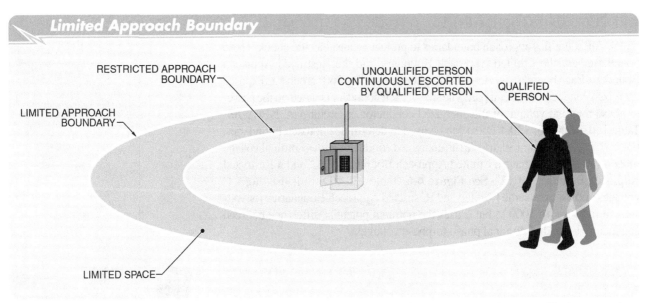

Figure 4-7. An unqualified person must be advised of specific hazards and continuously escorted by a qualified person when entering the Limited Space.

This exception is provided because it may be necessary for work to be audited by an unqualified person or to train apprentices or others who may not be familiar with a work practice or a task. Under no circumstances may the unqualified person cross the Restricted Approach Boundary. If the escort leaves the Limited Space, the unqualified person must also leave. Unqualified persons, regardless of occupational title or purpose, may not cross the Limited Approach Boundary unescorted.

Unqualified Persons. If unqualified persons are working near the Limited Approach Boundary, NFPA 70E 130.4(C)(2) requires they be warned of the hazards and that they stay outside the Limited Approach Boundary. This includes employees such as painters, plumbers, apprentices, helpers, and laborers who may be required to perform their tasks in the area of exposed energized conductors or circuit parts. If the equipment or circuits are placed in an electrically safe work condition by a qualified person, unqualified persons can work on or near them for the purposes of cleaning, tightening, or other tasks appropriate for deenergized work.

Qualified Persons. Only a qualified person can cross the Limited Approach Boundary unescorted. The qualified person must wear appropriate PPE for shock and arc flash hazards. An energized electrical work permit must also be completed. An energized electrical work permit (EEWP) is a document that describes the job planning needed to perform energized electrical work safely. An EEWP is not required when performing the following tasks:
- testing, troubleshooting, and voltage measuring, including calibration
- thermography and visual inspections if the Restricted Approach Boundary is not crossed
- access to an area with energized electrical equipment when electrical work is not performed and the Restricted Approach Boundary is not crossed
- housekeeping and other general nonelectrical tasks if the Restricted Approach Boundary is not crossed

The last two bullet points may pertain to entry into an operating substation. The substation yard is almost always within the Arc Flash Boundary, but groundskeepers and other nonelectrical personnel must have access. These additional exceptions should still provide an appropriate level of safety, but they also allow these workers to perform their tasks without completing an EEWP. All requirements in the permit must still be followed. To ensure all the proper steps were followed in the planning for these tasks, the author recommends that employees fill out an EEWP with the exception of the required signatures.

Restricted Approach Boundary

The *Restricted Approach Boundary* is the distance from an exposed energized conductor or circuit part where an increased risk of electric shock exists due to the close proximity of the person to the energized conductor or circuit part. The *Restricted Space* is the area between the Restricted Approach Boundary and the exposed energized conductors or circuit parts. Only a qualified person can cross the Restricted Approach Boundary. **See Figure 4-8.** These distances are given in OSHA Tables S-5, and R-6 and NFPA 70E Tables 130.4(D)(a) and 130.4(D)(b). In order to cross the Restricted Approach Boundary, a qualified person must meet one of the following three conditions:

- The circuits or equipment must be insulated for the voltage.
- The person must wear PPE that is insulated for the voltage.
- The person is insulated from the energized circuits or parts by an insulated bucket truck or other means.

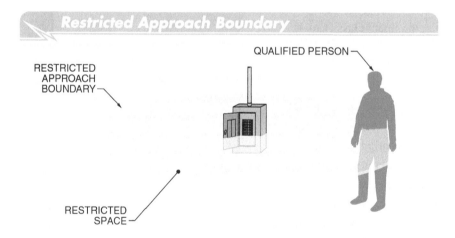

Figure 4-8. Only a qualified person can cross the Restricted Approach Boundary.

ARC FLASH BOUNDARY

The *Arc Flash Boundary* is the distance from exposed energized conductors or circuit parts where bare skin would receive the onset of a second-degree burn, equivalent to 1.2 cal/cm^2. According to OSHA, an employer must train employees about the dangers of being exposed to flames or electric arcs when they work on or near such hazards. This would include training employees how to determine the Arc Flash Boundary for specific tasks and circumstances.

SAFETY FACT

Several software packages are available to determine Arc Flash Boundary including EasyPower®, Millsoft®, and Electrical Transient Analyzer Program (ETAP®).

Arc Flash Boundary Per IEEE 1584

In 2002, the IEEE P1584 Work Group finished their work on IEEE 1584, *Guide for Performing Arc Flash Hazard Calculations.* It has since been amended as IEEE 1584b-2011. IEEE 1584 is based on several hundred staged arc flash tests performed at various distances and at various voltages and short-circuit currents. Equations were developed to provide guidance for calculating boundaries for arc flashes most likely to be experienced by industrial electrical employees. The following conditions apply to the equations developed by IEEE 1584:

- three-phase faults
- arc-in-a-box (enclosures) only
- 208 V to 15 kV systems
- 700 A to 106 kA short-circuit currents
- grounded and ungrounded electrical systems
- arc gaps between 13 mm (approximately 0.5″) to 152 mm (approximately 6.0″)

For higher voltages in open-air substations, IEEE 1584 recommends using the equations developed by Ralph Lee, PE, in his paper, "The Other Electrical Hazard: Electric Arc Blast Burns." In these circumstances, software is often used that is specifically designed to calculate the Arc Flash Boundary for overhead power lines, which are usually single phase-to-ground, open-air faults.

IEEE 1584 is designed to be used by engineers and experienced technicians. **See Figure 4-9.** The following items are shown on the IEEE 1584 calculations page:

- nominal voltage of bus (in kV)
- bolted fault current (in kA)
- total arcing fault current (in kA)
- equipment class (1 to 4)
- type of grounding (1 or 2)
- protective device type (0 to 14)
- gap (in mm)
- working distance (in mm)
- protective device arc fault current (kA)
- arc clearing time (in sec)
- reduced current clearing time (in sec)
- incident energy (in J/cm^2)
- incident energy (in cal/cm^2)
- Arc Flash Boundary (in mm)
- PPE per NFPA 70E (category)

It is important for the engineer performing the arc flash hazard analysis to be familiar with the characteristics of the electrical power system when performing these calculations. Certain conditions can change the operational characteristics of an electrical power system. For example, if a circuit breaker with an instantaneous trip function is used to feed a device or circuit several hundred feet away, the impedance of the cable may limit the available short-circuit current flowing to the fault. If the cable impedance is great enough, the short-circuit current could be below the set-point of the instantaneous trip function and the breaker will operate in long-time delay function. This may result in extended operating times, even though the circuit breaker has an instantaneous trip function.

SAFETY FACT

Even though arc-rated PPE and clothing may not be mandated by Table 130.7(C)(15)(A)(a), this does not mean that it may not be needed. Each worker must evaluate the equipment he or she is about to work on and determine if arc-rated PPE and clothing should be worn. In many cases, the table may indicate that no arc flash PPE is required, but the likelihood of injury may be higher than that worker would want to accept. Electrical workers should consider wearing arc-rated daily wear to ensure clothing does not ignite, which is the most likely cause of severe burns and death.

IEEE 1584 Calculations Page

Site: _____   Normal Operation

Results of arc flash calculations for determination of PPE and flash boundary

Bus Information Name of Location	kV of bus kV	Bolted Fault Current kA	Total Arcing Fault Current kA	Equipment Class 1 to 4	Grounding 1 or 2	Protective Device Type 0 to 14	Gap mm	Working Distance mm	Protective Device Arc Fault Current kA	Arc Clearing Time sec	Reduced Current Clearing Time sec	Incident Energy J/cm^2	Incident Energy cal/cm^2	Arc Flash Boundary mm	PPE per NFPA 70E Category
BFP-Ifp	13.8	11	10.66	1	2	0	153	1825	10.66	0.26	0.00	1.87	0.4	1113	0
BFP-Ifs	0.48	37.8	19.06	3	1	0	32	610	19.06	0.26	1.48	219.61	52.5	7930	X
BFP-Ifc2	0.48	13.9	7.28	1	1	0	32	610	7.28	0.26	0.35	10.63	2.5	888	1
Utility Example	68.8	5.37	5.37	1	1	0	153	1825	5.37	0.26	0.00	61.74	14.8	6400	3
KEMA Test 8244	4.16	5.44	5.33	1	2	0	102	483	5.33	0.26	0.00	11.08	2.6	717	1
KEMA Test 8283	4.16	40.43	38.32	3	2	0	102	610	38.32	0.26	0.00	101.02	24.1	13339	3
SqD 28397	0.4	103.34	38.49	4	1	0	25	610	38.49	0.26	0.06	96.59	23.1	3697	3
SqD F28	0.4	53.6	22.62	4	1	0	25	610	22.62	0.26	0.11	54.38	13.0	2605	3
Kinectrics 2338	0.59	3.35	2.46	1	1	0	32	610	2.46	0.26	0.21	2.91	0.7	464	0
250AMCCB	0.48	20	11.85	4	1	9	25	457	11.85	0.26	0.00	43.43	10.4	1702	3
2000A	0.48	50	25.93	4	1	12	25	457	25.93	0.26	0.00	25.40	6.1	1227	2
1000A	0.48	50	25.93	4	1	10	25	457	25.93	0.26	0.00	12.68	3.0	804	1
1000LI	0.48	50	25.93	4	1	11	25	457	25.93	0.00	0.00	20.18	4.8	1067	1
1200LI	0.48	50	25.93	4	1	13	25	457	25.93	0.00	0.00	35.06	8.4	1494	3
1200A Cl L	0.48	30	13.82	1	1	0	32	610	13.82	0.26	0.64	38.75	9.3	1695	3
600A RK1	0.48	15	7.76	1	1	0	32	610	7.76	0.26	0.59	19.20	4.6	1193	1
enter location	0.48	20	9.86	1	1	0	32	610	9.86	0.26	0.00	13.07	3.1	984	1
enter location	0.48	20	9.86	1	1	0	32	610	9.86	0.26	0.00	13.07	3.1	984	1
enter location	0.48	20	9.86	1	1	0	32	610	9.86	0.26	0.00	13.07	3.1	984	1
enter location	0.48	20	9.86	1	1	0	32	610	9.86	0.26	0.00	13.07	3.1	984	1
enter location	0.48	20	9.86	1	1	0	32	610	9.86	0.26	0.00	13.07	3.1	984	1
enter location	0.48	20	9.86	1	1	0	32	610	9.86	0.26	0.00	13.07	3.1	984	1
enter location	0.48	20	9.86	1	1	0	32	610	9.86	0.26	0.00	13.07	3.1	984	1
enter location	0.48	20	9.86	1	1	0	32	610	9.86	0.26	0.00	13.07	3.1	984	1
enter location	0.48	20	9.86	1	1	0	32	610	9.86	0.26	0.00	13.07	3.1	984	1
enter location	0.48	20	9.86	1	1	0	32	610	9.86	0.26	0.00	13.07	3.1	984	1
enter location	0.48	20	9.86	1	1	0	32	610	9.86	0.26	0.00	13.07	3.1	984	1
enter location	0.48	20	9.86	1	1	0	32	610	9.86	0.26	0.00	13.07	3.1	984	1
enter location	0.48	20	9.86	1	1	0	32	610	9.86	0.26	0.00	13.07	3.1	984	1
enter location	0.48	20	9.86	1	1	0	32	610	9.86	0.26	0.00	13.07	3.1	984	1
enter location	0.48	20	9.86	1	1	0	32	610	9.86	0.26	0.00	13.07	3.1	984	1
enter location	0.48	20	9.86	1	1	0	32	610	9.86	0.26	0.00	13.07	3.1	984	1

Figure 4-9. The IEEE 1584 calculations page can be used to determine the level of incident energy and the Arc Flash Boundary.

Arc Flash Boundary Per OSHA

OSHA, in its revisions to 1910.269 and 1926 Subpart V, requires employers covered by these regulations to make reasonable estimates of the incident energy an employee would receive if an arc flash were to occur. In OSHA 1910.269, Appendix E, OSHA provides guidance for employers to determine which method provides a reasonable estimate. Table 2 of Appendix E—Methods of Calculating Incident Heat Energy from an Electric Arc shows the currently available methods. Table 3 of Appendix E—Selecting a Reasonable Incident-Energy Calculation Method shows OSHA's evaluation of these methods. **See Figure 4-10.** IEEE 1584 and NFPA 70E are not appropriate above 15 kV. When the voltage is above 15 kV, OSHA found only the ARCPRO software provides reasonable estimates of incident energy.

Arc Flash Boundary Per NFPA 70E

Calculating the Arc Flash Boundary is one component of an arc flash risk assessment. In the 2015 edition of NFPA 70E, the term "hazard/risk analysis" was replaced with "risk assessment" in order to harmonize it with other standards.

Arc flash risk assessment is covered in 130.5 and includes determining if a hazard exists and the degree and extent of that hazard. The risk assessment identifies hazards, estimates the severity of injury or death, estimates the likelihood of an occurrence, and determines what protective measures, such as PPE, may be required. An arc flash risk assessment and shock risk assessment are part of the risk assessment process.

OSHA 1910.269 Appendix E, Tables 2 and 3

Table 2 – Methods of Calculating Incident Heat Energy from an Electric Arc

1. *Standard for Electrical Safety Requirements for Employee Workplaces,* NFPA 70E-2012, Annex D, "Sample Calculations of Flash Protection Boundary."

2. Doughty, T.E., Neal, T.E., and Floyd II, H.L., "Predicting Incident Energy to Better Manage the Electric Arc Hazard on 600 V Power Distribution Systems," *Record of Conference Papers IEEE IAS 45th Annual Petroleum and Chemical Industry Conference,* September 28–30, 1998.

3. *Guide for Performing Arc-Flash Hazard Calculations,* IEEE Std 1584-2002, 1584a-2004 (Amendment 1 to IEEE Std 1584-2002), and 1584b-2011 (Amendment 2: Changes to Clause 4 of IEEE Std 1584-2002).*

4. ARCPRO, a commercially available software program developed by Kinectics, Toronto, ON, CA.

* This appendix refers to IEEE Std 1584-2002 with both amendments as IEEE Std 1584b-2011

Table 3 – Selecting a Reasonable Incident-Energy Calculation Method†

Incident-Energy Calculation Method	600 V and Less‡			601 V to 15 kV∥			More than 15 kV		
	1ϕ	3ϕa	3ϕb	1ϕ	3ϕa	3ϕb	1ϕ	3ϕa	3ϕb
NFPA 70E-2012 Annex D (Lee Equation)	Y-C	Y	N	Y-C	Y-C	N	N§	N§	N§
Doughty, Neal, and Floyd	Y-C	Y	Y	N	N	N	N	N	N
IEEE Std 1584b-2011	Y	Y	Y	Y	Y	Y	N	N	N
ARCPRO	Y	N	N	Y	N	N	Y	Y∥	Y∥

† Although the Occupational Safety and Health Administration will consider these methods reasonable for enforcement purposes when employers use the methods in accordance with this table, employers should be aware that the listed methods do not necessarily result in estimates that will provide full protection from internal faults in transformers and similar equipment or from arcs in underground manholes or vaults.

‡ At these voltages, the presumption is that the arc is three-phase unless the employer can demonstrate that only one phase is present or that the spacing of the phases is sufficient to prevent a multiphase arc from occurring.

§ Although the Occupational Safety and Health Administration will consider this method acceptable for purposes of assessing whether incident energy exceeds 2.0 cal/cm², the results at voltages of more than 15 kilovolts are extremely conservative and unrealistic.

∥ The Occupational Safety and Health Administration will deem the results of this method reasonable when the employer adjusts them using the conversion factors for three-phase arcs in open air or in an enclosure, as indicated in the program's instructions.

Key

1ϕ: Single-phase arc in open air

3ϕa: Three-phase arc in open air

3ϕb: Three-phase arc in an enclosure (box)

Y: Acceptable; produces a reasonable estimate of incident heat energy from this type of electric arc

N: Not acceptable; does not produce a reasonable estimate of incident heat energy from this type of electric arc

Y-C: Acceptable, produces a reasonable, but conservative, estimate of incident heat energy from this type of electric arc.

OSHA

Figure 4-10. OSHA 1910.269, Appendix E, provides methods for calculating incident energy.

The Arc Flash Boundary was first identified as the Arc Flash Protection Boundary in the 1995 edition of NFPA 70E. The 2000 edition of NFPA 70E included the first tables for selecting arc flash PPE and clothing, Tables 130.7(C)(9) and 130.7(C)(10). These tables have been expanded and improved in each edition since.

In the 2015 cycle, David Wallis, then OSHA's representative to the NFPA 70E Committee, recommended a "clean sheet" approach to the tables. His recommendation was based on how OSHA determines whether PPE is necessary or not. In the 2015 edition of NFPA 70E, Table 130.7(C)(15)(A)(a) lists various tasks that may require arc-rated PPE and clothing. For most of these tasks, if the equipment is properly installed, properly maintained, has all covers and doors installed and properly secured, and there is no evidence of impending failure, then no arc-rated PPE or clothing is mandated. If any one of the conditions is not met, or if the technician about to perform the task cannot say with some degree of certainty that these conditions are met, arc-rated PPE and clothing are required.

Arc-rated PPE and clothing is selected using Table 130.7(C)(15)(A)(b). Table 130.7(C)(15)(A)(b) lists the categories of installed equipment, the limits of the table for each equipment category, and the Arc Flash Boundary. Table 130.7(C)(16) lists the PPE and clothing required for each arc flash PPE category.

Hazard/risk categories (HRCs) are no longer used in NFPA 70E and have been replaced with arc flash PPE categories. This is because the table method no longer bases the selection of arc-rated PPE and clothing on risk. The limits for each equipment category, as well as the arc flash PPE categories, have the same values as the pervious tables. See Table 130.7(C)(15)(A)(b) in NFPA 70E 2015 edition.

Unlike the tables in previous editions, which reduced the level of protection by one, two, or three numbers based on perceived risk, the new tables allow only one choice of arc-rated PPE and clothing. No reduction of protection is allowed. If there is an arc flash hazard, full-rated arc flash PPE and clothing must be worn. The NFPA 70E Committee believes that this new table method will be easier to use in the field and provide better protection when arc-rated PPE and clothing is necessary.

The short-circuit current and clearing time determine the incident energy for systems less than 600 V. Any product of the operating time of the OCPD and short circuit current that is less than the product of the limits given in the equipment categories of Table 130.7(C)(15)(A)(b) should provide a safe margin. For example, the limits for 600 V class motor control centers (MCCs) are a maximum short-circuit current of 65 kA and an operating time of 0.03 sec (2 cycles). This provides a product of 130 kA cycles. The recommended PPE for these limits should also be adequate for any combination of short-circuit current and operating time that does not exceed 130 kA cycles, as long as the voltage does not exceed 600 V.

Electrical Equipment Properly Installed, Properly Maintained, and Evidence of Impending Failure. Table 130.7(C)(15)(A)(a) has a note that defines what is meant by properly installed, properly maintained, and evidence of impending failure. Electrical equipment and systems that have been designed and approved by a professional engineer (PE) and inspected to ensure they meet local and national codes and standards, such as the NEC® or NESC, are examples of properly installed.

The most difficult condition for a technician to determine is probably whether the equipment has been properly maintained, especially if the technician does

SAFETY FACT

NFPA 70E is a standard. Standards represent the minimum acceptable requirements, not the best requirements. Per 130.7(A), Informational Note No. 1, the goal of NFPA 70E is to make an arc flash incident survivable, not to eliminate injury.

not work at that facility. In most cases, it is not practical to review test sheets from the last maintenance cycle, even when they are available. Larger facilities may have thousands of test results that may not be organized in a manner that would allow their review. NFPA 70E 205.3, Information Note No. 1, recommends the use of local indication to assist in determining condition of maintenance. Local indication refers to a labeling system, such as the system outlined in NFPA 70B, *Recommended Practice for Electrical Equipment Maintenance* in Section 11.27, Test or Calibration Decal System.

This is the system used by many NETA-member companies to alert and track equipment condition during shutdowns and turnarounds. There are three labels that could be affixed to the equipment once maintenance is complete. **See Figure 4-11.** If the equipment is ready for continued service and has no defects, a white label is placed on it. Equipment with minor defects that do not affect the operation or safety of the equipment is indicated with a yellow label. A red label is used for equipment that needs immediate repair and should be removed from service.

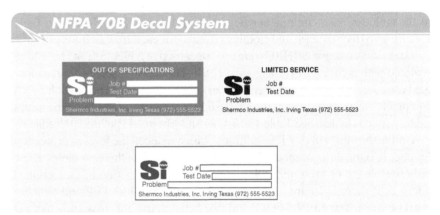

Figure 4-11. Using a labeling system such as the one described in NFPA 70B enhances safety by allowing field service technicians to immediately determine the condition of maintenance.

These labels only indicate the condition of maintenance at the time of test or maintenance. As time progresses, the accuracy of the labels may become questionable. After three years, the labels should be viewed more cautiously, as the equipment is due for another maintenance cycle. The best safe work practice is to evaluate the equipment for the likelihood of failure, even if it is the day after maintenance was performed on it. A new note for Table 130.7(C)(15)(A)(b) allows the arc flash PPE category to be reduced by one number if the circuit or equipment is rated 600 V or less and is protected by upstream current-limiting fuses or circuit breakers sized 200 A or less.

Evidence of impending failure is somewhat subjective. It is up to the technician performing the task to determine if there is any evidence of impending failure. Evidence of impending failure may include extremely degraded condition of the equipment, any evidence of improper operation or insulation failure, or the smell of ozone.

Working on medium-voltage equipment may require PPE rated higher than 40 cal/cm^2, but guidance for such tasks is not provided by Tables 130.7(C)(15)(A)(a) and (b). It is recommended by the author that an employee wear protective clothing and PPE with a minimum arc rating of 65 cal/cm² when working on medium-voltage equipment or where PPE Category 4 is recommended by the table method. PPE with an arc rating of 65 cal/cm² that weighs about the same as PPE with an arc rating of 40 cal/cm² is available for such work.

Arc Flash Boundary Equations. The Arc Flash Boundary can be calculated using the equations in NFPA 70E Informative Annex D. The Arc Flash Boundary is dependent on the available short-circuit current, maximum total clearing time of the OCPD, the voltage of the circuit, and a standard factor that varies if the actual short-circuit current is known or not known. The incident energy received by an employee at the Arc Flash Boundary is not considered life threatening but can still cause pain and a severe first-degree burn with possible blistering.

The circuit voltage has less effect on the Arc Flash Boundary than the short-circuit current or operating time of an OCPD. The Arc Flash Boundary must be calculated for every bus in the electrical system and can vary greatly depending on the short-circuit current and clearing time of the OCPD. For example, a 480 V motor control can have an Arc Flash Boundary of 10″ or 60″, depending on the short-circuit current and operating time of the OCPD. The Arc Flash Boundary is the distance at which the incident energy equals 1.2 cal/cm² for all voltages. **See Figure 4-12.**

> **SAFETY FACT**
>
> *When using the table method to select arc-rated PPE and clothing, be certain the circuit or system does not exceed the limits given in Table 130.7(C)(15)(A)(b) for short circuit available current or operating time of the OCPD. If either limit is exceeded, an incident energy analysis must be performed to determine the incident energy exposure of the worker and the table method cannot be used.*

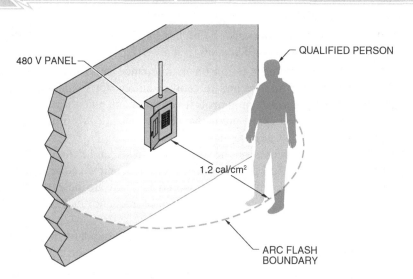

Figure 4-12. The Arc Flash Boundary is always where the calculated incident energy is equal to 1.2 cal/cm².

Maintenance and Testing of Overcurrent Protective Devices (OCPDs). Informational Note 1 of NFPA 70E 130.5 states that improper or inadequate maintenance of OCPDs can cause longer than normal operating times, which increases the incident energy received by the employee. Section 130.5(D) of NFPA 70E requires that equipment such as switchboards, panelboards, industrial control panels, meter socket enclosures, and motor control centers that are in other than dwelling units (homes, manufactured housing, and the like) and are likely to require examination, adjustment, servicing, or maintenance while

energized are to be field-marked with a label. The label must contain the nominal system voltage, the Arc Flash Boundary, and at least one of the following:
- either the available incident energy with the corresponding working distance or the arc flash PPE category (using the table method) for the equipment, but not both
- the minimum arc rating of clothing
- the site-specific level of PPE

This requirement was added because many employees may not have the necessary information to determine the required arc-rated protective clothing and PPE to perform specific tasks at facilities. Only equipment such as switchboards, panelboards, industrial control panels, meter socket enclosures, and MCCs that may require inspection, maintenance, or servicing while energized require labels. An exception to the labeling requirements allows for labels applied before September 30, 2011 as long as they provide the available incident energy or required level of PPE.

Section 130.5 Informational Note No. 4 refers the user to Chapter 2, Safety-Related Maintenance Requirements. Section 205.3 requires electrical equipment to be maintained in accordance with the manufacturer's recommendations or industry consensus standards. Section 200.1 Informational Note refers the user of NFPA 70E to one of the three following industry consensus standards:
- NFPA 70B, *Recommended Practice for Electrical Equipment Maintenance*
- ANSI/NETA MTS, *Standard for Maintenance Testing Specifications for Electrical Power Distribution Equipment and Systems*
- IEEE 3007.2, *IEEE Recommended Practice for the Maintenance of Industrial and Commercial Power Systems*

NFPA 70B provides information on maintenance philosophies, forms, maintenance frequency tables, and recommendations for maintaining and testing all types of electrical equipment. It is a complete guide for setting up a preventive maintenance (PM), predictive maintenance (PdM) or reliability-centered maintenance (RCM) program.

ANSI/NETA MTS focuses on the inspections, tests, and maintenance required by common electrical power system equipment and what the results of those inspections, tests, and maintenance should be. There are tables providing results for almost any test, as well as a maintenance frequency matrix in Annex B. It is a great companion standard to NFPA 70B.

IEEE 3007.2 is more of an overview maintenance standard. It provides useful information for maintaining electrical power systems and equipment.

Incident Energy Analysis. An *incident energy analysis* is part of an arc flash risk assessment and determines the incident energy a worker would receive at a specific working distance. It also determines the Arc Flash Boundary and the required arc-rated PPE and clothing. *Incident energy* is a measurement of thermal energy (in cal/cm^2) at a working distance from an arc fault. Factors that go into determining incident energy include system voltage, available short-circuit current, arc current, and the time required for OCPDs to open.

Incident energy is directly proportional to time. If the time of exposure increases, the incident energy received by a surface (usually a person) will increase in proportion. For example, a well-maintained circuit breaker may have a 0.03 second operating time for its instantaneous function. Depending on short circuit available current, the arc flash warning label may indicate an incident energy at an 18″ working distance of 8 cal/cm^2. If the circuit breaker

---

SAFETY FACT

Do not use the earlier editions of NFPA 70E. NFPA 70E is a safe work practices standard that is sometimes changed to reflect the most recent information available. The newest edition of NFPA 70E always supersedes the previous edition.

does not operate in 0.03 seconds, but slows down to 0.06 seconds due to poor maintenance practices, the incident energy received by a worker would increase from 8 cal/cm^2 to 16 cal/cm^2. If that same worker were wearing 10 cal/cm^2 arc-rated PPE and clothing, which would be a logical choice for that estimated incident energy exposure, there is a substantial likelihood he or she would be underprotected.

Using the same example, if the circuit breaker failed to operate, the next device upstream would be required to clear the fault. Depending on the settings of the OCPD upstream, it may operate using a long time-delay function, instead of an instantaneous function. In this scenario, the operating time could be several seconds and the incident energy received by the worker would drastically increase and the likelihood of burns would increase.

The NFPA 70E Committee is aware of the relationship between operating time and incident energy and how condition of maintenance will affect operating time. NFPA 70E Section 130.5 requires that the arc flash risk assessment consider the design, opening time, and condition of maintenance of the OCPD.

ENERGIZED ELECTRICAL WORK PERMITS

An *energized electrical work permit (EEWP)* is a document that describes the job planning needed to perform energized electrical work safely. It includes a risk assessment and specifies PPE to be used and other measures that may be required to protect both the employee and others who may be in the area. According to NFPA 70E 130.2(B), an EEWP is required to perform energized electrical work. An example of an EEWP is located in NFPA 70E Informative Annex J. An EEWP is made up of three main parts. **See Figure 4-13.**

According to NFPA 70E 130.2(B), an EEWP is required when a qualified person is performing tasks within the Restricted Approach Boundary or interacting with equipment in a manner that could cause failure, such as racking circuit breakers or inserting or removing MCC buckets. The EEWP is part of the documentation and approval process required for a risk assessment.

The risk assessment includes an arc flash risk assessment, a shock risk assessment, and other safe work practices. The arc flash risk assessment determines incident energy exposure to the employee, the Arc Flash Boundary, and the arc-rated PPE and clothing required. The shock risk assessment determines the Limited and Restricted Approach Boundaries and the type of rubber insulating gloves, insulated tools, or other shock protective equipment and devices needed. As part of the risk assessment, additional safe work practices would also be determined and documented. These practices may include setting up barriers or barricades, using attendants to limit access to the work area, marking look-alike equipment to prevent workers from working on adjacent equipment, or using fall arrest equipment.

SAFETY FACT

The EEWP in NFPA 70E Informative Annex J is not mandatory. It can be modified to fit the specific needs of the company using it.

Energized Electrical Work Permit (EEWP)

ENERGIZED ELECTRICAL WORK PERMIT
Greater than 50 Volts and less than 600 Volts

PART I
- Location: _____ Job #: _____ Date: _____
- Equipment/circuit: _____ Task: _____
- Justification of why the equipment/circuit cannot be de-energized:
  - ☐ Absence of voltage testing   ☐ Operational Limits   ☐ Other _____
  - ☐ Troubleshooting / Testing   ☐ Process or life support equipment
- Signature of Requestor _____ Title _____ Date _____

PART II

1. A. **SHOCK HAZARD ANALYSIS:** (After voltage test, cover all exposed energized conductors and circuit parts if possible)
   - Voltage: _____ ☐ DC ☐ AC ☐ Use appropriate voltage detector: _____
   - Working distance: _____ feet _____ inches ☐ from body ☐ from hand
   - Specific shock hazards: _____

   B. **SHOCK PROTECTION BOUNDARIES AND PPE:** Reference Table 130.4(D)(a) and 130.4(D)(b) in NFPA 70E
   - ☐ Glove class _____ ☐ Limited Approach _____ in. ☐ Restricted Approach _____ in.
   - ☐ Prohibited Approach _____ in.

   C. **ARC FLASH HAZARD ANALYSIS:** Follow the sequence below and check all items that apply
   - ☐ Metal-clad equipment   ☐ Open-air equipment
   - ☐ There are Labels, Incident energy at working distance _____ cal/cm^2. Be certain the arc rating of PPE exceeds this number.
   - ☐ No labels, use the Tasks tables 130.7(C)(15)(A)(a) and 130.7(C)(15)(A)(b) in NFPA 70E 2015 to determine the Hazard Risk Category: HRC ☐ 1 ☐ 2 ☐ 3 ☐ 4
   - ☐ Arc Flash Protection Boundary _____ in.
   - ☐ Isolating an upstream device? This may be safer than opening the 480-volt Main circuit breaker.
   - ☐ Where at all possible use a remote switching device to operate circuit breakers or switches.
   - ☐ Where at all possible use a remote racking device for inserting or removing draw out circuit breaker.
   - ☐ Safe Work Zone barrier to be placed at _____ in.
     - ☐ All conductive objects removed from person and placed outside Safe Work Zone
   - Short Circuit Amps _____ Fuses/ C.B. / Relay _____ Clearing Time Cycles/Sec: _____
   - Working Distance: ☐ 18"   ☐ other _____ feet _____ inches _____ cal/cm^3
   - Condition of equipment   ☐ Clean/Excellent ☐ Medium/Good ☐ Dirty/Poor   Last Calibration Date _____
   - ☐ If equipment is poor condition, if there are no recent calibration stickers or there are other extenuating situations, the best decision may be to not proceed with the current work plan but to find a safer way to do the job.

2. **ADDITIONAL PPE BASED ON ABOVE ANALYSIS & JOB SCOPE:** (See JHA for minimum PPE)
   - ☐ Insulated tools   ☐ Floor mat   ☐ 40 cal flash suit w/ hood   ☐ Ground cluster
   - ☐ Arc flash face shield ☐ balaclava   ☐ Rubber blanket   ☐ Arc flash blanket
   - ☐ Other _____

3. **GROUND CLUSTERS:** (Also see Grounding for Personal Protection Policy & Procedure and the LOTO Form)
   - ☐ Use longest hot stick reasonable   ☐ Use body position to keep the body outside of flash and blast direction (stand to one side)

PART III

I UNDERSTAND THE HAZARDS ABOVE AND THE WORK CAN PROCEED SAFELY *(PRINT NAME CLEARLY BELOW)*

_____ _____   _____ _____
TECHNICIAN   Date   ADDITIONAL QUALIFIED PERSON   Date

_____ _____
SUPERVISOR   Date

Figure 4-13. An energized electrical work permit (EEWP) documents the steps necessary to protect workers when hazardous electrical work is performed.

Energized Electrical Work Permit — Part I

Part I of an EEWP permit should be completed by the requester. The requester is usually the manager or supervisor who decides the necessity of performing energized electrical work. Justification for the work is required, as well as a signature and date. When a requester signs an EEWP, they must understand that they are assuming some of the responsibility. The requestor should be cognizant of the hazards that could be encountered and the repercussions of an accident. The justification of energized electrical work and its repercussions are factors for the requester to consider before signing off on an energized work permit.

Energized Electrical Work Permit — Part II

Part II of an EEWP can be used for job planning and includes a risk assessment. This part should be completed by the qualified person doing the work. By completing Part II, the qualified person will have completed some of the planning necessary for performing the work. As the qualified person plans the task, a list of items that are required for the specific tasks involved in the work can be compiled. These items may include safety barricade tape, cones, stanchions, or PPE and other equipment as indicated by the risk assessment.

Energized Electrical Work Permit — Part III

Part III of an EEWP consists of spaces for the required signatures that are to be completed by the management of a company. One of the key elements of the EEWP is that it must be approved and signed by different managerial and technical persons. This is to ensure no unauthorized energized electrical work is performed and that the work is performed safely. The management must approve all energized work before it is started. There may be variations as to whose signatures are actually needed and when. The management is responsible for controlling who performs the work, when the work is performed, and how the work is performed. The management is also responsible if an accident were to occur.

Exemptions to Energized Electrical Work Permits

The 2015 edition of NFPA 70E has additional tasks in the list of tasks that are exempt from an EEWP. In previous editions, only voltage testing and visual inspections (if the Restricted Approach Boundary is not crossed) were exempt from an EEWP. The 2015 edition exempts the following activities:

- Testing, troubleshooting, and voltage measuring—This includes calibration of instruments.
- Thermography and visual inspections if the restricted approach boundary is not crossed—Thermographers that have others remove panel covers and doors also do not have to wear arc flash protective PPE or clothing if they remain outside the Restricted Approach Boundary and do not break the plane of the enclosure. If they are removing covers, they have to wear arc-rated PPE and clothing adequate for the hazard.
- Access to and egress from an area with energized electrical equipment if no electrical work is performed and the Restricted Approach Boundary is not crossed—This applies to equipment and machine rooms where energized electrical equipment is cycling on and off.
- General housekeeping and miscellaneous nonelectrical tasks if the Restricted Approach Boundary is not crossed—Groundskeepers and others who perform nonelectrical maintenance tasks within outdoor substations will benefit from this new exception.

SAFETY FACT

When a task is exempt from an EEWP, it does not mean safety is discarded. A job safety analysis (JSA)/job hazard analysis (JHA) still needs to be completed, a job briefing may need to be conducted, and all job preplanning still needs to occur, along with wearing the appropriate PPE. The only thing that is not required is the actual permit.

FIELD NOTES:
COMPANY RESPONSIBILITY

I was at a facility teaching a 3-day electrical safety class. At the end of the third day, one of the electricians came up to me and said, "You know, my supervisors really like me." I thought that was a pretty strange statement, so I said, "Really, why is that?" He responded, "Because I'll do this work energized and the other guys won't." I said, "Haven't you seen the videos and pictures of people who thought accidents won't happen to them?" He said, "I know what I'm doing, it's not a problem."

I spoke with the electrical supervisor before I left and told him the facility had a major problem when shift supervisors encourage a 21-year-old worker to perform energized work when the equipment should be placed into an electrically safe work condition. He agreed and said he would look into it.

If that worker was seriously injured or killed, lawyers would have a field day knowing that the company, despite its written policies, actually encouraged energized work. Companies must control energized work carefully and ensure it is done to the highest safety standards.

SAFE WORK ZONES

A *safe work zone* is a barrier that is established to protect unqualified or unaware persons from entering an area that has electrical hazards. The phrase "safe work zone" is not specifically called for in OSHA regulations or NFPA 70E. However, it is implied because both documents require restricting access to areas where unqualified or qualified persons unaware of the hazardous work being performed may enter.

According to OSHA 1910.335(b), *"Alerting techniques. The following alerting techniques shall be used to warn and protect employees from hazards which could cause injury due to electric shock, burns, or failure of electric equipment parts: (1) Safety signs and tags. Safety signs, safety symbols, or accident prevention tags shall be used where necessary to warn employees about electrical hazards which may endanger them, as required by 1910.145. (2) Barricades. Barricades shall be used in conjunction with safety signs where it is necessary to prevent or limit employee access to work areas exposing employees to uninsulated energized conductors or circuit parts. Conductive barricades may not be used where they might cause an electrical contact hazard. (3) Attendants. If signs and barricades do not provide sufficient warning and protection from electrical hazards, an attendant shall be stationed to warn and protect employees."*

Similar requirements related to alerting techniques are provided in NFPA 70E 130.7(E). This section also includes an additional requirement for look-alike equipment. One of the three alerting techniques must be used with installations where there are similar types of equipment and devices in the same area as the one being placed in an electrically safe work condition. An example of look-alike equipment is a motor control center. There can be dozens, even hundreds, of motor control centers in the same location. It is likely that an employee may mistake one motor control center for another, causing a possibly hazardous situation.

An employee can also be put in a dangerous situation while working on circuit breakers in an outdoor substation. Circuit breakers do not make noise or give any indication that they are energized. It is possible that an employee may mistake an energized circuit breaker with a deenergized circuit breaker. Barricading the energized equipment and identifying it would help reduce the possibility of accidents.

Determining Safe Work Zones

Safe work zones are established at the Limited Approach Boundary or the Arc Flash Boundary, whichever is the greatest distance from the exposed energized conductors or circuit parts. For example, a 480 VAC circuit has a Limited Approach Boundary of 42″ and an Arc Flash Boundary of 31″. This means the safe work zone barrier would be set according to the Limited Approach Boundary because it is the greater distance from the electrical hazard. If the Limited Approach Boundary is 42″ and the Arc Flash Boundary is 48″, the safe work zone barrier would be set according to the Arc Flash Boundary because the arc flash hazard presents the greatest danger by having a longer distance than the Limited Approach Boundary. **See Figure 4-14.** Per NFPA 70E 130.7(E)(2), in instances where the Arc Flash Boundary is greater than the Limited Approach Boundary, barricades must not be located closer to the equipment than the Arc Flash Boundary.

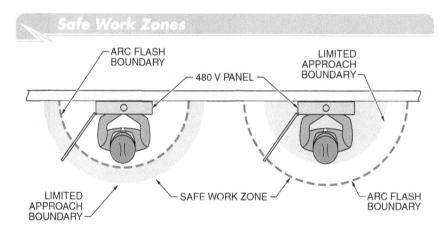

Figure 4-14. The Limited Approach Boundary or the Arc Flash Boundary, whichever is the greatest distance from the electrical hazard, determines safe work zones.

Establishing Safe Work Zones

Safe work zones are established by using safety barricade tape that is placed approximately waist high and supported by cones or stanchions. Hard barriers or attendants may be needed to restrict access if safety barricade tape is judged to be inadequate. Signs may also be needed to alert employees in the area of the hazards that may be present.

Safety Barricade Tape. The type of safety barricade tape used can be important, depending on the normal practices of a given company. Standard yellow and black safety barricade tape with the word "CAUTION" means "enter when aware of the hazards." However, this type of safety barricade tape is not recommended for work involving electrical hazards.

Typically, red is the most recognizable color to demonstrate that there is dangerous electrical work taking place. Red and black or red and white safety barricade tape with the word "DANGER" means "enter only if qualified and authorized." Orange and black safety barricade tape with the words "WARNING – ENERGIZED CIRCUITS – KEEP OUT" is also highly effective in communicating that dangerous electrical work is taking place. Most qualified and unqualified persons will avoid an area if they know there are exposed energized conductors or circuit parts located behind a barricade. **See Figure 4-15.**

Figure 4-15. Different types of safety barricade tape can be used to establish safe work zones, although standard yellow and black safety barricade tape with the word "CAUTION" is not recommended for work involving electrical hazards.

SUMMARY

- Approach boundaries must be defined to warn both qualified and unqualified persons that they are entering an area that contains electrical hazards.
- Safe work zones provide safe distances to protect unqualified or unaware persons from shock and arc flash hazards.
- Safe work zones are established by placing safety barricade tape, nonconductive chains or rope, or other barricades at either the Limited Approach Boundary or the Arc Flash Boundary, whichever is farther from the exposed energized conductor or circuit part.
- NFPA 70E specifies that an unqualified person must be continuously escorted by a qualified person while inside the Limited Approach Space and that both persons must wear the appropriate PPE.
- Determining shock and arc flash approach distances are a critical part of being a qualified person.
- Shock and arc flash can cause severe injuries and death for thousands of employees each year. Many of these injuries and fatalities are to unqualified persons, which is why OSHA and NFPA 70E both place so much emphasis on keeping unqualified persons out of dangerous situations.

Digital Learner Resources
ATPeResources.com/QuickLinks
Access Code: 705798

APPROACH BOUNDARIES FOR SHOCK AND ARC FLASH HAZARDS

Review Questions

Name _____ Date _____

_____ 1. An unqualified person may cross the ___ only if they are advised of the specific hazards and are continuously escorted by a qualified person.
   A. Prohibited Approach Boundary
   B. Restricted Approach Boundary
   C. Limited Approach Boundary
   D. Arc Flash Boundary

_____ 2. The Limited Approach Boundary for a 480 V circuit is ___″.
   A. 36
   B. 42
   C. 48
   D. 60

_____ 3. Per OSHA, clearance (safe approach) distances for qualified employees working on or near exposed energized electrical conductors or circuit parts can be determined by using Table ___.
   A. R-4
   B. R-5
   C. S-5
   D. S-6

_____ 4. Approach boundaries for DC voltages are determined by using NFPA 70E Table ___.
   A. 130.4(D)(a)
   B. 130.4(D)(b)
   C. 130.7(D)(15)(a)
   D. 130.7(D)(15)(b)

_____ 5. The primary factor that determines the Arc Flash Boundary is the ___.
   A. type of latch on the enclosure
   B. size of the enclosure
   C. maximum total clearing time of the OCPD
   D. method of system grounding

_____ 6. The short-ciruit current and ___ determine the incident energy for systems less than 600 V.
   A. voltage
   B. arc rating
   C. type of enclosure
   D. clearing time

T   F   7. The approach distances listed in OSHA Table S-5 are only used for energized overhead power lines.

T   F   8. The Limited Approach Boundary is the distance from an exposed energized conductor or circuit part at which a person can get an electric shock and is the closest distance an unqualified person can approach.

_____ 9. The ___ is the distance from exposed energized conductors or circuit parts where bare skin would receive the equivalent to 1.2 cal/cm².
   A. Arc Flash Boundary
   B. Limited Approach Boundary
   C. Prohibited Approach Boundary
   D. Restricted Approach Boundary

_____ 10. An incident energy analysis is used to determine ___.
   A. system voltage
   B. PPE for a specific task
   C. thermal energy at a working distance from an arc fault
   D. none of the above

_____ 11. An example of an energized electrical work permit is located in ___.
   A. NFPA 70E Informative Annex J
   B. NFPA 70E Informative Annex H
   C. NFPA 70E Informative Annex D
   D. NFPA 70E Informative Annex F

_____ 12. A factor that goes into determining incident energy is ___.
   A. system voltage
   B. available short-circuit current
   C. arc current
   D. all of the above

_____ 13. An energized electrical work permit is required by NFPA 70E for ___.
   A. performing energized electrical work
   B. tasks that are required to be performed by a certified electrical worker
   C. work on equipment rated less than 50 V
   D. troubleshooting electrical circuits

_____ 14. A(n) ___ is a document that describes the job planning needed to perform energized electrical work safely.
   A. energized electrical work permit
   B. risk assessment
   C. accident report
   D. incident energy review

_____ 15. The Arc Flash Boundary is the distance at which the incident energy equals ___ for all voltages.
   A. 1 cal/cm²
   B. 1.2 cal/cm²
   C. 1.5 cal/cm²
   D. 1.8 cal/cm²

T   F   **16.** Rubber insulating gloves are not required for voltages less than 300 V when there is a possibility of contact with energized conductors or circuit parts.

T   F   **17.** A safe work zone is established to exclude people who may be unqualified or unaware of a hazard.

_____ **18.** The Restricted Approach Boundary for a 4160 V circuit is ___″.
    A. 12
    B. 26
    C. 32
    D. 60

_____ **19.** The boundary used to determine the safe work zone is the ___.
    A. Arc Flash Boundary
    B. Prohibited Approach Boundary
    C. Limited Approach Boundary
    D. Both A and C

PERFORMING A RISK ASSESSMENT

NFPA 70E is mainly concerned with electrical hazards and requires that a risk assessment be performed when working on or near exposed energized conductors or circuit parts. A risk assessment includes a shock risk assessment and an arc flash risk assessment. OSHA is concerned with anything that may cause a hazard in the workplace. OSHA requires a risk assessment that covers all risks involved with a task, but this text only covers electrical risk assessments. Both OSHA and NFPA 70E call for a risk assessment, which is also called a job hazard analysis (JHA) or a job safety analysis (JSA), to protect workers from hazards encountered on the job.

OBJECTIVES

- Define risk assessment.
- Identify the recognized electrical hazards.
- Define the type of hazards OSHA refers to in Section 5(a), General Duty Clause.
- List the items OSHA directs a company to identify as part of a risk assessment.
- Describe the importance of maintaining overcurrent protective devices.
- List the items that should be considered when assessing the risk involved in a particular task.
- Explain the different methods of performing a risk assessment.

RISK ASSESSMENT

A *risk assessment* is a field document that accounts for all hazards. In previous editions of NFPA 70E, a risk assessment was referred to as a hazard/risk analysis. This was changed to harmonize with other standards. A shock risk assessment and an arc flash risk assessment are parts of a risk assessment. NFPA 70E 110.1(G) covers risk assessment requirements. Informative Annex F covers the risk assessment procedure in more detail.

A risk assessment will identify the hazards associated with the task to be performed, assess the degree of the risks, and list controls that can be implemented as outlined in ANSI/AIHA/ASSE Z10 and the ANSI/ASSE/ISO Risk Management Standards Package. In actual practice, a risk assessment accomplishes the same tasks as a hazard/risk analysis. OSHA regulations and NFPA 70E require that the risks and hazards be identified and eliminated or steps be taken to ensure they do not cause injury or death. Deenergizing the electrical equipment is required unless it would cause additional or increased hazards or is infeasible. When working on energized electrical equipment, it is necessary that the appropriate PPE, work methods, tools, equipment, and safe work practices are chosen and the residual risk is assessed once all the methods of control have been implemented. If the residual risk is still too great, the task cannot be performed as planned.

Risk Assessment—NFPA 70E 130.3

If the task a worker is about to perform will expose him or her to electrical hazards, safe work practices must be determined. In order to determine the safe work practices needed, a shock risk assessment and an arc flash risk assessment must be performed as part of the job planning. These risk assessments will determine the type and extent of the hazards, the approach boundaries, the

SAFETY FACT

NFPA 70E 130.5 requires that the arc flash risk assessment be updated at least every five years or whenever major modifications that could affect the results of the arc flash risk assessment occur.

degree of risk associated with the task, and the appropriate PPE. Once these have been determined, the appropriate safe work practices, such as using appropriate tools, work methods, alerting techniques, and other safeguards for personnel protection, can be determined.

NFPA 70E 130.3 also requires that the safe work practices used be consistent with the electrical hazards and the associated risk. Only qualified persons can work on exposed energized electrical conductors or circuit parts.

Electrical Hazards

Per OSHA's General Duty Clause, Section 5(a)(1), *"each employer shall furnish to each of his employees employment and a place of employment which are free from recognized hazards that are causing or are likely to cause death or serious physical harm to his employees."* A recognized hazard is a hazard that is recognized by the industry and OSHA, and it should be recognized by the employer. The recognized hazards of electricity are shock, arc flash, arc blast, and ignition of flammable clothing.

The magnitude of the shock hazard is determined by the phase-to-phase nominal voltage. The arc flash hazard is determined by the incident energy created by an electrical arc. Incident energy is expressed in cal/cm^2 at a specified working distance from the arc source. An incident energy analysis is conducted on an electrical power system to determine the incident energy at the working distance, the Arc Flash Boundary, and the appropriate arc-rated clothing and PPE. An arc flash hazard warning label is affixed to electrical equipment to provide this information to qualified workers. **See Figure 5-1.**

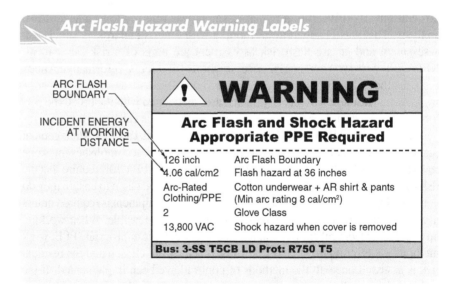

Figure 5-1. Arc flash hazard warning labels list the arc flash boundary, incident energy at the working distance, and the nominal phase-to-phase voltage. It may also list the Limited and Restricted Approach Boundaries and specific items of required PPE.

NFPA 70E 130.5(D) states that arc flash warning labels may include an incident energy analysis to determine PPE or the table method may be used, but not both. In the past, arc flash software companies have defaulted to the table method (hazard/risk categories) and have listed the HRC on the label to identify the arc-rated clothing and PPE required. This is not acceptable based on NFPA 70E requirements. The owner of the equipment is responsible for the accuracy of arc flash warning labels. As newer versions of software allow the correct information to be entered, it is the responsibility of the equipment owner to specify what information is entered on the label instead of the PPE category. Table H.3(b) in Informative Annex H should be used to select the appropriate PPE when an incident energy analysis is performed.

Nominal Voltage. Nominal voltage is the electrical system design voltage. This is the voltage that appears on nameplates and data plates of electrical equipment as well as on single-line diagrams and schematics. The nominal phase-to-phase voltage is used to determine the shock hazard. **See Figure 5-2.**

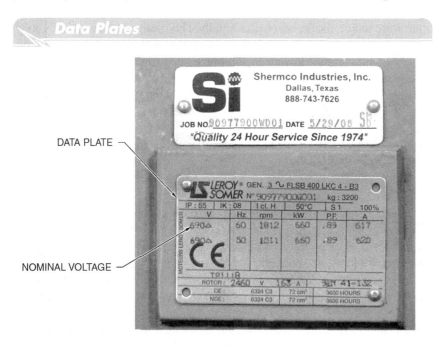

Shermco Industries

Figure 5-2. Nameplates, data plates, or single-line diagrams can be used to determine the nominal voltage of a circuit or system.

According to OSHA 1910.332(b)(3), *"Qualified persons (i.e. those permitted to work on or near exposed energized parts) shall, at a minimum, be trained in and familiar with the following: . . . (ii), the skills and techniques necessary to determine the nominal voltage of exposed live parts."*

OSHA uses nominal voltage instead of actual voltage because the accuracy of the voltage is not as much of a concern as the sequence that is followed when a qualified person is about to perform voltage measurements or other similar activities. If OSHA used actual voltage instead of nominal voltage, a worker would have to choose the voltage detector and rubber insulating PPE to be used before the voltage of the electrical system is known. This may lead to accidents if the worker chooses incorrectly. By using the phrase nominal voltage, OSHA is directing the qualified worker to determine the system's design or operating voltage before choosing the voltage detector and rubber insulating PPE.

A typical nameplate will list the nominal voltage and other ratings. The nominal voltage in a given part of an electrical power system does not change. The approach distances for a specific range of nominal voltages also always remain the same. For example, Class 00 rubber insulating gloves would be appropriate for a 480 V shock exposure. They would be an appropriate choice for any 480 V electrical system or equipment. As long as the phase-to-phase voltage exposure does not exceed 480 V nominal, this set of gloves can be used. If the nominal voltage exceeds the maximum use voltage specified for Class 00 gloves, Class 0 gloves may be required.

Arc Flash Hazards

One of the basic arc flash rules is that incident energy decreases by the inverse square of the distance from an arc source. In other words, as a worker moves away from an arc source, the heat decreases rapidly. This means that the worker's body position is important when performing tasks on energized conductors or circuit parts. If a worker moves toward an arc source, the incident energy will increase by the square of the distance, meaning it will increase rapidly.

The recommended working distance from a potential arc source is provided in IEEE 1584. The working distance for most low-voltage tasks is 18″. This distance is measured from an arc source to the chest and face area of a worker with arms extended a comfortable length in front while using hand tools. A distance of 24″ is typically used for racking low-voltage power circuit breakers in and out of their cubicles.

A working distance of 36″ is typically used for medium-voltage systems up to 15 kV. This distance is measured from a potential arc source to the worker's chest and face area while using a live-line tool. Typically, a worker using a 36″ live-line tool will not be 36″ away from the arc source because his or her hand position on the live-line tool will use up approximately 12″ to 18″. A 48″ live-line tool is required to move the worker 36″ away from the arc source. The best practice is to use the actual working distance to calculate incident energy, especially if it is known that the distance will be greater than or less than the typical 36″.

Incident energy is proportional to time. For example, if the time of exposure is doubled, the incident energy also doubles. This is why worker safety is a major concern when dealing with the maintenance of OCPDs. If the recommended maintenance is not performed on OCPDs, they are less likely to perform in accordance with the manufacturer specifications or the time-current characteristic curves. The equations for arc flash hazard calculations are based upon the OCPD operating within the manufacturer specifications. The lack of maintenance can cause the operating time to greatly increase. In some cases, the OCPD may not operate at all.

Condition of maintenance, especially the condition of maintenance of OCPDs, has been a continuing concern among NFPA 70E Committee members. Each cycle, the committee adds more direction on this subject. Section 130.5 Informational Note No. 1 states that inadequate or improper maintenance can cause the operating time of an OCPD to increase, which increases the incident energy exposure to the worker. In the 2015 edition, an additional caution was added to Informational Note No. 1 stating that improperly installed or maintained equipment may cause the selected arc-rated PPE and clothing to be inadequate. NFPA 70E 130.5 Informational Note No. 4 refers the user of the standard to Chapter 2, Safety-Related Maintenance Requirements, for direction on maintaining electrical equipment and OCPDs.

SAFETY FACT

An arc flash hazard is expressed as incident energy. About half of the heat from an arc flash is radiated heat, and half is convection heat. Cal/cm² is used because it is a unit of measure that can be easily related to skin burn models to prevent permanent burns.

NFPA 70E 130.7(A) Informational Note No. 2 states that the committee believes that equipment that has been properly installed and maintained exposes the worker to no increased risk when operated normally. This also implies that if the equipment is not properly installed and maintained, then it is not safe to operate. The notes for Table 130.7(C)(15)(A)(a) were added in the 2015 cycle, along with a new table, and define what is meant by "properly installed," "properly maintained," and "evidence of impending failure." For most tasks listed in Table 130.7(C)(15)(A)(a), arc flash PPE is not mandated if the equipment is properly installed and properly maintained, all covers and doors are installed and secured, and there is no evidence of impending failure.

NFPA 70E 200.1 Informational Note refers the user of NFPA 70E to three industry consensus standards for maintenance: NFPA 70B, *Recommended Practice for Electrical Equipment Maintenance,* ANSI/NETA MTS-2011, *Standard for Maintenance Testing Specifications for Electrical Power Distribution Equipment and Systems,* and IEEE 3007.2-2010, *IEEE Recommended Practice for the Maintenance of Industrial and Commercial Power Systems.* Section 205.3 states that electrical equipment is to be maintained in accordance with the manufacturer's recommendations or industry consensus standards. Section 205.4 states the same thing for OCPDs and adds that inspections, tests, and maintenance are to be documented.

One of the discussions the NFPA 70E Committee has had on a continual basis is that if the requirements of Chapter 2 for maintenance are not being met, it would not be possible to meet the requirements of Chapter 1. If an OCPD and associated electrical equipment are not maintained, there is no method to determine how long it will take the OCPD to operate.

FIELD NOTES:
CIRCUIT BREAKER OPERATING TIME

A technician pressed the TRIP button on a circuit breaker and nothing happened. He went to his toolbox to get tools and when he was on his way back to the circuit breaker, it tripped open. This delay probably lasted 45 to 90 seconds. The effect this delay would have on the incident energy exposure to a worker would have been tremendous and there would be no way to determine what arc-rated clothing and PPE would be adequate for that level of hazard. OCPDs and associated equipment must be properly maintained in order to determine appropriate arc-rated clothing and PPE.

Assessing Hazards

According to OSHA 1910.132(d)(1), *"The employer shall assess the workplace to determine if hazards are present, or are likely to be present, which necessitate the use of personal protective equipment (PPE). If such hazards are present, or likely to be present, the employer shall: (i) Select, and have each affected employee use, the types of PPE that will protect the affected employee from the hazards identified in the hazard assessment; (ii) Communicate selection decisions to each affected employee; and, (iii) Select PPE that properly fits each affected employee. Note: Non-mandatory Appendix B contains an example of procedures that would comply with the requirement for a hazard assessment."*

SAFETY FACT

Since November 15, 2007, OSHA has required employers to pay for all PPE that is required for employees during any job. Prior to this final ruling, OSHA specified that the employer must pay for only some PPE. The final ruling gives employers specific direction to pay for all PPE, since some employees are not able to afford the costs or may choose incorrect PPE based on the costs.

OSHA 1910.132 describes how employers are responsible for the proper PPE that employees use. This regulation directs employers to complete the following:
- Perform a hazard assessment if there are hazards requiring the use of PPE.
- Select and provide PPE that is appropriate for the hazard.
- Ensure that employees are using the PPE they have received.
- Inform each employee of the required PPE for the hazards that they will encounter.
- Ensure that the PPE fits each employee properly.
- Refer to OSHA 1910 Subpart I Appendix B for example procedures for a hazard assessment.

In order to choose the proper PPE and equipment, a hazard assessment (risk assessment per NFPA 70E) must be performed. If the magnitude of the hazard is not known, appropriate PPE cannot be chosen. OSHA 1910.132 applies to all tasks requiring PPE, not only electrical tasks. However, OSHA 1910.132 is still closely associated with OSHA 1910.335.

According to OSHA 1910.335(a)(1)(i), *"Employees working in areas where there are potential electrical hazards shall be provided with, and shall use, electrical protective equipment that is appropriate for the specific parts of the body to be protected and for the work to be performed."* The key word in OSHA 1910.335(a)(1)(i) is "appropriate." In order to select the appropriate PPE, a risk assessment, also known as a job hazard analysis (JHA) or job safety analysis (JSA), must be performed.

General Duty Clause

According to the General Duty Clause, Section 5(a)(1) of the Occupational Safety and Health (OSH) Act, *"Each employer—(1) shall furnish to each of his employees employment and a place of employment, which are free from recognized hazards that are causing or are likely to cause death or serious physical harm to his employees; (2) shall comply with occupational safety and health standards promulgated under this Act."*

The General Duty Clause is a general regulation that has proven to be open to interpretation by federal courts. Often, it cannot be used alone to confirm a citation. A complement to the General Duty Clause is NFPA 70E. Since NFPA 70E is prescriptive, OSHA compliance officers can use it to determine the specific work practices or PPE that should be used for abatement of a citation.

Written Hazard Assessment

According to OSHA 1910.132(d)(2), *"The employer shall verify that the required workplace hazard assessment has been performed through a written certification that identifies the workplace evaluated; the person certifying that the evaluation has been performed; the date(s) of the hazard assessment; and, which identifies the document as a certification of hazard assessment."*

OSHA 1910.132(d)(2) requires a written hazard assessment that is certified. The four items that OSHA directs an employer to identify as part of a hazard assessment include the following:
- the workplace that has been evaluated
- the person that has certified the evaluation
- the date(s) the assessment was performed
- the document as a hazard assessment

SAFETY FACT

The terminology in NFPA 70E and OSHA regulations is sometimes different. This is because NFPA 70E is updated every three years, whereas the OSHA regulations are not. For example, OSHA 1910.331 through .335 have been unchanged since 1990.

These OSHA requirements clearly identify the need for a hazard/risk assessment. The assessment is required initially, as well as each time any type of hazardous work is to be performed. The initial assessment should be broad, determining the PPE and procedures that may be required for all work and areas within the job site. The assessment must also cover the hazards and risks involved when performing work on energized conductors and circuit parts.

Maintaining Overcurrent Protective Devices

The routine maintenance of OCPDs can be critically important for employee safety. If the recommended maintenance is not performed on OCPDs, they may not perform in accordance with the manufacturer specifications or the time-current characteristic curves. Lack of maintenance on an OCPD can cause its operating time to increase. In some cases, the OCPD may not operate at all. This can create a serious increase in the incident energy. If the incident energy is increased, PPE chosen based on the arc flash hazard warning label may be insufficient.

Lack of Lubrication. According to several studies from NETA, IEEE, and Shell Oil Company, inadequate lubrication is the number one cause of OCPD failure. Inadequate lubrication is an example of not meeting manufacturer specifications. Components within circuit breaker contact assemblies are lubricated by the manufacturer. Over time this lubrication dries out. **See Figure 5-3.**

Shermco Industries

Figure 5-3. Over time, the lubrication applied to circuit breakers and switches dries out. Equipment can be damaged if circuit breakers are not properly maintained.

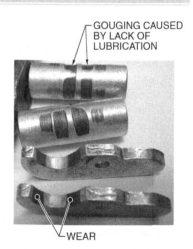

Shermco Industries

Figure 5-4. A lack of maintenance can damage critical circuit breaker components such as the moving contact assembly of a low-voltage circuit breaker. This can cause a circuit breaker to operate slowly and occasionally seize during operation.

Initially, the lubricants become thicker, causing the circuit breaker to slow down. If allowed to progress, the lubricants will dry completely and flake off the components, and metal-to-metal wear will begin. **See Figure 5-4.** At some point the circuit breaker can seize, and the contacts will no longer open. This can create a safety issue for workers operating the circuit breaker or any equipment being fed by it. The lubricants dry out due to the current flow through the contact assemblies. As current flows through a conductor, heat is produced. The greater the flow of current through the conductor, the more heat that is generated.

The study conducted by NETA involved 340,000 responses from NETA-member companies from around the U.S. The study showed that 22% of the circuit breakers tested in the field had defective OCPDs, 43% had mechanical issues, and 11% would not function at all. Most of the circuit breakers in the mechanical failure and DNF categories failed due to lack of lubrication.

Shell Oil Company did a system-wide study on their circuit breakers. In the study, they found that after a circuit breaker had been in service for three to five years without being operated, 30% of them did not meet manufacturer specifications when tested. After being in service for seven to ten years, 50% of the circuit breakers did not meet manufacturer specifications. If the circuit breakers had been in service for 17 to 20 years, the failure rate was above 90%. These were circuit breakers that had not been operated or tripped during that time period.

IEEE 493, *IEEE Recommended Practice for the Design of Reliable Industrial and Commercial Power Systems* (Gold Book), contains the results of a survey on equipment failure versus time in service. The study found that circuit breakers with less than 12 months of service after maintenance had a failure rate of 12.5%. Circuit breakers between 12 months and 24 months of service after maintenance had a failure rate of 19%. After 24 months, the failure rate was almost 78%. The longer a circuit breaker is in service without operating or having maintenance performed on it, the greater the probability of failure.

Corrosion. As the load on a circuit breaker cycles, the constant heating and cooling will produce a small amount of corrosion within the breaker. This corrosion can affect the release of the trip latch of molded- and insulated-case circuit breakers. The trip latch is what causes the circuit breaker to trip to the OPEN position.

FIELD NOTES: LACK OF MAINTENANCE

The circuit breaker repair shop at Shermco Industries sees an astonishing number of circuit breakers and switches of all voltage ratings suffering from a lack of lubrication, which is the number one cause of failure to operate. Many equipment owners do not realize that the conductive (current-carrying) paths of circuit breakers and many switches are lubricated when they are built by the manufacturer and that the lubrication dries out over time, even when the breaker (or switch) is not operated.

Each year, Shermco technicians calibrated the protective relays at a customer's job site, but they did not perform maintenance on the circuit breakers. When asked about the circuit breakers, the customer usually responded, "Those breakers never operate. They're like brand new." Several in-depth explanations about the problems caused by drying lubrication failed to change the customer's stance, so no breaker maintenance was performed for this company.

Several months after one of the times their protective relays were calibrated, a 13.8 kV underground feeder cable failed and the fault cascaded through six breakers before it was finally interrupted. All the relay flags were dropped, which means the relays saw the problem but the circuit breaker's operating mechanisms had frozen and did not operate. The entire section of switchgear had to be replaced at a substantial cost in excess of $750,000. The customer's insurance carrier would not pay for this lost revenue since the circuit breakers were not maintained.

Effect of Time on Incident Energy. Suppose a panel is protected by an upstream 800 A circuit breaker that has a short-time delay setting of 6 cycles. The 6 cycle delay is an intentional delay feature of the circuit breaker that allows it to selectively coordinate with the circuit breakers in the panel it is supplying. If the circuit breaker has been maintained and operates in accordance with the manufacturer specifications, the incident energy would be 5.8 cal/cm^2 at 18″. The Arc Flash Boundary would be 47″. A worker wearing arc-rated PPE for a maximum incident energy of 10 cal/cm^2 should be protected if the circuit breaker has been properly maintained.

However, if the circuit breaker has not been properly maintained, it may take longer to interrupt an arcing fault. The longer the circuit breaker takes to open, the greater the actual incident energy would be. If the circuit breaker clears in 30 cycles rather than 6 cycles due to improper maintenance, the actual incident energy would be 29 cal/cm^2 at 18″. The Arc Flash Boundary would increase to 125″. **See Figure 5-5.** A worker wearing arc-rated PPE for a maximum incident energy of 10 cal/cm^2 would be underprotected and would suffer serious burn injuries if an arc flash were to occur. The change in incident energy from 5.8 cal/cm^2 to 29 cal/cm^2 would break open the protective garment and expose the worker and any underlayers, causing possible ignition if the underlayers are not arc rated. Typically, an incident energy of 29 cal/cm^2 will break open two arc-rated layers rated for 10 cal/cm^2.

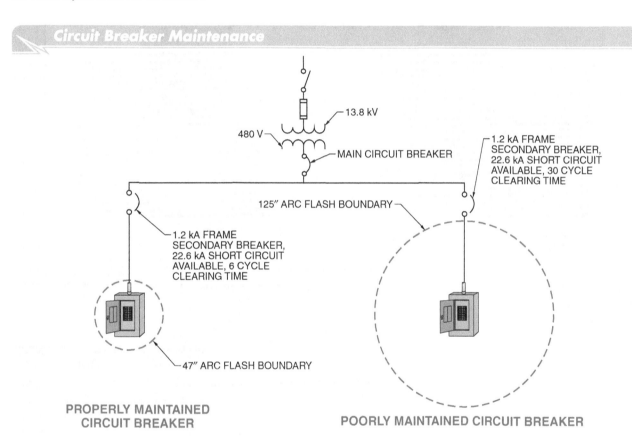

Figure 5-5. A circuit breaker that is not properly maintained may take longer to interrupt an arcing fault, which in turn creates a greater Arc Flash Boundary.

If there is any question about the ability of an OCPD to function properly, it is safer as part of the risk assessment to assume that the OCPD will not function at all. It is safer because the incident energy and Arc Flash Boundary cannot be accurately estimated. In this case, it is best to deenergize the circuit. The time spent in a planned outage would be far less than what would be required to repair and replace damaged equipment, and deenergizing the circuit would create a safer work environment for the workers. A planned outage for maintenance may also be less costly than the repairs necessary to fix extensive damage to the circuit breakers. **See Figure 5-6.**

Damage Due to Improper Circuit Breaker Maintenance

DAMAGED MEDIUM-VOLTAGE CIRCUIT BREAKER DUE TO SEIZED MECHANICAL COMPONENTS

DAMAGED CHAIN OF CIRCUIT BREAKER LIFT MECHANISM

DAMAGED SECTION OF SWITCHGEAR

Figure 5-6. Repairs to equipment due to improper maintenance may be far more expensive than the cost to properly maintain an electrical system.

If it is decided that the circuit or equipment cannot be placed in an electrically safe work condition, the next device upstream that is deemed reliable has to be considered as the operational protective device. For example, if an OCPD is not maintained, the next properly maintained OCPD upstream is used to determine the arc flash hazard. It is probable that, due to the increase in operating time, the incident energy will be substantially higher. If the next OCPD upstream is not reliable, the device that is upstream from it is used to assess the arc flash hazard. This may mean going upstream to the utility OCPD.

Short-Time Delay. Another factor in determining the arc flash hazard is whether the electrical system uses circuit breakers that have a short-time delay (STD) function. The STD function is used to provide an intentional delay for a short-circuit condition for the purpose of selective tripping. Selective tripping, also known as protective devices coordination, is used to ensure that the OCPD closest to the load will operate first. If the closest OCPD does not operate, the next OCPD upstream from it will.

One of the basic principles of protective device coordination is that two devices with instantaneous functions placed in series cannot be coordinated. If the short circuit available current is high enough, both will trip. **See Figure 5-7.**

Power System Using Circuit Breakers with an Instantaneous Trip Function

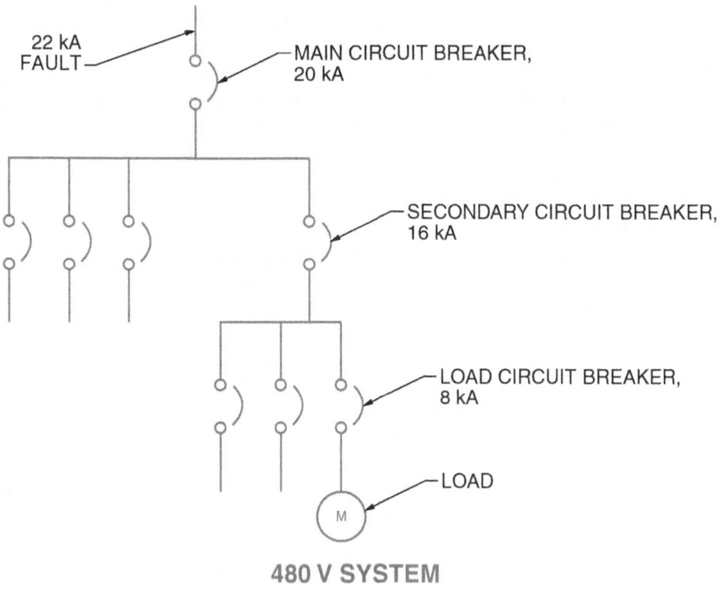

[Diagram: 480 V SYSTEM showing 22 kA FAULT feeding into MAIN CIRCUIT BREAKER (20 kA), branching to SECONDARY CIRCUIT BREAKER (16 kA), then to LOAD CIRCUIT BREAKER (8 kA), feeding a motor LOAD.]

Figure 5-7. An entire electrical system is at risk of shutting down if a fault occurs on a load in a system using instantaneous trip devices.

In this example, a main circuit breaker, a secondary circuit breaker, and a load circuit breaker are essentially in series, and all three have instantaneous functions. The main circuit breaker has the instantaneous minimum pickup set at 20 kA, the secondary circuit breaker is set to 16 kA, and the load circuit breaker is set to 8 kA.

It may seem that the load circuit breaker would be more likely to operate than either of the others, but that may not be true. If the load being fed by the load circuit breaker fails and draws 22 kA, the short circuit current will flow through the main circuit breaker first. Since the instantaneous function has no intentional time delay, it would immediately unlatch and begin to trip. The short circuit current may then flow through the secondary circuit breaker, which also has no intentional time delay, and unlatch. There is a very good chance that the short circuit current will be interrupted before it reaches the load circuit breaker. This is exactly what is not wanted, because at this point the entire facility would shut down.

When an STD function is used on the main incoming and secondary circuit breakers, a small time delay is required before they trip. As short-circuit current flows through the main incoming and secondary circuit breakers, they begin to respond but cannot trip until they time out. The load breaker, which has an instantaneous trip, operates without any intentional time delay and interrupts the short circuit. Then, the main and secondary circuit breakers both reset. **See Figure 5-8.**

In the event that the load breaker does not operate, the secondary circuit breaker will operate after a 0.18 sec nominal delay. The 0.18 sec nominal delay is located at the center of the time band for the trip device and the actual tripping time could vary. If the secondary circuit breaker does not operate, then the main incoming circuit breaker will operate after a 0.33 sec nominal delay. As the fault cascades into the system, more and more of the power system is shut down. The entire power system may need to be shut down to interrupt the fault.

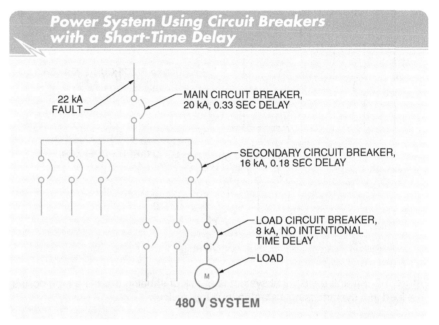

Figure 5-8. Short-time delay circuit breakers are used as main and secondary circuit breakers to prevent an entire system from shutting down when a fault on the load occurs.

Although an STD allows proper selective tripping for a power system, there is a disadvantage. If the tripping is delayed for the main incoming and secondary circuit breakers, a fault will last longer than the normal 4 cycles (0.067 sec) for the instantaneous function on the circuit breaker in this example. Because incident energy is proportional to time, the incident energy will be increased in proportion to the delay time of the STD function. This causes safety concerns for personnel working on the part of the power system that is protected by these devices.

The safety concerns caused by the STD function can be somewhat resolved by the use of a maintenance switch. A maintenance switch temporarily allows an instantaneous function to override the STD function, providing no intentional time delay. There are two primary drawbacks to a maintenance switch: the technicians have to remember to switch it into service and they have to remember to switch it to the normal STD condition. If these conditions are not met, the system could be subject to nuisance tripping. There are also other methods available to make systems safer, such as arc flash detection systems, differential protective relaying, and zone selective interlocking (ZSI).

Determining the Hazard

A hazard is created by a combination of the short circuit available current, the operating time of the OCPD, and the voltage of the electrical power system. When an electrical power system is energized, the hazard never changes. The amount of risk involved in performing a task, however, does change.

For example, reading the meter on a 15 kV switchgear door and racking a circuit breaker out of the same piece of equipment presents the same hazard. The difference between the two is the amount of risk involved with each task. Arc flash hazard warning labels provide information on the hazard, but they do not indicate the level of risk. This is why a risk assessment is an important element whenever work is to be performed on energized electrical equipment or devices.

Determining Risk

Sometimes determining the hazard can be easier than determining the risk. There are no calculations for determining risk, but NFPA 70E Informative Annex F contains a matrix that may assist in determining risk. Risk assessment is primarily based on judgment.

The conditions at a specific job site must be addressed at the time the work is being performed. When assessing the risk associated with a task, some of the issues that need to be considered include the following:

- The condition of the equipment being worked on
- The operating environment—Equipment that is in a positive-pressure, filtered, and air-conditioned area is going to deteriorate much slower than weather-exposed equipment.
- The last time the equipment was maintained and calibrated—Protective relays and circuit breakers that have calibration stickers older than three years generally cannot be relied on to function properly.
- The age of the equipment and its general design—The design of the venting may unintentionally direct hot gases toward personnel. Low-voltage power circuit breakers and insulated-case circuit breakers can be designed as drawout-type circuit breakers. They are usually inserted or removed using a screw-drive racking mechanism, but they can use a levering-type mechanism. Circuit breakers using a screw-drive racking mechanism can often be inserted or removed remotely, while the levering-type mechanism requires the worker to be positioned directly in front of the circuit breaker.
- The performance history of the equipment—Any equipment problems or breakdowns, as well as technical bulletins from the manufacturer, must be noted.
- The history of failure of the equipment—Past incidents involving equipment failure or other severe issues must be assessed.
- The potential for backfed or induced voltages
- The loading of the equipment—Due to the heat generated from current, heavily loaded equipment deteriorates more quickly than lightly loaded equipment. Also, on some electrical power systems, one tie circuit breaker must be closed before the other is opened. If both tie circuit breakers are closed at the same time, the incident energy could be increased greatly.
- The design of the power system—The power system may be a ring bus, radial bus, double-ended sub, or main-tie-main. Properly isolating each type of power system requires careful consideration to ensure the equipment to be worked on is in an electrically safe work condition.
- The skill level of the personnel doing the work—Highly experienced and qualified workers should be performing the tasks. They should be familiar with the equipment manufacturer's operating instructions.

SAFETY FACT
Regardless of the method used to determine PPE, a risk assessment must always be performed prior to the start of each task.

When the risk associated with performing a specific task is properly evaluated, a qualified person may find that additional PPE is often necessary. The issues listed above must be considered every time energized electrical work is performed. When using arc flash hazard warning labels, it may be possible to lower the amount of required PPE based on the risk assessment.

There are situations where the hazard may be high but the risk may be low, such as when performing a thermographic infrared (IR) scan. Typically, IR scans are noninvasive and are often performed at a distance of 3′ or more from the equipment being scanned. A worker is typically outside the Restricted Approach Boundary of the switchgear during IR scans, except when removing the panels. Even though the 2015 edition of NFPA 70E allows a thermographer to not wear arc-rated clothing and PPE, the risk of injury as a result of an arc is still present. Arc-rated PPE of some level is recommended by the author. **See Figure 5-9.**

NFPA 70E Table 130.7(C)(15)(A)(a) shows that performing the actual infrared scan does not require arc-rated PPE or clothing, provided the thermographer does not remove the panels or covers and does not cross the Restricted Approach Boundary. It may still be prudent to install IR windows in areas that have limited egress or tight quarters.

For example, a piece of equipment has an arc flash hazard warning label that indicates an incident energy exposure of 18.7 cal/cm^2 at an 18″ working distance. The qualified person performing an IR scan would typically wear 25 cal/cm^2 arc-rated protective clothing and PPE. However, due to the working distances involved with the IR task, the qualified person can be outside the Restricted Approach Boundary. Eliminating arc-rated PPE for the IR scan operator could be an option, based on the risk assessment. However, the qualified person removing the equipment panels would be at a higher level of risk due to proximity to the exposed parts and would have to wear full arc-rated PPE and clothing that is appropriate for the hazard. Also, there is the risk of dropping tools into the equipment. Whether the NFPA 70E tables are used or an incident energy analysis is performed, a risk assessment is always performed.

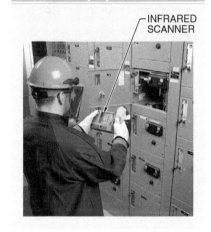

Fluke Corporation

Figure 5-9. IR scanning often occurs at a greater distance than the working distance specified on the arc flash hazard label.

METHODS FOR PERFORMING A RISK ASSESSMENT

Three methods can be used to perform a risk assessment. One method used to perform a risk assessment is to use the risk assessment procedure located in NFPA 70E Informative Annex F. The second method involves completing forms. The third method uses Appendix E in ANSI/AIHA/ASSE Z10.

NFPA 70E Informative Annex F

A risk assessment can be performed by using the risk assessment procedures and flow charts in NFPA 70E Informative Annex F. NFPA 70E 110.1(G) requires an electrical safety program (ESP) to identify the risk assessment procedure to be used before working within the Limited Approach Boundary. According to NFPA 110.1(G) Informational Note No. 2, the risk assessment procedure may identify when a second person is required to act as a safety backup, as well as the training and PPE or other equipment that person may require. NFPA 70E Informative Annex F provides an example of a risk assessment procedure and flow chart.

FIELD NOTES:
RISK ASSESSMENT PROCEDURE

Studies have shown that workers in companies with written electrical safety programs (ESPs) were more likely to wear the proper PPE than workers in companies that do not have written ESPs. In 2005, Michael McCann performed a study entitled *Live Work Practices Among Construction Workers*. In cooperation with the IBEW, the study surveyed 5000 electrical workers about their work practices and electrical safety programs. The study found that only 48% of electrical workers had a written ESP, and of those who did, many did not specify what PPE was to be used or when it should be worn. A risk assessment procedure ensures that the hazards associated with performing the task are properly identified and evaluated and that appropriate steps are taken to reduce the risk to a level where workers will not be seriously injured or killed.

Informative Annex F has evolved into more of a risk assessment guide than a risk assessment evaluation procedure. Risk is difficult to quantify because it is largely subjective. The process outlined in Informative Annex F will provide guidance to assessing the risks associated with each task.

The Risk Assessment Process flow chart (Figure F.1(a) in NFPA 70E) lists the typical steps for performing a risk assessment. Each step in the flow chart references the appropriate NFPA 70E or NEC® section requiring and explaining the need for it. The Risk Assessment Process flow chart contains the following elements:

- hazard identification
- identifying the tasks to be performed
- documenting the hazards associated with each task
- estimating the risk for each task
- determining protective measures and equipment needed to reduce the risk for each task
- assessing the residual risk
- determining whether it is safe to perform the task

NFPA 70E describes four risk factors that include the following:

- Severity (Se)—Severity involves estimating the extent of possible injuries and has four levels ranging from death to minor bruising or lacerations.
- Frequency (Fr)—Frequency is how often a hazardous task is performed and has five levels ranging from ≤ 1 per hour to > 1 per year.
- Probability (Pr)—Probability involves estimating how likely an injury would be if an accident were to occur and has five levels ranging from very high to negligible.
- Probability of Avoidance (Av)—Probability of Avoidance is how likely an individual is to avoid or limit harm during performance of a task. There are three levels of avoidance: impossible, rare, and probable.

These values are entered into the appropriate columns in Table F.2.5 as part of the initial risk assessment. Unlike the 2009 edition of NFPA 70E, the 2015 edition does not use the values except to provide a general sense of severity for the evaluation of the risk factors given. Informative Annex F also provides several risk-reduction strategies that include engineering controls such as barriers, awareness devices such as signs and alarms, written procedures for specific tasks, training in accordance with OSHA 1910.332 through .333 and NFPA 70E 110.2, and PPE appropriate for the task and specific body parts to be protected. Each of these strategies should be considered and implemented for each task being assessed.

Section F.4 of Informative Annex F provides detailed information on risk evaluation. Section F.5 covers risk reduction verification. Once the risk reduction steps have been implemented, it is important to gauge their effectiveness. Verification of their effectiveness can be accomplished through auditing. Auditing must be performed prior to starting the work to ensure no person is placed at undue risk. An example audit is given in Figure F.5.2. Known or possible hazards are listed in the left-hand column, the corrective action is provided in the center column, and confirmation of implementation is checked off in the right-hand column. Even after every step has been taken to reduce risk, there will always be some residual risk because no task is risk-free. A determination has to be made at that time as to whether it is safe to proceed with the task or if additional measures need to be implemented.

Informative Annex F was completely rewritten in the 2012 edition and should be reviewed carefully. There is considerably more detail on performing risk assessments and the risk assessment process. The new wording should clarify some areas that were not fully detailed in the 2009 edition.

SAFETY FACT

The forms located in the appendix were created by Tony Demaria and Gary Donner of Tony Demaria Electric, Inc., Los Angeles, CA. They are examples of forms that can be used to perform a risk assessment.

Forms for Performing a Risk Assessment

Forms may be developed as an alternate method for performing a risk assessment. **See Appendix.** The forms in the Appendix include a Risk Assessment Form, Energized Electrical Work Permit, Lockout/Tagout Switching and Grounding Procedure, Risk Assessment Matrix, and Grounding of Equipment for Personal Safety Policy and Procedure. Most of these forms are in checklist format to reduce the amount of handwriting. If the task or other information is not listed, open lines allow specific information to be entered. *Note:* The forms in the Appendix are examples that were developed based on a specific type of facility and may not be applicable for all facilities.

Risk Assessment Form. Although most risk assessment forms contain similar information, the specific needs for the type of work being performed in a facility can be met by adding or removing information. **See Figure 5-10.** This figure shows an example risk assessment form for a typical petrochemical facility. The following items are to be completed on the risk assessment form:
- Enter a brief description of the work scope, which is required at the top of the form. Mental planning and the planning process begin with this item. As the different tasks are entered, planning begins for determining what tools, equipment, and PPE will be needed at each step.

- Determine if an energized electrical work permit is needed. It is important to note that any of the tasks listed below must have an energized electrical work permit completed.
- Determine if lockout/tagout is required.
- Identify the common tasks that are performed on a regular basis. Beside each task is a listing of known hazards, as well as a reminder to the worker about the consequences of performing that task. The last column provides guidance based on NFPA 70E for typical PPE requirements or other means of minimizing or eliminating the hazard.
- Receive the various signatures that are required at the bottom of the form and list all of the workers involved in the task. The signatures required may vary according to the needs or requirements of the customer. At a minimum, however, the signatures included in the example risk assessment form must be completed.

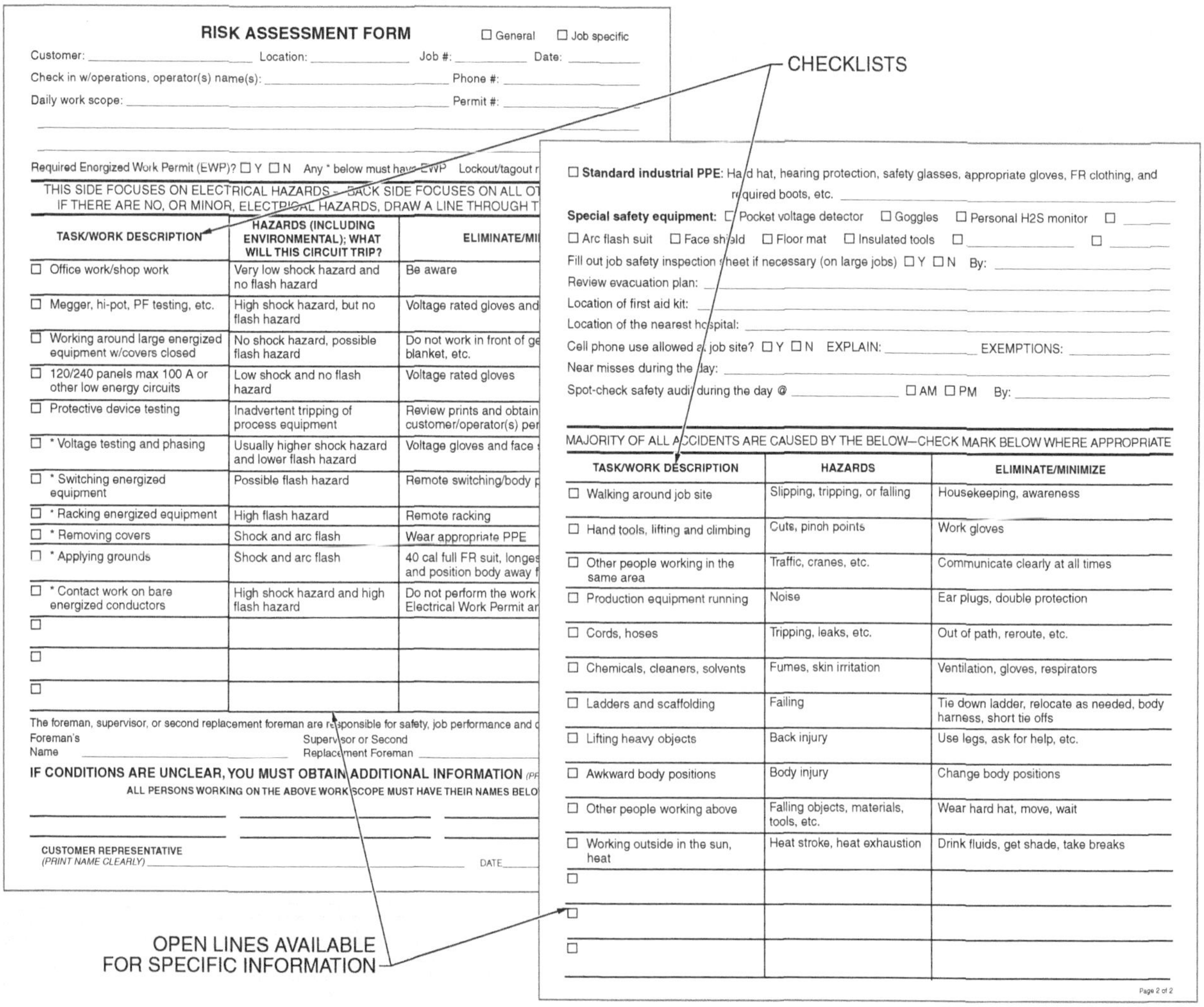

Figure 5-10. Risk assessment forms should be in a checklist format and, if possible, limited to one page (front and back) to make it faster and easier for employees to complete.

The other side of the example risk assessment form covers most common hazards at a typical petrochemical facility. At the top is a list of standard PPE that is required at all times in the petrochemical facility. The next item down lists special safety equipment for electrical work. Other types of facilities may require information to be added or removed to suit their needs. Extra spaces allow the user to make the form job-specific. Other requirements listed on this side include the following:

- Provide a review of the evacuation plan. If it becomes necessary to leave the area in an emergency, the recommended route as well as assembly areas will be important.
- Provide the location of a first aid kit. The safety kit needs to meet the requirements of OSHA 1910.151 and .269(b).
- Provide the location of the nearest hospital and note any special phone numbers or other directions required for emergency aid.

Energized Electrical Work Permits. An *energized electrical work permit* is a document that describes the job planning needed to perform energized electrical work safely. **See Appendix.** The Energized Electrical Work Permit form is an example form developed for use by electrical test and maintenance companies and may require modification to fit specific requirements depending on the company. This permit is used for systems less than 600 V. Informative Annex J in the NFPA 70E provides a more generic energized electrical work permit.

The top of the example energized electrical work permit is where general information is provided, such as the name of the customer, the location, job number, date, and the equipment or circuit being worked on. For example, the note beneath the general information may read, "The goal is to perform work deenergized. When deenergizing use remote switching and racking, and when grounding use longest hot stick as practical." Employees should take advantage of every opportunity to put more distance between themselves and the potential arc source. Other items on the energized electrical work permit form include the following:

- Justification why the equipment/circuit cannot be deenergized—Listed in this section are some common reasons why it may not be feasible to deenergize. If the listed reasons do not fully explain why the energized electrical work must be performed, the box marked "Other" can be checked off and a full explanation provided. However, the reasons that attempt to justify work on energized conductors or circuit parts are often unacceptable by OSHA regulations. Work on energized conductors and circuit parts should be performed only as a last resort.
- Minimum required PPE—Listed in this section is the PPE that is required for energized electrical work.
- Shock hazard analysis—The nominal voltage, type of voltage detector or tester, and working distance are required in this section. The employee must also check to see if there are other possible shock hazards nearby.
- Shock Protection Boundary and PPE—Gloves are required between 50 V and 1000 V. The box for the appropriate glove class should be checked off. If there is a reason for a different glove class, the box marked "Other" can be checked off and the appropriate glove class listed in the blank space provided.
- Arc flash hazard analysis—A minimum 40 cal/cm^2 flash suit is required when racking circuit breakers, removing or replacing covers, inserting

or removing motor control center (MCC) buckets, or applying personal protective grounds. The requirements for PPE follow the minimum requirements of NFPA 70E. It should be noted that when a risk analysis is performed, it might require that PPE with a higher arc rating than the minimum be used.

SAFETY FACT

Emergency responders may experience difficulty finding their way around a large facility or a facility with multiple buildings. To achieve the quickest response possible, workers may be required to call a special emergency number instead of 911. This number is usually used to contact security personnel and provide them with information such as a building number or doorway. Security personnel can then contact emergency responders and quickly lead them to the specific site of an emergency.

Arc flash protective clothing and PPE of 8 cal/cm^2 is required for control circuits, heater circuits, lighting circuits, or testing. If the equipment does not have an arc flash hazard warning label or does not meet the labeling requirement of NFPA 70E 130.5(D), the available short-circuit current from the nearest transformer feeding that circuit or device must be calculated. This calculation will probably result in a greater short-circuit current than what is actually at the work location, but a conservative estimate is more beneficial to the safety of the worker.

Two default times are given for the OCPD protecting the equipment or circuit being worked on. If the OCPD uses a short-time delay, the delay time is entered. Working distance is usually 18″ for low voltages and 36″ for medium voltages. If the actual working distance is different, it can be entered on the form.

Determining the condition of equipment is also part of the risk assessment. The appropriate box should be checked off. The date of the last calibration should also be noted and entered in the blank space provided. If the last calibration date is more than three years old, the ability of the OCPD to function properly may be uncertain and additional precautions might be necessary.

- Arc Flash Boundary and PPE—The only way to determine Arc Flash Boundary is to calculate the distance where 1.2 cal/cm^2 incident energy would be received to the face and chest areas.
- Additional PPE based on the above analysis and job scope—The additional tools and equipment that are required should be checked off.
- Ground clusters—Personal protective grounds, or ground clusters, must be inspected before each use. If they are found to be defective, they should be turned in for repair or disposal according to ASTM F2249. OSHA 1910.269(n)(4)(i) requires that personal protective grounds be sized to allow the available short-circuit current to flow for the necessary time to clear the fault without fusing. OSHA 1910.269(n)(6) requires that personal protective grounds be applied or removed using live-line tools. A tag should show when the personal protective grounds were last tested. When personal protective grounds are placed, there is a separate form that must also be completed.
- Signatures—Signatures are required from the technician performing the work, supervisor, and an authorized customer representative.

FIELD NOTES:
CUSTOMIZED COMPANY APPROACH BOUNDARIES

Companies may decide to exceed the minimum safety requirements as established by OSHA and NFPA 70E. For example, many companies use Shock Approach Boundaries that may be different than those given in NFPA 70E Table 130.4(D)(a) and Table 130.4(D)(b). A company may require a Restricted Approach Boundary of 12" for all low-voltage work, while another company may have an 18" boundary. When a company uses boundaries that are more conservative than OSHA or NFPA 70E regulations, OSHA will hold that company to its written safety policies rather than OSHA regulations. A company must enforce its written policies as stringently as OSHA regulations.

SAFETY FACT

Reliable, up-to-date equipment information is an important resource used to maintain equipment. Years of equipment maintenance and modification information are normally stored for future maintenance work. Employees must have easy access to historical information, drawings, and modification documentation when performing maintenance work and troubleshooting equipment.

Calculating the Short-Circuit Current from a Transformer. All electrical workers must understand how to calculate short-circuit current from a transformer. This knowledge is one of the foundations for electrical safety. Available short-circuit current is calculated using the full load current (FLC) rating and the impedance of a transformer. The FLC rating is listed on the nameplate of a transformer. This rating is often several thousand amperes. The impedance number is also on the nameplate in the form of a percentage. The impedance of industrial transformers can range from 2% to 10%, but most are in the 5% to 7% range. To calculate available short-circuit current, apply the following equation:

$$I = \frac{FLC}{\left(\frac{Z}{100}\right)}$$

where
I = available short-circuit current (in A)
FLC = full load current (in A)
Z = impedance (in %)
100 = constant

For example, what is the available short-circuit current for a transformer that has a full load current rating of 2400 A and a nameplate impedance of 5.5%?

$$I = \frac{FLC}{\left(\frac{Z}{100}\right)}$$

$$I = \frac{2400}{\left(\frac{5.5\%}{100}\right)}$$

$$I = \frac{2400}{0.055}$$

$$I = 43{,}636 \text{ A or } 43.6 \text{ kA}$$

A shortcut method can also be used to calculate available short-circuit current. The shortcut method involves multiplying the FLC of the nearest upstream transformer by a multiplier. The multiplier is determined from the impedance of the transformer. **See Figure 5-11.** For example, if a transformer has an impedance of 4.8%, the multiplier 25 is used.

Multipliers Used for Calculating Transformer Short-Circuit Current

Transformer Impedance*	Multiplier
2–2.99	50
3–3.99	33
4–4.99	25
5–5.99	20
6–6.99	17
7–7.99	14
8–8.99	13
9–9.99	11
10–10.99	10

* in %

Figure 5-11. The multiplier used to calculate transformer short-circuit current is determined by the impedance of the transformer. The impedance of industrial transformers can range from 2% to 10%.

If a transformer has an impedance of 5%, the multiplier 20 is used. To calculate available short-circuit current using the shortcut method, apply the following equation:

$I = FLC \times M$

where

I = available short-circuit current (in A)

FLC = full load current (in A)

M = multiplier

For example, using the shortcut method, what is the available short-circuit current for a transformer that has an FLC rating of 2405 A and a nameplate impedance of 5.75%?

$I = FLC \times M$

$I = 2405 \times 20$

$I = $ **48,100 A** or **48.1 kA**

The important point to remember when using the shortcut method is that short-circuit current increases as impedance decreases. The shortcut method gives a more conservative calculation. This allows an electrical worker to overestimate the short-circuit current. Generally, it is better to overestimate the short-circuit current than to underestimate it. An electrical worker that has overestimated the short-circuit current would be better protected from injury because he or she would be wearing more PPE than may be required if the actual incident energy were to be calculated. Underestimating the short-circuit current can create a hazardous situation because an electrical worker may be wearing PPE that has too low of an arc rating.

Lockout/Tagout Switching and Grounding Procedure. The Lockout/Tagout Switching and Grounding Procedure form is an example form that can be modified to meet the needs of a specific electrical system. **See Appendix.** The top of the example form shows three notes, which include the following:

- "A good way to achieve an electrically safe work condition is the example below." The example form can be modified to fit a specific circumstance and system.
- "It is always best to use an updated single-line diagram." An updated single-line diagram is critically important for worker safety. Without it, an electrical worker cannot be sure that the electrical power system devices have been isolated. The single-line diagram for the actual system may be configured completely differently from the single-line diagram on the example form. For example, it may be a double-ended substation, main-tie-main power system, or loop system. **See Figure 5-12.** The current single-line diagram should always be used for the specific system being worked on. Per NFPA 70E 205.2, all single-line diagrams must be kept in a legible condition and up to date. This only applies when single-line diagrams exist for the electrical system.
- "Whenever possible, visually verify that all blades of the disconnecting devices are fully open or drawout-type circuit breakers are withdrawn to the fully disconnected position." Medium-voltage air break switches are often neglected and receive no maintenance, even when the circuit breakers at a facility are well maintained. If the arcing contact does not open, it is typically caused by a lack of lubrication at the contact pivot point. **See Figure 5-13.** This could mislead an electrical worker into thinking the circuit was safely isolated, when the phase would still be energized.

SAFETY FACT

When using the shortcut method to calculate available short-circuit current, the multiplier can be determined for any transformer by using simple math. To determine the multiplier, divide 100 by the impedance of the transformer. For example, a transformer with an impedance of 2% would have a multiplier of 50 (100 ÷ 2 = 50).

Single-Line Diagrams

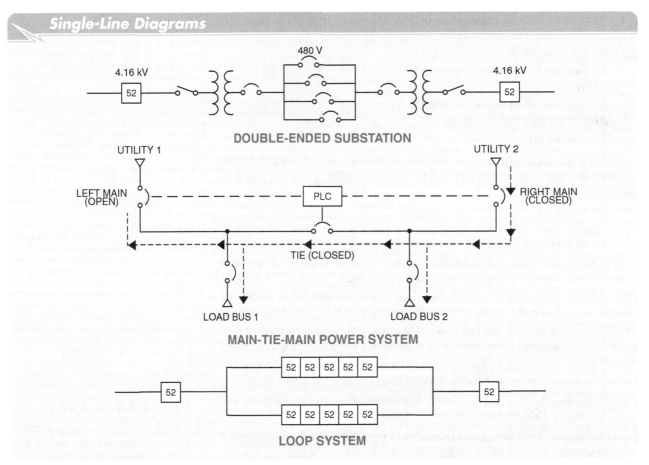

Figure 5-12. An updated single-line diagram should always be used to develop a switching procedure that is specific to the system. This is because a power system, such as a double-ended substation, a main-tie-main power system, loop system, or other scheme, may have specific hazards while operating.

Medium-Voltage Air Break Switches

Power & Generation Testing, Inc.

Figure 5-13. It is important to visually verify that all blades of the disconnecting devices are fully open or that drawout-type circuit breakers are withdrawn to the fully DISCONNECTED position.

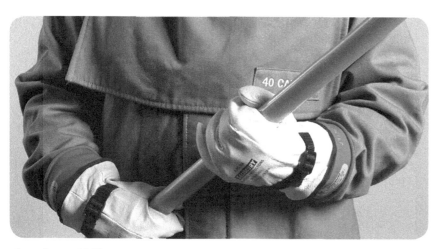

According to OSHA 1910.269(n)(6), live-line tools are used to apply and remove personal protective grounds.

An updated single-line diagram should always be used to develop a switching procedure that is specific to the system. Each step is checked off as it is performed according to the sequence on the form. Additional blanks are included on the form if changes must be made to the procedure.

It is critically important that all steps of the switching procedure be accurate. If there are any discrepancies during the switching sequence, the procedure must be stopped immediately and an accurate switching procedure developed before proceeding. The bottom of the form is typically used for the signatures of the technician, supervisor, and authorized customer representative.

Risk Assessment Matrix. The risk assessment matrix is a method used to assign a number value to specific worksite conditions using three risk factors. The Risk Assessment Matrix form is an example form that is intended as a general guide and should only be used to augment a full risk assessment. **See Appendix.** The Risk Assessment Matrix example form is divided into the following three sections:

- A. Frequency of Performing Task—Place a check next to the number representing the frequency the task is performed. The number 1 represents tasks performed very infrequently. The number 5 represents tasks that are performed constantly/several times a day.
- B. Probability of Being Harmed—Place a check next to the number representing the probability of being harmed while performing the task. The number 1 represents no probability of being harmed. The number 5 represents the probability of being harmed is certain.
- C. Possible Degree of Harm—Place a check next to the number representing the severity of an accident if it were to occur. The number 1 represents no degree of harm. The number 5 represents death.

After the three sections are complete, the ratings are then added together to provide a value for the total risk involved with the task. This number is compared to the risk assessment rating matrix located on the lower right of the form. **See Figure 5-14.** The values in the matrix indicate the following measures:

- 1–5, Proceed with the task—This indicates that a task is fairly safe with little risk involved.
- 6–9, Proceed with caution—This indicates that a task has some risk, and steps must be taken to reduce the hazard.
- 10–11, Reassess the plan to see if there is a better plan—This indicates that a task may result in serious injury. There is a significant possibility that an accident could occur. If there are other ways to perform the work, they should be investigated.
- 12–15, Make a new plan—This indicates that a task must not be performed. Alternate methods of performing the task must be considered.

Risk Assessment Matrix

RISK ASSESSMENT MATRIX

Customer: ______________ Date: ______________

Location: ______________ Job #: ______________

Task: ______________

ASSIGN A, B, AND C BELOW WITH A NUMBER 1 THROUGH 5; 1 BEING THE LOWEST AND 5 BEING THE HIGHEST

A. *FREQUENCY OF PERFORMING TASK*

☐ 1. Very infrequently ______________
☐ 2. ______________
☐ 3. ______________
☐ 4. ______________
☐ 5. Constantly/several times per day ______________

FREQUENCY OF PERFORMING TASK

B. *PROBABILITY OF BEING HARMED*

☐ 1. No probability of being harmed ______________
☐ 2. ______________
☐ 3. ______________
☐ 4. ______________
☐ 5. You will be harmed ______________

PROBABILITY OF BEING HARMED

C. *POSSIBLE DEGREE OF HARM*

☐ 1. No degree of harm ______________
☐ 2. ______________
☐ 3. ______________
☐ 4. ______________
☐ 5. Death ______________

POSSIBLE DEGREE OF HARM

RISK ASSESSMENT = **A** + **B** + **C** = ______

RISK ASSESSMENT = ______ + ______ + ______ = ______

COMMENTS: ______________

RISK ASSESSMENT RATING MATRIX

☐ **1-5** Proceed with the task
☐ **6-9** Proceed with caution
☐ **10-11** Reassess the plan to see if there is a better plan
☐ **12-15** Make a new plan

RISK ASSESSMENT RATING MATRIX

TOTAL RISK = ______________

(PRINT NAME CLEARLY BELOW) *DATE*

TECHNICIAN: ______________

FOREMAN: ______________

DATE

AUTHORIZED CUSTOMER REP: ______________

OTHER: ______________

Figure 5-14. The risk assessment matrix is used to rate the risk involved for a particular task.

Grounding of Equipment for Personal Safety Policy and Procedure. The Grounding of Equipment for Personal Safety Policy and Procedure form is an example form that is used as a guide to ensure proper and safe grounding of electrical equipment. **See Appendix.** It is important for an electrical worker to notice the following two sentences that appear near the top of the form:

- Any circuits over 600 V or where possible stored electrical energy may exist must be grounded before any direct contact.
- Circuits less than 600 V may be grounded if deemed necessary. Such circuits include but are not limited to uninterruptible power supplies, variable frequency drives, or any other equipment that may contain large capacitors, circuits where backfed or induced voltages may be present, or circuits that contain automatic transfer switches.

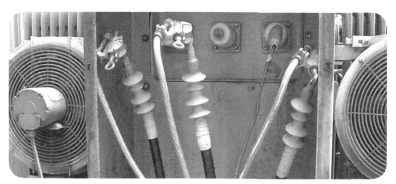

Shermco Industries

Grounding equipment must be installed at each point where work is being performed on deenergized equipment.

An electrical worker should follow the steps provided on the form. Each step is checked off as it is completed. This form can be modified to meet the specific needs of equipment.

ANSI/AIHA/ASSE Z10—Appendix E

Another method that can be used to perform a risk assessment is to incorporate the requirements from ANSI/AIHA/ASSE Z10, *American National Standard for Occupational Safety and Health Management Systems*. Appendix E, Assessment and Prioritization, provides a procedure for conducting a risk assessment. A generalized matrix is also provided. This matrix can be applied to all hazards not just electrical hazards.

The method that is used will not eliminate all risk. There will always be some risk remaining, especially when electrical systems are involved. The remaining risk must be evaluated and a decision made as to whether it is acceptable to proceed with the task. If the remaining risk is not acceptable, additional steps must be taken or an alternate method must be used to perform the task.

SUMMARY

- A risk assessment is critically important for safe work practices on or near exposed energized electrical conductors and circuit parts.
- Electrical hazards include shock, arc flash, arc blast, and ignition of flammable clothing.
- A risk assessment must be performed in order to select the appropriate PPE.
- The nominal voltage of an electrical device determines the shock hazard.
- As a worker moves away from an arc source, heat decreases rapidly.
- Improper circuit breaker maintenance can damage equipment and create hazards.
- The three methods for performing a risk assessment include using the procedures and forms of NFPA 70E Informative Annex F and Appendix E of ANSI/AIHA Z10.

Digital Learner Resources
ATPeResources.com/QuickLinks
Access Code: 705798

PERFORMING A RISK ASSESSMENT

Review Questions

Name _____   Date _____

_____ 1. The General Duty Clause in Section 5(a) of the OSHA regulations states that each employer shall furnish to each of his employees employment and a place of employment, which are free from ___ hazards.
   A. actual
   B. serious
   C. recognized
   D. all

_____ 2. A risk assessment is a field document that accounts for ___ hazards.
   A. chemical
   B. noise
   C. electrical
   D. all

T   F   3. Nominal voltage can be found on the nameplates of electrical equipment.

T   F   4. According to OSHA, a written hazard assessment does not need to be certified.

_____ 5. Per OSHA 1910.332, a qualified person must be able to determine ___ voltage.
   A. actual
   B. peak
   C. nominal
   D. RMS

_____ 6. Incident energy decreases by the ___ of the distance from an arc source.
   A. inverse square
   B. square
   C. hypotenuse
   D. root mean square

_____ 7. If the time of exposure to an arc flash is doubled, the incident energy is ___.
   A. halved
   B. tripled
   C. quadrupled
   D. doubled

_____ 8. A common weakness with circuit breakers and switches is that ___.
   A. they are too large
   B. the arc chutes will not interrupt the arc quickly enough
   C. the contacts must travel too far to extinguish an arc
   D. the contact pivot point binds up

_____ 9. The main cause of OCPD failure is ___.
   A. exposure to high temperatures
   B. lack of lubrication
   C. corrosion
   D. overvoltage

_____ 10. If a 13,800 V transformer has a full-load current of 3,250 A and a 3% nameplate impedance, the available short-circuit current is ___ A.
   A. 76,350
   B. 108,300
   C. 121,600
   D. 170,879

_____ 11. OSHA 1910.335 directs employers to select and have each employee use PPE that is ___.
   A. adequate
   B. appropriate
   C. cost effective
   D. stain-resistant and attractive

T   F   12. If OCPDs are not properly maintained, they may not perform in accordance with manufacturer specifications.

T   F   13. When assessing the risk associated with an electrical task, the age of the electrical equipment being worked on is not a factor.

_____ 14. What is the most common reason for circuit breaker misoperation?
   A. They have large springs and contacts that take up a lot of room.
   B. The arc must travel up the arc chute, cool off, deionize, and be interrupted when it tries to alternate at a zero crossover point.
   C. The contacts must have a minimum air gap in order to extinguish the arc.
   D. The pivot point lubrication dries out due to current flow through the circuit breaker.

_____ 15. What is the purpose of a short-time delay function on circuit breakers?
   A. It allows extra time for transient overloads, such as in air compressors, to cycle and prevent nuisance tripping.
   B. It allows the ground fault circuit interrupter function to operate first.
   C. It allows the device closest to the load to operate first (selective tripping).
   D. It is required by the NEC® for solidly grounded systems above 1200 A.

_____ 16. According to NFPA 70E ___, one-line diagrams are required to be kept in a legible condition and updated.
   A. 205.2
   B. 210.1
   C. 200.2
   D. 203.3

_____ 17. Which of the following is not a recognized hazard of electricity?
   A. arc flash
   B. shock
   C. asphyxiation
   D. arc blast

T   F   18. A risk assessment analysis is a field document that accounts for only electrical hazards.

ESTABLISHING AN ELECTRICALLY SAFE WORK CONDITION

NFPA 70E 120 uses the phrase "electrically safe work condition" to describe the deenergized state of conductors and circuit parts. The procedure for deenergizing conductors and circuit parts is referred to as electrical lockout/tagout. OSHA and NFPA 70E follow the same line of reasoning with electrical lockout/tagout. The lockout/tagout process ensures power is removed from equipment to ensure the safety of personnel working with the equipment.

OBJECTIVES

- Identify the OSHA regulations that cover electrical lockout/tagout.
- Explain the difference between induced voltage and backfed voltage.
- Explain how to perform absence-of-voltage testing.
- List the three types of test instruments that are commonly used to verify the absence of voltage.
- Explain simple and complex lockout/tagout procedures.
- Identify the NFPA 70E standards for training.
- List the equipment needed for proper lockout/tagout.
- List the items that must be addressed and the steps that must be taken while planning a lockout/tagout procedure.
- Explain the elements of control that should be included in a lockout/tagout procedure.
- Explain the standards concerning temporary protective grounding equipment per NFPA 70E 120.3.
- List the safety precautions that must be followed when using temporary protective grounding equipment.

DEENERGIZED PARTS

According to OSHA 1910.333(a)(1), *"Live parts to which an employee may be exposed shall be deenergized before the employee works on or near them, unless the employer can demonstrate that deenergizing introduces additional or increased hazards or is infeasible due to equipment design or operational limitations. Live parts that operate at less than 50 volts to ground need not be deenergized if there will be no increased exposure to electrical burns or to explosion due to electric arcs."*

It is clear from this OSHA regulation that workers should not work on energized electrical parts. However, the exception that OSHA has included after the comma in the first sentence may create confusion among some readers. When words such as "unless" and "but" are used, readers may mistakenly assume that what follows is the most important part of the sentence. This is not the case. For example, with the wording of this regulation, a company may attempt to rationalize working on energized conductors and circuit parts as a result of misinterpreting the sentence that includes the exception. This is poor practice because it puts the employee and employer at risk, which may lead to willful citations.

FIELD NOTES: EMPLOYEE TRAINING

It is not always the company that justifies working on energized conductors and circuit parts. Sometimes employees decide to perform energized work even though they are not properly trained to place equipment in an electrically safe work condition.

Case in point, a wind power generating site has a policy that no energized work is to be performed and that there must be two people present if work is performed near energized conductors or circuit parts for certain tasks. When worker A clocked in at the site, he found that his coworker was out for the day. Worker A's task was to perform routine maintenance on electrical equipment at the base of a generating tower, which included tightening all connections. According the procedures at this site, worker A should not have performed this task alone.

Since his partner was not in, two other employees invited worker A to join them. Worker A declined and proceeded to finish the checklist of tasks he had started. Worker A opened the main circuit breaker from the tower-mounted generator and tagged it out. The pad-mounted transformer, however, was still energized and caused the load side of the circuit breaker to remain energized.

Worker A was attempting to remove the uninterruptible power supply (UPS) when it came into contact with the energized bus connected to the main circuit breaker. The resulting fault created an arc flash, damaging the electrical control panel inside the generating tower to such an extent that it could not be reasonably repaired. Worker A received second- and third-degree burns on approximately 15% of his body. He later died while confined to the hospital due to complications from his injuries.

Worker A would not have been considered a qualified person according to the OSHA definition. He either could not choose the appropriate PPE or he chose not to wear it, he did not plan out the job safely due to misinterpreting the single-line diagram, and he chose to violate his company's safety policies. After an accident of this type, OSHA reviews employee training records to see if the training was adequate for the risk involved in the job duties. The most common citations are for unqualified persons performing tasks, lack of or inadequate training, and not wearing PPE.

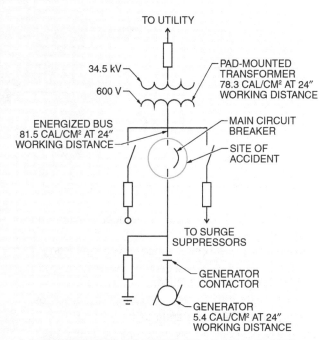

SINGLE-LINE DIAGRAM OF ACCIDENT SITE

Shermco Industries

DAMAGED ELECTRICAL CONTROL PANEL

SAFETY FACT

Electrical power must be removed when electrical equipment is serviced or repaired. To ensure the safety of employees working with the equipment, power is removed and equipment is locked out and tagged out.

PLACING ELECTRICAL EQUIPMENT IN AN ELECTRICALLY SAFE WORK CONDITION

According to OSHA 1910.333(b)(1), *"This paragraph applies to work on exposed deenergized parts or near enough to them to expose the employee to any electrical hazard they present. Conductors and parts of electric equipment that have been deenergized but have not been locked out or tagged in accordance with paragraph (b) of this section must be treated as energized parts, and paragraph (c) of this section applies to work on or near them."*

OSHA recognizes that electrical equipment may not be deenergized, even though lockout/tagout has been performed. It is necessary to comply with all of the regulations under OSHA 1910.333(b) and to follow them in their given order, otherwise the equipment will not be electrically safe. The regulations under OSHA 1910.333(b) can be viewed as steps that are used to place electrical equipment in an electrically safe work condition.

Electrical Lockout/Tagout

According to OSHA 1910.333(b)(2), *"While any employee is exposed to contact with parts of fixed electric equipment or circuits which have been deenergized, the circuits energizing the parts shall be locked out or tagged or both in accordance with the requirements of this paragraph. The requirements shall be followed in the order in which they are presented. (i.e., paragraph (b)(2)(i) first, then paragraph (b)(2)(ii), etc.)."*

The primary method that OSHA requires for protecting qualified electrical workers from electrical hazards is electrical lockout/tagout. *Lockout* is the process of removing the source of electrical power and installing a lock to prevent the power from being turned ON. *Tagout* is the process of placing a danger tag on a source of electrical power, which indicates that the equipment may not be operated until the danger tag is removed. Lockout/tagout is necessary to provide for the safety of employees working on electrical equipment.

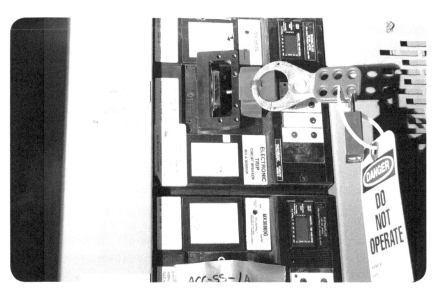

Equipment must be locked out or tagged out or both before installation, preventive maintenance, or servicing is performed.

Procedures for Deenergizing Conductors and Circuit Parts

According to OSHA 1910.333(b)(2)(ii)(A), *"Safe procedures for deenergizing circuits and equipment shall be determined before circuits or equipment are deenergized."*

Before deenergizing equipment, a qualified person must determine and write the lockout/tagout procedure to be performed. A mental list of procedural steps is unacceptable. There are two primary reasons for having a written procedure. The first reason is that another qualified person can verify the procedure and review it for errors or omissions. The second reason is to have a known, accurate procedure in place so other workers can use it in the future.

Disconnecting Conductors and Circuit Parts from All Energy Sources

According to OSHA 1910.333(b)(2)(ii)(B), *"The circuits and equipment to be worked on shall be disconnected from all electric energy sources. Control circuit devices, such as push buttons, selector switches, and interlocks, may not be used as the sole means for deenergizing circuits or equipment. Interlocks for electric equipment may not be used as a substitute for lockout and tagging procedures."*

Circuit breakers, switches, and fused disconnects are appropriate devices for isolating electrical equipment for lockout/tagout. Pushbutton (STOP/START) stations or other control devices are not appropriate devices for isolating electrical equipment for lockout/tagout. If desired, they can be used in conjunction with circuit breakers, switches, and fused disconnects, but cannot be the primary means of isolation.

Releasing Stored Electric Energy

According to OSHA 1910.333(b)(2)(ii)(C), *"Stored electric energy which might endanger personnel shall be released. Capacitors shall be discharged and high capacitance elements shall be short-circuited and grounded, if the stored electric energy might endanger personnel. Note: If the capacitors or associated equipment are handled in meeting this requirement, they shall be treated as energized."*

OSHA requires that stored electrical energy in circuit parts be released. Capacitors are specifically mentioned in this paragraph, as well as any lines or equipment that can also store a capacitive charge. Uninterruptible power supplies (UPSs), variable frequency drives (VFDs), and other types of equipment can have large capacitors that require discharging and grounding.

Preventing Other Devices from Energizing Conductors and Circuit Parts

According to OSHA 1910.333(b)(2)(ii)(D), *"Stored non-electrical energy in devices that could reenergize electric circuit parts shall be blocked or relieved to the extent that the circuit parts could not be accidentally energized by the device."*

An automatic transfer switch (ATS) is an example of a device capable of reenergizing electric circuit parts. Once a circuit is deenergized, the ATS is designed to reenergize the circuit to provide emergency power, usually through the use of a standby generator. The generator may be on the opposite side of the building where it could start and not be heard. An ATS can be dangerous if a worker does not know that it is connected to an electrical system. A thorough understanding of the electrical single-line diagram is required to properly and safely ensure power systems and their equipment are deenergized and safe to work on.

SAFETY FACT

Long runs of power cable can build and hold a capacitive charge that can be lethal. This charge can actually rebuild, even after the circuit conductors have been grounded. Be certain to always test for the absence of voltage when testing or handling power cable to ensure it is completely discharged.

Attaching Locks and Tags

According to OSHA 1910.333(b)(2)(iii)(A), *"A lock and a tag shall be placed on each disconnecting means used to deenergize circuits and equipment on which work is to be performed, except as provided in paragraphs (b)(2)(iii)(C) and (b)(2)(iii)(E) of this section. The lock shall be attached so as to prevent persons from operating the disconnecting means unless they resort to undue force or the use of tools. (B) Each tag shall contain a statement prohibiting unauthorized operation of the disconnecting means and removal of the tag."*

There are a variety of tags available with different statements on them. Most commercially available tags include the appropriate statements that meet OSHA regulations. **See Figure 6-1.** In addition to the words "DANGER" or "WARNING," contact information for the worker who placed the lock and tag is to be included. The worker's name and phone number are usually enough to meet this requirement. The expected date of completion is also included on many tags to provide information on the length of time that the equipment is expected to be out of service.

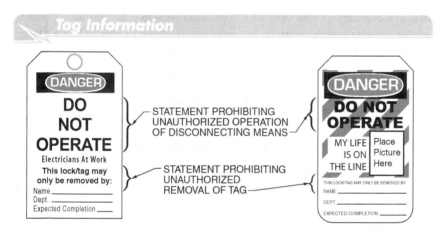

Figure 6-1. Most commercially available tags contain a statement prohibiting unauthorized operation of the disconnecting means and removal of the tag.

According to OSHA 1910.333(b)(2)(iii)(C), *"If a lock cannot be applied, or if the employer can demonstrate that tagging procedures will provide a level of safety equivalent to that obtained by the use of a lock, a tag may be used without a lock."*

Some equipment manufactured prior to 1970 is incapable of fitting a lock. Some types of equipment can be retrofitted to provide a means to lock it out, while other types cannot. A tag can be used without a lock only on equipment that has no means of attaching a lock.

Demonstrating that tagging procedures will provide a level of safety equivalent to that obtained by the use of a lock is a difficult requirement to meet. A lock provides physical security against tampering and limits access to equipment for everyone except the person who placed the lock. A tag-only system has to be used under very controlled circumstances and requires a lot of training and supervision to accomplish the level of safety required.

According to OSHA 1910.333(b)(2)(iii)(D), *"A tag used without a lock, as permitted by paragraph (b)(2)(iii)(C) of this section, shall be supplemented by at least one additional safety measure that provides a level of safety equivalent to that obtained by use of a lock. Examples of additional safety measures include the removal of an isolating circuit element, blocking of a controlling switch, or opening of an extra disconnecting device."*

If tagout-only procedures must be used, it is then necessary to provide an additional measure of safety. Any of the three examples listed by OSHA should be considered when isolating electrical equipment. Removing an isolating circuit element, blocking a controlling switch, and opening an extra disconnecting device are all effective safety measures that can be used together with tagout-only procedures.

According to OSHA 1910.333(b)(2)(iii)(E), *"A lock may be placed without a tag only under the following conditions: (1) Only one circuit or piece of equipment is deenergized, and (2) The lockout period does not extend beyond the work shift, and (3) Employees exposed to the hazards associated with reenergizing the circuit or equipment are familiar with this procedure."*

Lockout-only procedures are more secure than tagout-only procedures, but their disadvantage is that contact information may not be available to other employees who need to work on the same system or equipment. OSHA places strong restrictions on lockout-only procedures and limits their use to only one piece of equipment. When lockout-only procedures are implemented, employees and everyone in the area who could be exposed to the hazards associated with reenergizing the equipment must be trained in those porcedures.

Verifying Conductors and Circuit Parts Are Deenergized

According to OSHA 1910.333(b)(2)(iv), *"The requirements of this paragraph shall be met before any circuits or equipment can be considered and worked as deenergized."*

Applying locks and/or tags does not make electrical equipment safe. Equipment must be verified as deenergized before it can be considered in an electrically safe work condition. It is necessary to follow all OSHA regulations to avoid backfed voltages, induced voltages, and other hazards.

According to OSHA 1910.333(b)(2)(iv)(A), *"A qualified person shall operate the equipment operating controls or otherwise verify that the equipment cannot be restarted."*

OSHA further requires that the equipment be test-operated as a final check to verify that it will not start. This may not apply to every situation, such as power system work, but it must be performed when appropriate.

According to OSHA 1910.333(b)(2)(iv)(B), *"A qualified person shall use test equipment to test the circuit elements and electrical parts of equipment to which employees will be exposed and shall verify that the circuit elements and equipment parts are deenergized. The test shall also determine whether any energized condition exists as a result of inadvertently induced voltage or unrelated voltage backfeed even though specific parts of the circuit have been deenergized and presumed to be safe. If the circuit to be tested is over 600 volts, nominal, the test equipment shall be checked for proper operation immediately after this test."*

OSHA repeatedly uses the term "qualified person" in this section. Only a qualified person may test electrical circuits, and not everyone is a qualified person. This is important to remember because electrical circuits and devices may not be in a deenergized state even if they appear to be. Conductors or circuit parts should be considered energized until an absence-of-voltage test is completed. Rather than testing at the breaker, switch, or disconnect, the equipment must be electrically tested at the point of contact or where the work will take place. If desired, additional testing can be performed at the source disconnect.

An employee performing an absence-of-voltage test must be qualified to perform the work.

Ghost Voltages

There may be situations when a piece of equipment still contains voltage despite all appropriate lockout/tagout procedures being taken to properly deenergize it. This unexpected voltage is often referred to as a ghost voltage, which is an induced or backfed voltage. When a ghost voltage is present, it must first be verified that the correct piece of equipment is being worked on. Employees must be certain they are working on the correct piece of equipment and not the adjacent equipment. This is a common enough problem that the NFPA 70E Committee included it in NFPA 70E 130.7(E)(4), a section that addresses look-alike equipment.

It must also be verified that the equipment is actually placed in the OFF position. If the circuit breaker or fuse feeding the circuit is not clearly marked or if a molded-case circuit breaker has tripped, electric shock can occur. If a molded-case circuit breaker has tripped, it does not necessarily mean the contacts are completely open. Always set a tripped circuit breaker to the full OFF position before working on it.

Induced Voltage. An *induced voltage* is an unexpected voltage that is created by placing a deenergized conductor in the magnetic field surrounding a conductor that is energized and carrying a load. The expanding and contracting magnetic field around the energized conductor cuts the deenergized conductor, inducing a voltage. This is sometimes referred to as transformer action or induction.

The biggest danger from induced voltages is in outdoor, high-voltage substations. However, low-voltage circuits routed through cable trays can also induce a voltage into deenergized cables that are in the same cable tray. **See Figure 6-2.**

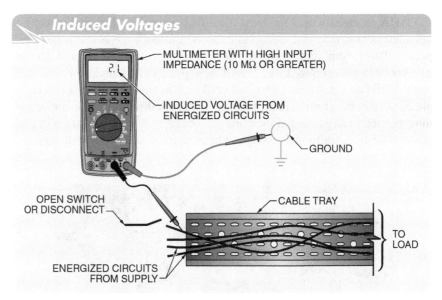

Figure 6-2. Low-voltage circuits routed through cable trays can induce a voltage into deenergized cables that are in the same cable tray.

When there is an induced voltage present, the equipment or circuit will have to be grounded in order to perform deenergized work. The ground will discharge the induced voltage. As long as the ground is in place, it will be safe to work on the equipment.

Backfed Voltage. A *backfed voltage,* or backfeed, is an unexpected voltage from another circuit and is backfed through other electrical devices. Possible sources of backfed voltages include control power transformers (CPTs), indicating lights, and circuits that are not part of the system. It may take some time and effort to determine from where a backfed voltage is coming. However, this must be determined before disconnecting and deenergizing the equipment.

It is important to determine whether the voltage is induced or backfed. If there is backfed voltage coming through control circuits or any other devices, grounding will cause an arc flash because the circuit being grounded is connected to a power source. Even though backfed voltages are lower than nominal voltage, the arc may be substantial enough to damage equipment or cause injury. Furthermore, causing an arc is an unsafe act that may result in disciplinary actions. It is best to avoid any situation that could cause an arc. Induced voltages are unable to supply short-circuit current and are not an arc flash hazard, but they are a shock hazard when woking on overhead powerlines and can cause serious injury or death.

Determining Induced or Backfed Voltages. To determine whether a voltage is induced or backfed, a combination of test instruments is used to test the circuit and verify the results. After initially measuring the voltage on a circuit to ground using a high-input impedance voltage tester, a low-input impedance voltage tester is used. Some manufacturers have voltage testers that have both a high-input impedance metering circuit and a low-input impedance metering circuit in one instrument. **See Figure 6-3.** This type of voltage tester reduces the number of test instruments required to perform the testing described. A low-input impedance voltage tester will either indicate voltage, which is the usual indication of a backfed voltage, or it will indicate zero voltage, which is the usual indication of an induced voltage. Either indication must be verified.

Voltage Testers

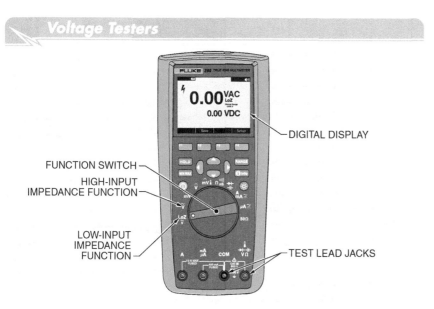

Figure 6-3. Some models of voltage testers contain both low- and high-input impedance functions.

To verify an initial test result, a low-voltage proximity tester can be used to measure along the circuit being tested while the low-impedance voltage tester remains connected. If the proximity tester indicates no voltage and no voltage is displayed on the low-input impedance tester, then the induced voltage is verified and a ground may be applied.

If the low-input impedance voltage tester is measuring a voltage (even though it may only be less than 100 V) and the proximity tester indicates the presence of voltage, the voltage on the circuit is probably backfed. **See Figure 6-4.** In this case, the worker must locate its source before proceeding with the work because applying a ground on this circuit would result in an arc.

FIELD NOTES:
ADVANTAGES OF USING HIGH-QUALITY VOLTAGE TESTERS

High-quality voltage testers typically have high-input impedance. Behind the input terminal of every voltage tester is a resistor. The resistor determines whether a voltage tester has high-input impedance or low-input impedance. The higher the resistance value, the less current will flow through the voltage tester to ground. This reduces the possibility of causing electrical components to unintentionally operate when being tested while energized.

While using a low-input impedance voltage tester, I was troubleshooting a 9,000 ton chiller that had an intermittent problem. I connected the test probe to one side of a coil and then touched the other probe to ground. When the probe contacted ground, enough current flowed through the voltage tester to ground to cause the coil being tested to operate, tripping the chiller off-line. The voltage tester being used had an input impedance of only a few thousand ohms. A better quality voltage tester with high-input impedance would not have allowed enough current through it to cause the coil to operate. This also could have averted a costly equipment shutdown.

A high-input impedance voltage tester will also detect induced voltages, while a low-input impedance voltage tester will not. Connecting a low-input impedance voltage tester to a conductor or circuit part that is energized with an induced voltage to ground allows enough current to flow through the meter to dissipate the voltage. This essentially grounds the circuit. Because an induced voltage is not connected to a generation source, connecting it to ground will effectively deenergize the circuit. Induced voltages usually present a shock hazard but not an arc flash hazard because there is a very small amount of current available.

SAFETY FACT

Digital multimeters (DMMs) that have both low- and high-input impedance functions are perfect for testing for induced and backfed voltages and eliminate the difficulty of carrying around multiple testers.

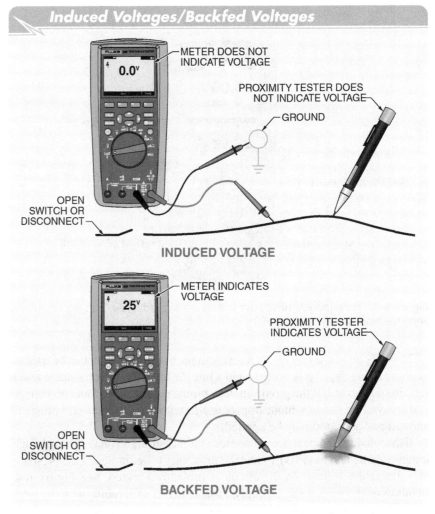

Determining Induced and Backfed Voltages

Figure 6-4. A low-voltage proximity tester and a low-input impedance voltage tester are used to determine induced and backfed voltages.

Reenergizing Conductors and Circuit Parts

According to OSHA 1910.333(b)(2)(v), *"These requirements shall be met, in the order given, before circuits or equipment are reenergized, even temporarily."*

The requirements mentioned in OSHA 1910.333(b)(2)(v) refer to the steps that are involved before reenergizing equipment. This OSHA regulation requires that all steps be completed in the given order. Failure to complete a step or completing the steps in different order may cause accidents. During the course of setting up large electrical motors or other components, it is necessary to "bump" the motor to test rotation direction or shaft alignment. The natural tendency for employees in these situations is to take shortcuts, which can cause accidents, OSHA requires all steps be completed, even when bumping a motor.

According to OSHA 1910.333(b)(2)(v)(A), *"A qualified person shall conduct tests and visual inspections, as necessary, to verify that all tools, electrical jumpers, shorts, grounds, and other such devices have been removed, so that the circuits and equipment can be safely energized."*

Reenergizing an electrical system or component that is grounded or jumpered can have immediate and dangerous consequences. A visual inspection is all that is required, unless there are parts of the system or equipment that cannot be seen. If parts of the system or equipment cannot be seen, additional electrical tests may be needed to verify that the system or equipment is safe to reenergize. Control power transformers will often show up as a ground when a megohmmeter is used. However, using an ohmmeter can verify that there are no short circuits without requiring every control power transformer to be disconnected from the system.

According to OSHA 1910.333(b)(2)(v)(B), *"Employees exposed to the hazards associated with reenergizing the circuit or equipment shall be warned to stay clear of circuits and equipment."*

Employees are advised to keep a reasonable distance from the equipment when it is being reenergized. If an explosion were to occur, an employee too close to the equipment may be hit by flying debris. By staying clear of equipment during reenergizing, an employee can reduce the chances of an accident.

According to OSHA 1910.333(b)(2)(v)(C), *"Each lock and tag shall be removed by the employee who applied it or under his or her direct supervision. However, if this employee is absent from the workplace, then the lock or tag may be removed by a qualified person designated to perform this task provided that: (1) The employer ensures that the employee who applied the lock or tag is not available at the workplace, and (2) The employer ensures that the employee is aware that the lock or tag has been removed before he or she resumes work at that workplace."*

To remove a lock or tag for an absent employee according to OSHA 1910.333(b)(2)(v)(C), it must be determined that the employee has left the job site and is not simply in a location where he or she cannot be located immediately. A reasonable attempt must be made to contact the missing employee, usually by mobile phone. If the employee cannot be contacted, the lock may be removed. Each company may have a different procedure for removing locks. When the employee returns, the lock is returned and an explanation is given as to why it was removed.

According to OSHA 1910.333(b)(2)(v)(D), *"There shall be a visual determination that all employees are clear of the circuits and equipment."*

This regulation is similar to OSHA 1910.333(b)(2)(v)(B), but is intended for a different reason. It is used to verify that employees will not be injured or killed if any undetected issues exist that could cause unexpected equipment failure.

FIELD NOTES: FOLLOWING OSHA REGULATIONS

A machine shutdown procedure was in process at a manufacturing facility. An employee who was working on switchgear prepared for his scheduled lunch break by placing his tool bag on the deenergized bus he was working on and closed the cubicle door. During the lunch break, the employee left the facility to attend to an emergency at home. As he quickly left the job site, he left his tool bag in the closed cubicle.

Later that day as the work was finishing up, the foreman observed that all of the cubicle doors were secured. He mistakenly assumed that the job had been completed by his employee and began to reenergize the equipment. When the section of switchgear that the employee had been working on was reenergized, the tool bag caused a three-phase short circuit. This caused an explosion that destroyed the cubicle that contained the tool bag, as well as the adjacent cubicle. Fortunately, there were no injuries or deaths because it had been verified that all employees were clear of the circuits and equipment.

OSHA 1910.333(b)(2)(v)(D) is used to visually verify that employees will not be in danger if an unexpected equipment failure were to occur. Undetected problems may cause hazardous situations, especially when reenergizing equipment.

TESTING FOR ABSENCE OF VOLTAGE

An *absence-of-voltage test* is a test performed to verify that a conductor or circuit part contains no voltage. NFPA 70E uses the phrase "absence of voltage test" instead of voltage testing because the system or equipment is expected to be in a deenergized state. The expectation that equipment is deenergized affects how testing is performed and determines the type of test instrument used.

Testing is made safer and easier to perform through proper planning and preparation. To properly plan and prepare for testing, certain questions must be addressed before any measurements are taken. These questions include the following:
- Is it necessary to perform the work energized?
- Is the task considered troubleshooting or absence-of-voltage testing?
- What is the nominal voltage of the system?
- What test instrument will be used to verify that the system is deenergized?
- How are the measurements from the test instrument interpreted?
- What are the limitations of the test instrument?
- Is the test instrument and are its leads and accessories functioning properly?
- Is the required lockout/tagout procedure complete?
- What types of PPE will be required?

NFPA 70E 110.4(A)(5) requires that the test instruments used for absence-of-voltage testing be verified for proper operation before and after each absence-of-voltage test, regardless of the circuit voltage (low, medium, or high voltage). There are three types of test instruments that are commonly used to verify the absence of voltage. These include noncontact voltage testers, solenoid voltage testers, and digital multimeters (DMMs).

Noncontact Voltage Testers

A *noncontact voltage tester* is a test instrument that indicates the presence of voltage when the test tip is near an energized conductor or circuit part. **See Figure 6-5.** Because they can indicate the presence of voltage by being near it, noncontact voltage testers are also known as proximity voltage testers. When a noncontact voltage tester tip glows, it indicates that voltage is present. A noncontact voltage tester can also be used to provide an early warning that voltage is present when it is used on medium-voltage equipment. If the medium-voltage equipment is energized, the tester will usually indicate a voltage as soon as the cubicle door is opened.

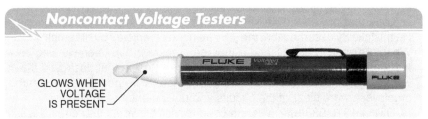

Fluke Corporation

Figure 6-5. A noncontact voltage tester glows to indicate the presence of voltage when the test tip is near an energized conductor or circuit part.

If a noncontact voltage tester does not indicate a voltage, it does not necessarily mean voltage is absent. There is a possibility that the tester is providing a false negative indication. Because of the possibility of a false negative indication, noncontact voltage testers should only be used as a quick precheck to determine whether a conductor or circuit part is energized. A low-voltage proximity test should always be followed by a test that uses a direct-contact meter. This should be done because the expanding and contracting magnetic field around low-voltage conductors is not strong enough to provide a positive indication under all circumstances.

If the conductor or circuit part is energized, a low-voltage proximity tester may indicate a false negative indication under the following conditions:
- if the insulated test point touches grounded metal, the cable being tested is partially buried, or the user is isolated from ground
- if the instrument is used inside of a metal enclosure
- if the conductor is partially buried or is wet inside the jacket

No proximity tester will indicate voltage when testing shielded cable. Shielded cable is usually used for medium- and high-voltage applications, but it is also used for signal and communications circuits. A high-voltage proximity tester is typically used for voltages above 600 V. The expanding and contracting magnetic field that surrounds a medium- or high-voltage energized conductor is much stronger and can be detected reliably. In addition, in order to comply with OSHA 1910.333(b)(2)(iv)(B), phase-to-phase and phase-to-ground tests must be conducted, which is something a proximity tester cannot be used for. This is not a requirement for higher voltage circuits, where noncontact voltage detectors are commonly used.

Solenoid Voltage Testers

A *solenoid voltage tester* is a test instrument that indicates approximate voltage level and type (AC or DC) by the movement and vibration of a pointer on a scale. **See Figure 6-6.** Solenoid voltage testers have several disadvantages. For example, a solenoid tester can wear out over time and its voltage scale may become difficult to read. If the solenoid is so weak that its pointer does not discernibly vibrate and the voltage indicator cannot be read, it must be replaced because it is no longer reliable.

Some types of solenoid voltage testers may not indicate voltage if it drops below 70 V to 90 V. Also, voltage testers with neon indicator lights may stop indicating at about 30 V. If a person receives an electric shock, it may not be life threatening, but it could cause them to contact electrical equipment that gives them a life-threatening shock. Some solenoid testers are not fused and do not comply with CAT safety rating requirements. If a transient voltage hits the system while it is connected, there is nothing to protect the user from serious injury. If a solenoid-type tester is used, it must be CAT-rated and fused.

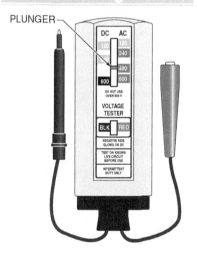

Figure 6-6. A solenoid voltage tester indicates approximate voltage level and type (AC or DC) by the movement and vibration of a pointer on a scale.

Digital Multimeters

A *digital multimeter (DMM)* is a test instrument that can measure two or more electrical properties and displays the measured properties as numerical values. **See Figure 6-7.** Safety issues arise when the employee using a DMM is fatigued or careless. One of the most common mistakes when using a DMM is turning the dial to the wrong function. For example, the dial may be turned to amps instead of volts. Also, older DMMs that do not include an autoranging feature may be mistakenly put into a range that is too high. This may indicate a voltage that is lower than what it actually is. Using a modern autoranging DMM resolves this issue.

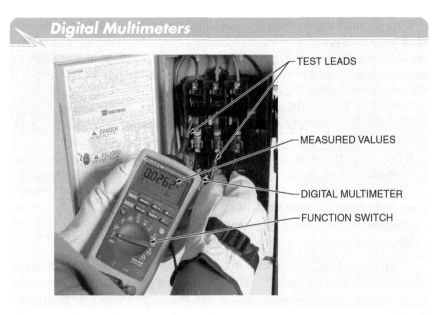

Figure 6-7. Digital multimeters (DMMs) are used to measure two or more electrical properties.

Any direct-contact meter, such as a DMM, can be dangerous if connected to a circuit with a voltage that exceeds its rating. For example, a technician troubleshooting a 2.3 kV or 4.16 kV motor starter control circuit using a 600 V voltage tester is at risk of injury or death if the test leads make contact with the high-voltage circuits. The control power transformer for the motor starter control circuit is often mounted on the side of the drawout unit and the terminals cannot be seen clearly. **See Figure 6-8.** The control power transformer will have the higher voltage on one side (2.3 kV or 4.16 kV) and the control voltage on the other side (typically 208 V or 480 V). During troubleshooting, the technician may attempt to test the 480 V circuit and accidentally come into contact with a higher voltage circuit, often at the control power transformer. The resulting short-circuit current usually causes an arc flash that would burn or electrocute the technician.

Shermco Industries

Figure 6-8. The control power transformer or motor starter control circuit is often mounted on the side of a drawout unit, which can make it difficult to distinguish the high-voltage terminals from the low-voltage terminals.

ESTABLISHING AN ELECTRICALLY SAFE WORK CONDITION PER NFPA 70E

NFPA 70E 120 includes a combination of the requirements from OSHA 1910.147 and 1910.333(b) to provide a more complete procedure for establishing an electrically safe work condition. Among other requirements, NFPA 70E 120 requires that the employer must establish lockout/tagout procedures, provide the training and necessary equipment for performing the lockout/tagout, audit employees for compliance, and audit the procedures to verify that they are current and accurate. These requirements are specified in NFPA 70E 120.2(D) and (E).

Simple Lockout/Tagout Procedure—NFPA 70E 120.2(D)(1)

Lockout

The simple lockout/tagout procedure is performed when lockout/tagout tasks do not fall under the complex lockout/tagout procedures. For a task to be considered simple lockout/tagout, only one set of conductors or one source is required to be deenergized in order to achieve an electrically safe work condition. Also, the person using the simple lockout/tagout procedure must be qualified. The simple lockout/tagout procedure is not required to be in writing for every device covered by it. One general lockout/tagout procedure is adequate.

Complex Lockout/Tagout Procedure— NFPA 70E 120.2(D)(2)

The complex lockout/tagout procedure follows the requirements of OSHA 1910.147 and 1910.333(b)(2). A complex lockout/tagout procedure is required under any of the following conditions:
- There are multiple energy sources.
- There are multiple work crews at the job site.
- More than one type of tradesworker is performing the work.
- Work is taking place at multiple locations.
- Multiple employers have personnel performing the work.
- There are multiple disconnecting means and/or multiple types of disconnecting means.
- There is a specific sequence for lockout/tagout.
- The time required to complete the job will take longer than allowed by a single shift.

NFPA 70E 120.2(D)(2) also lists several requirements that employees must fulfill when performing a complex lockout/tagout procedure. The requirements for a complex lockout/tagout procedure include the following:
- A written procedure that specifies the person in charge of the lockout/tagout is required. In a group lockout/tagout situation, there must be a single person in charge of implementing the lockout/tagout who is responsible for the safe and proper execution of the lockout/tagout procedure.
- A method of accounting for all employees who may be exposed to the electrical hazards when performing the lockout/tagout procedure is required.
- Each employee is responsible for placing a lock and tag to the group lockout device and must remove them when work is completed. **See Figure 6-9.**
- Every concern of the employees affected by the lockout/tagout procedure must be addressed.

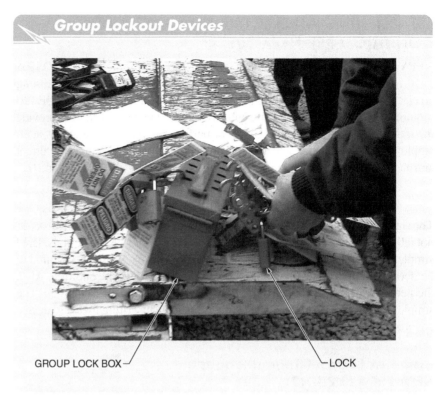

Figure 6-9. Group lock boxes are used to ensure that no one worker has access to the locked-out equipment unless all workers have removed their locks from the group lock box.

Coordination—NFPA 70E 120.2(D)(3)

NFPA 70E 120.2(D)(3) addresses the efforts that an employer must make to coordinate lockout/tagout procedures. The lockout/tagout procedures must be coordinated with all other facility lockout procedures and must be specific to the job site where the work is taking place. If there are other procedures that are to be used, they must also be coordinated to ensure that they will be based on the same concepts. Any time an employee may be exposed to electrical hazards, the procedures must include absence-of-voltage testing.

The requirement for lockout/tagout procedures is also included in OSHA 1910.333(b). Locks and tags used for electrical lockout/tagout may be similar to devices for other types of lockout/tagout, such as those used for hydraulic flow control valves, but they must be used only on electrical equipment. The locks must be a different color if the same type is used. OSHA states that the locks must be uniquely identified.

Training, Retraining, and Training Documentation—NFPA 70E 120.2(B)(2), (3), and (4)

NFPA 70E 120.2(B)(2), (3), and (4) addresses the training and retraining of employees and training documentation. Section 120.2(B)(2) states that training is needed by all persons that could be affected by a lockout/tagout. They must be trained to understand the procedures and what their responsibilities are under those procedures.

NFPA 70E 120.2(B)(3) covers retraining. Retraining is required any time a procedure is revised or at three-year intervals and for employees who have been reassigned and may not understand the lockout/tagout requirements of their new location or assignment. Lockout/tagout procedures may vary, even within a single facility, based on the specific types and configurations of the installed equipment and the power system. A ring bus system would have different lockout/tagout requirements than a double-ended substation, but both might be found in one facility or, in a larger company, one campus where there are several buildings on corporate property.

NFPA 70E 120.2(B)(4) covers the documentation required for lockout/tagout training and retraining. Once an employee has demonstrated proficiency, the training can be documented and the employee is considered qualified to perform that task. "Demonstration" means the employee performs the task, either in a classroom, a controlled situation, or the field, while being observed. A written test is not adequate by itself. The Informational Note provides some examples of the required documentation, such as an outline, a syllabus, training objectives, or a table of contents.

Equipment—NFPA 70E 120.2(E)

NFPA 70E 120.2(E) addresses the necessary equipment for proper lockout/tagout. Lockout/tagout equipment includes locks, keys, hasps, tags as well as their attachment means (usually nylon cable ties), and other devices needed to safely secure a device from being inadvertently energized.

Lock Application. All equipment or devices used after January 2, 1990 for lockout/tagout must have a provision for attaching a lock. This requirement was not in effect before that time, and much of the equipment manufactured prior to the 1970s did not have a provision for locking the equipment out. At the time, it was assumed that no one would reenergize circuits or equipment unless they had authorization to do so.

Lockout/Tagout Devices. NFPA 70E 120.2(E)(2) states that the employer must provide all equipment needed for lockout/tagout and that the employees must use the employer-supplied equipment. It is unacceptable for employees to use their own personal locks for lockout/tagout. Locks and tags must be unique and readily identifiable as lockout/tagout devices and may not be used for any other purpose.

> **SAFETY FACT**
> Locks used for lockout/tagout have to be uniquely identified, but one type of identifier can be used for all lockout/tagout locks. Some companies use a different color lock for each trade or shop (red for electrical, blue for mechanical, and so on), but this is not mandated by OSHA or NFPA 70E. The identifier for the lockout/tagout-specific lock can only be used for that one purpose.

Several lockout attachments are available to provide a means to lock out the various types of control devices.

SAFETY FACT

Lockout devices prevent operation unless tools or undue force is used. If a person is truly determined to remove a lock, he or she will find a way to remove it using tools or undue force. However, NFPA 70E and OSHA both require that lockout devices prevent operation under normal circumstances.

Lockout Devices. NFPA 70E 120.2(E)(3) describes the requirements for lockout devices. These requirements are also provided in OSHA 1910.147. The requirements for a lockout device include the following:
- The lock must be either a key or combination lock.
- The device must include a means of identifying the worker who placed the lock.
- The device may be only a lock, as long as it is easily identified as a lockout device and it identifies the employee who placed it.
- Lockout devices must prevent operation unless tools or undue force are used.
- Tags, when used with locks, are required to have a statement prohibiting the unauthorized removal of the lock or operation of the equipment. **See Figure 6-10.**

Figure 6-10. A hasp, lock, and tag are used to prevent the operation of equipment.

- Lockout devices must be suitable for the environment they are to be subjected to for the length of time they are to be used in that environment. For example, locks that are not suitable for outdoor use can rust and tags can fade over long periods of time to the point where they cannot be read. Locks and tags used outdoors must be designed for outdoor use.
- The worker placing the lock must have control of the key or the combination. In a letter of interpretation dated February 28th, 2000, OSHA stated that "*the 'one person, one lock, one key' practice is the preferred means and is accepted across industry lines, but it is not the only method to meet the language of the standard. However, prior to the use of the master key method, specific procedures and training, meeting the 1910.147(e)(3) exception, must be developed, documented, and incorporated into your energy control program. Among the features essential to a compliant master key procedure is a reliable method to ensure that access to the master key will be carefully controlled by the employer such that only those persons authorized and trained to use the master key in accordance with the employer's program can gain access.*" Employers are cautioned to use the master key method carefully and meet all the above requirements.

Tagout Devices. NFPA 70E 120.2(E)(4) describes the requirements for tagout devices. The requirements for a tagout device include the following:
- The tagout device must have a tag and a means of attachment.

- The tagout device must be suitable for the environment and the length of time it is to be exposed to the environment and must be easily identified as a tagout device.
- The tagout device attachment means must be capable of withstanding a force of 50 lb. The tagout device attachment means must be nonreusable, manually attachable, self-locking, and nonreleasable. It also must have the equivalent resistance and environmental suitability of a nylon cable tie.
- The tagout device must provide a statement prohibiting the unauthorized removal of the lock or operation of the equipment.

Note: An exception to the first three requirements allows both a hold card tagging tool on an overhead conductor and a live-line tool used to place a tag on a disconnect that is isolated from workers.

Electrical Circuit Interlocks. NFPA 70E 120.2(E)(5) requires up-to-date diagrammatic drawings be consulted to verify that circuits or equipment cannot be reenergized by operating an electrical circuit interlock. An electrical circuit interlock, such as an ATS, automatically switches power from a primary source to a secondary source when a power loss is detected. The secondary power source, such as an emergency generator, is usually located in an isolated area and a worker may be unaware that a switchover has occurred.

Control Devices. A disconnecting means must be used to lock and tag devices. Disconnecting means include circuit breakers, switches, fuses, and disconnect switches. Pushbuttons, selector switches, and other control devices cannot be used as the primary lockout means. Control stations can be used to lock devices, but the primary disconnect must also be locked out in order for the equipment to be in an electrically safe work condition.

Procedures—NFPA 70E 120.2(F)

NFPA 70E 120.2(F) addresses the procedures that must be followed to achieve an electrically safe work condition. The employer or facility manager must keep a copy of the written lockout/tagout procedures in an accessible location. The procedure must also be available to all employees.

Planning. NFPA 70E 120.2(F)(1) requires that the lockout/tagout procedure be planned. Specific items and steps that must be addressed while planning the lockout/tagout procedure include the following:

- Locating sources—Up-to-date single-line diagrams are the preferred primary reference for determining the location of energy sources. If up-to-date single-line diagrams are unavailable, an equally effective means of determining energy source locations is required. However, a company should exercise caution in these cases because the phrases "equally effective" and "providing a level of protection equal to" used by NFPA 70E can be misleading. Even though other methods are permitted, a company risks willful citations if an accident occurs or if OSHA is performing an audit.

    Maintaining single-line diagrams is equally as important as having them available. A single-line diagram must be maintained to reflect any changes made to a system. For example, when a private contractor installs new equipment, it can alter systems designs. Per NFPA 70E 205.2, when single-line diagrams are provided, they must be kept current and in legible condition. An up-to-date single-line diagram allows employees to identify energy sources, possible sources of backfed voltage, and the locations of isolating devices.

If a single-line diagram is unavailable, employees are only able to depend on their memory or another unreliable resource. This section does not require companies to have single-line diagrams. It only requires companies to maintain them if they have them.

- Exposed persons—If other employees may be exposed to hazards during lockout/tagout, those employees must be identified. This includes employees working in areas that may be affected by the lockout/tagout. For example, if the loss of power causes lights to go out or ventilation equipment to turn off, the employees in those areas and any other employees that may be exposed to hazards during, or as a result of, the lockout/tagout must be identified. Also, all PPE that is required by the employees performing the lockout/tagout must be specified.
- Person in charge—When performing multiple lockout/tagout procedures in the same facility or location, an employee is placed in charge of the lockout/tagout. This employee has overall responsibility for the performance of the lockout/tagout and must verify that all steps required are complete. The person in charge is to be identified in the procedure, as well as their responsibilities during lockout/tagout.
- Simple lockout/tagout—Simple lockout/tagout procedures must meet the requirements of NFPA 70E 120.2(D)(1).
- Complex lockout/tagout—Complex lockout/tagout procedures must meet the requirements of NFPA 70E 120.2(D)(2).

Elements of Control—NFPA 70E 120.2(F)(2)

NFPA 70E 120.2(F)(2) provides a list of specific elements of control. The list describes what should be provided in the lockout/tagout procedure. These elements of control identify the method in which the energy sources are to be locked out.

Deenergizing Equipment (Shutdown). The lockout/tagout procedure must identify the person performing the actual isolation of the equipment to be shut down. The procedure must also identify the location of the equipment, such as operating circuit breakers, switches, or disconnects, and how to deenergize the load.

Stored Energy. Both OSHA and NFPA 70E require that all stored energy that may endanger workers be released or restrained. The lockout/tagout procedure must identify the stored energy that is to be released or restrained, where the energy is located, and how it is to be released or restrained. The NFPA 70E Committee also included a list of items to test to ensure that all energy sources are properly released or restrained. **See Figure 6-11.**

Releasing Stored Electric or Mechanical Energy

- ☑ Capacitors and high capacitance elements must be discharged, short circuited, and grounded.
- ☑ Automatic transfer switches must be blocked and restrained.
- ☑ All sources of induced voltages or backfeeds must be located and isolated.
- ☑ Springs must be released or restrained.
- ☑ Hydraulic and/or pneumatic equipment must be immobilized.
- ☑ Any other source of stored energy must be blocked, relieved, or restrained.

Figure 6-11. NFPA 70E includes a list of items that must be checked to ensure energy sources are properly released or restrained.

Disconnecting Means. The lockout/tagout procedure must explain the method of verifying that the system, circuit, or equipment is in an electrically safe work condition. NFPA 70E 120 can be used to verify that equipment is in an electrically safe work condition.

Responsibility. The person in charge must be identified during planning for the lockout/tagout procedure. The responsibilities of the person in charge include verifying that the lockout/tagout procedure is implemented and ensuring that the work is completed during lockout/tagout before the locks and tags are removed. When multiple jobs or tasks are being performed, or multiple companies are performing the work, specific directions on how the lockout/tagout is to be implemented and coordinated, as well as who is responsible for coordination, must be included in the lockout/tagout procedure.

Verification. The lockout/tagout procedure must contain a method for verifying that the equipment cannot be restarted. The normal method of starting the equipment must be applied in order to verify that the equipment cannot restart. If the normal method of starting the equipment is applied and the equipment starts, then the source of the energy must be determined. The energy may be from an induced or backfed voltage, either of which can cause injury or death.

Testing. Absence-of-voltage testing includes several requirements. The lockout/tagout procedure must establish requirements concerning absence-of-voltage testing, which include the following:

- The type of voltage tester to be used must be identified. Typically, it is up to the discretion of the employee to determine the type of voltage tester. However, some companies may determine the type of tester the employee must use based on the situation. There are several types of voltage testers and each has its advantages and disadvantages.

- The PPE that will be used by the employee during absence-of-voltage testing must be identified. The typical level of protection needed when applying grounds or removing bolted covers for medium-voltage circuits is PPE Category 4. Absence-of-voltage testing on 480 V equipment is usually a PPE Category 2 task, as long as the limits given in NFPA 70E Table 130.7(C)(15)(A)(b) are not exceeded. **See Figure 6-12.**

- The employee performing both the absence-of-voltage testing and the required live-dead-live testing must be identified. Live-dead-live testing is used to verify the proper operation of a test instrument. A known energized source is used to verify that the voltage tester is operating properly. Then tests are performed on the assumed deenergized circuits. Then the voltage tester is again verified against a known energized source. OSHA 1910.269 requires this method for electrical systems rated above 600 V, while NFPA 70E requires it for all voltages.

    NFPA 70E 110.4(A)(5) was modified for the 2015 edition of NFPA 70E to state that verification of proper operation of the test instrument must be performed on a known voltage source. This means that testing the voltage detector on an energized circuit, preferably of the same voltage as the circuit about to be tested, is required. The committee did not believe that using a self-tester or battery-powered tester was an adequate test for the test instrument. Testing on a similar voltage is not mandated, but for voltage detectors that have range switches, it will help ensure the switch is not defective in that range.

- The work area boundaries must be defined. These boundaries include the Limited and Restricted Approach Boundaries for the shock hazard, the Arc Flash Boundary, and the location of the safe work zone barriers. NFPA 70E 130.4(B) Informational Note states that the Arc Flash Boundary may be a greater distance from the equipment than the Limited Approach Boundary and that the two boundaries are independent of each other. Also, in the 2015 edition of NFPA 70E 130.7(E)(2), a new requirement was added that where the Arc Flash Boundary is greater than the Limited Approach Boundary, the safety barrier tape or barricades are not to be placed closer than the Arc Flash Boundary. This requirement was added to make clear that the safety barrier tape must be placed so that other workers are not subjected to either hazard.
- A test instrument must be used on the conductors or circuit parts before they are touched. Each conductor or circuit part must be tested. Some companies use the test-before-touch (TBT) program that was developed by the DuPont Company. The TBT program clearly implements NFPA 70E requirements for absence-of-voltage testing.
- Retesting is required when conditions change or the job site is left unattended. If an employee must leave the work area for any reason, the TBT program requires that the equipment or circuits be retested to verify that they remain in an electrically safe work condition.
- Alternate solutions must be determined for testing inaccessible areas. If the circuit or equipment is in an inaccessible location or is situated so that absence-of-voltage testing cannot be performed, extra planning is required to verify that the circuit or equipment is deenergized. This may include testing a different circuit from the same power supply that is in a more readily accessible location.

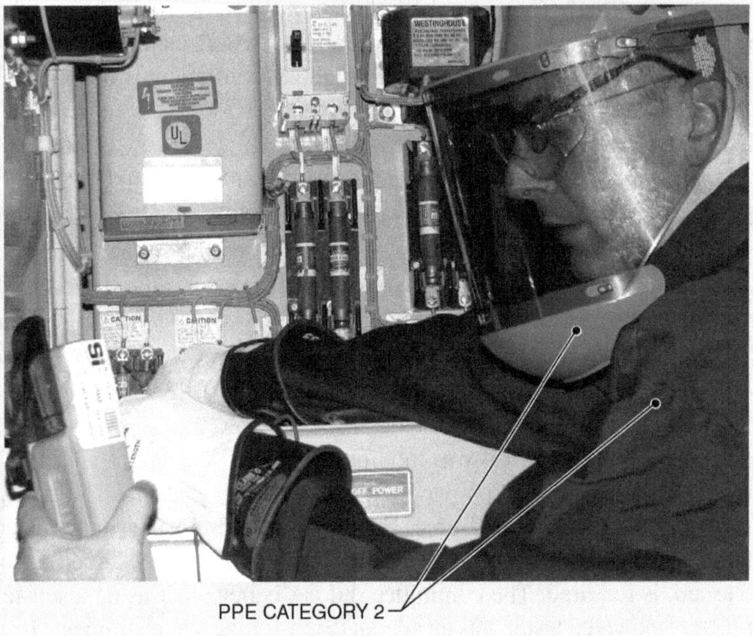

Figure 6-12. The typical level of protection needed when applying grounds or testing for the absence of voltage for low-voltage circuits is PPE Category 2.

Grounding. Grounding is always required for systems and equipment that are connected to UPS systems or VFDs because they have large capacitors. However, employees may sometimes forget to consider the necessity of grounding low-voltage equipment or circuits. Proper grounding is important because it is a sure method of preventing a system or circuit from being reenergized. A grounded system also helps protect against induced or backfed voltages that may appear while work is being performed. The requirements for grounding can be found in NFPA 70E 120.2(F)(2)(g) and 120.3. The lockout/tagout procedure must identify the following requirements for grounding:

- Grounding requirements must be established. Temporary personal protective grounds must meet the requirements for ASTM F855, *Standard Specifications for Temporary Protective Grounds to Be Used on De-energized Electric Power Lines and Equipment*. They also must be properly sized to carry the available short-circuit current for the amount of time needed to safely clear the fault. Ground clamps must be rated for the same amount of short-circuit current as the temporary personal protective grounds. The ground clamps also must be the proper type for the bus, such as for flat buses or tubular buses, or component being grounded.
- The length of time that the ground conductors are needed must be specified. Ground conductors may be installed temporarily or for the entirety of the task. During electrical testing, grounds often must be removed or relocated on the device being tested. The grounding sequence and required PPE should be addressed as well.
- A separate section on grounding is not required in the lockout/tagout procedure because grounding requirements may be covered in other parts of the procedure. Grounding requirements may also be addressed in other work rules.

Shift Change. If the work extends into another shift, a procedure must be established that specifies how to transfer responsibility for the lockout/tagout to the incoming work crew. OSHA refers to a change in work shift responsibility as a transfer of control. OSHA also requires that the entire lockout/tagout procedure be completed each time transfer of control is initiated. However, the entire lockout/tagout procedure is often not completed because the worker coming onto the job site assumes that the lockout/tagout was performed properly. This may not always be the case. The OSHA regulation ensures that all required steps were initially performed and the system or equipment is properly secured for each shift.

Coordination. How other tasks or jobs will be coordinated with the lockout/tagout must be specified in the procedure. This includes identifying tasks performed at remote locations and the person responsible for coordinating between them. Coordination also applies to work being performed by other companies, especially if that work involves the same equipment or parts of the electrical system being worked on by different companies. For example, if work were being performed on a 15 kV-class switch, proper coordination would prevent another work crew from testing the cables connected to that switch.

Accountability for Personnel. Workers who are or may be exposed to hazards created by the lockout/tagout or during the lockout/tagout must be accounted for. The lockout/tagout procedure must have a method to account for these workers. For example, if power to the exhaust ventilation in a hazardous location must be disconnected during lockout/tagout, the workers in that area must be notified and vacated from the area before the circuit is actually deenergized.

SAFETY FACT

Proper communication and coordination between multiple work crews can prevent accidents.

Lockout/Tagout Application. The lockout/tagout procedure must clearly identify when and where a lock is the only device applied, as well as when and where only a tag is used. Usually, this is necessary for older equipment that does not have any provisions for lockout/tagout. The lockout/tagout procedure must also establish the following:

- Lockout is placing a lock on the energy source of a piece of equipment, which disables it. A lock prevents the operation of the energy source and should not be removable through normal means. However, the lock may be forced open if tools or undue force is used.

- Tagout is placing a tag in the same location as the lockout device. A tag warns against operating the device, provides contact information for the person who has placed the locks and tags, and lists the equipment under lockout/tagout.

- If the equipment, circuit, or device cannot accept a lock, it is unacceptable as the primary means of disconnecting the energy source.

- The use of only a tag is permitted when the equipment design cannot accept a lockout device. Performing only tagout requires the implementation of a second means of isolating a circuit. The lockout/tagout procedure must establish when only tagout will be used, the person responsible for implementation, and how to account for workers who may be exposed to any hazards that are created. Secondary means of isolating a circuit include racking out circuit breakers, removing fuses, blocking control switches, and/or opening a second disconnecting device.

Removal of Lockout/Tagout Devices. A written lockout/tagout procedure must specify how to remove locks and tags. Also, when the person who placed the locks and tags is unavailable, the steps taken to remove them must be specified. These steps include the following:

1. An attempt must be made to contact the person who placed the locks and tags. OSHA regulations clarify that a reasonable attempt must be made. For example, if it is known that the person is not at the job site, a call is placed to his or her phone number on record and the time and circumstances are recorded.

2. The locks and/or tags are removed by the employer.

3. When the person who placed the locks returns, he or she must be informed that the locks and tags were removed.

Release for Return to Service. The necessary steps to return equipment or circuits to service must be identified in the lockout/tagout procedure. These steps include the following:

1. Visual examinations or tests must be performed as needed to verify that the equipment or circuits are safe to reenergize. This verifies that all jumpers, grounding devices, shorts, tools, or mechanical restraints have been removed before reenergizing. If the equipment cannot be visually inspected, electrical tests with test instruments must be performed.

2. Workers that are affected by reenergizing equipment or circuits must be notified. This includes machine operators, equipment operators, or any other workers in the immediate area.

3. Because they have the most knowledge about the equipment, machine or equipment operators who are qualified persons should be the employees that verify whether the equipment can be safely started up.

4. Nonessential items must be removed from the immediate area. The less equipment and materials that are in the immediate work area, the lower the chances for an accident.

5. All personnel must vacate the area that may be exposed to any hazards created during reenergizing of the equipment or circuits.

6. All components and equipment that were blocked and restrained must be unblocked and made ready for service.

Temporary Release for Testing/Positioning. The lockout/tagout procedure must identify the steps and qualified personnel required to temporarily reenergize equipment or circuits for repositioning or testing for other purposes. The responsibilities of the qualified personnel must be identified as well. The steps to temporarily reenergize equipment or circuits are the same steps required to return the equipment to service. An Informational Note to NFPA 70E 120.2(F)(2)(n) states that NFPA 70E 110.4(A) provides requirements for using test instruments and equipment.

TEMPORARY PROTECTIVE GROUNDING EQUIPMENT—NFPA 70E 120.3

NFPA 70E 120.3 provides a brief description of the required procedures for placing temporary protective grounds. This standard covers similar information that is contained in OSHA 1910.269(n). Often, the temporary protective grounding equipment used are personal protective grounds. The information concerning temporary protective grounding equipment per NFPA 70E 120.3 includes the following:

- Placement—Temporary protective grounds must be placed so that they do not expose employees to hazardous differences in potential. This is referred to as creating an equipotential zone. An *equipotential zone* is an area in which all conductive elements are bonded or otherwise connected together in a manner that prevents a difference of potential from developing within the area. A shock hazard exists when there is a difference in voltage between conductors or between conductors and ground. An equipotential zone places everything within the zone at the same voltage. There is no shock hazard when the potential is equal. A new requirement in the 2015 edition of NFPA 70E states that the location, sizing, and application of grounds must be part of the job planning.

- Capacity—Temporary protective grounds must have adequate ampacity to carry the short-circuit current for the amount of time necessary to safely clear a short circuit. **See Figure 6-13.** For example, if sizing is inadequate or the connection has high impedance due to poor connections or construction, a personal protective ground cluster can fuse or vaporize, causing an arc flash.

- Equipment approval—Temporary protective grounding equipment must meet ASTM F855 requirements. This is to prevent employees from using welding cables, welding clamps, automotive jumper cables, or any other similar type of equipment for protective grounds.

- Impedance—The impedance of the temporary protective grounding equipment must be low enough to allow the immediate operation of overcurrent protective devices in the event that the circuit is accidentally reenergized.

Temporary Protective Grounds

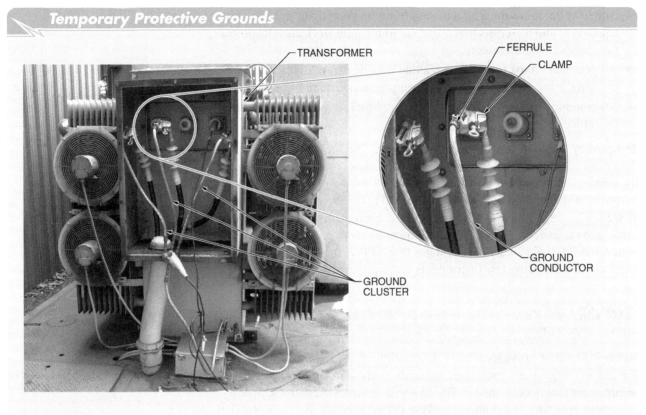

Shermco Industries

Figure 6-13. The clamps, ferrules, and conductors of a ground cluster must be rated for the amount of available short-circuit current that may flow for the period of time needed to clear the fault.

Additional Safety Precautions for Temporary Protective Grounding Equipment

The safety information in NFPA 70E related to temporary protective grounding equipment is brief. It does not expand upon the same type of information found in OSHA 1910.269(n). There are several additional safety precautions that should be followed when using temporary protective grounding equipment. These safety precautions include the following:

- There must be a method to track and control the placement of temporary protective grounding equipment. Several methods are available. One such method uses a bar code and a brass tag/clasp assembly. Each ground set is given a bar code label and a brass tag/clasp assembly. When the ground is placed in service, the clasp containing the brass tag is applied to the lockout box, while the second tag stays attached to the ground. The location of each ground, identified by the unique number on the tag, is listed on the lockout/tagout forms. Another method uses specialized tags. Green tags are placed on the locks applied during the lockout/tagout procedure. **See Figure 6-14.** The grounds are also listed as part of the lockout/tagout procedure.

- Never use components or devices for temporary protective grounds that are not sized or approved for that purpose. Items such as welding cables and clamps or automotive jumper cables may vaporize or explode off of the conductor when the short-circuit current flows through them. The resulting arc flash can damage equipment, create a fire hazard, and injure or kill nearby workers.

- Use as short a personal protective ground cluster as possible for temporary protective grounding. A shorter ground cluster will reduce the risk of damage if one of the conductors whips out when accidentally reenergized. In the ASTM F18 Committee meetings, a maximum of 20′ was discussed, because the length begins to affect impedance.
- Always test circuits for the absence of voltage before placing temporary protective grounds. There may be an induced voltage on the circuit or line to be grounded. This is usually seen in overhead power lines located in outdoor substations. The expanding and contracting magnetic field from other nearby power lines will often induce a voltage in a circuit that has been deenergized. This voltage can be lethal, even though it may be much less than the nominal voltage of the circuit. In situations where an induced voltage is present, the ground must be placed quickly to prevent unnecessary arcing. Once the ground is placed, the induced voltage will dissipate.
- Place temporary protective grounds close enough to protect workers, but not so close that they can injure other people nearby. When the short-circuit current hits the ground cluster it may whip around with tremendous force. This can cause people to fall off equipment, ladders, and platforms, which can cause injury or possibly death. However, the grounds must be positioned close enough to the work area to adequately protect people who may be working on the lines or equipment.

SAFETY FACT

Temporary protective grounds are used to provide protection against electric shock. Equipment is grounded to provide a direct path to the earth for unwanted current without causing harm to employees or equipment. Unwanted current may result from static buildup, faulted equipment, capacitors, electromagnetic coupling, high voltage testing, and backfed voltage.

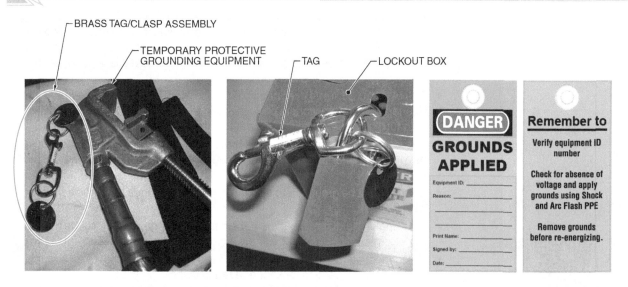

Figure 6-14. Several different methods can be used to track and control the placement of temporary protective grounding equipment.

- Temporary protective grounds must be placed using live-line tools. Never use hands, even with rubber insulating gloves, to place grounds. An arc flash may expose a worker to extremely high temperatures, well above what may be expected.
- Never come into contact with temporary protective grounds while they are being placed or after they are placed. The jacket on the grounds is used to protect the fine conductors of the ground cable from being damaged, not for protecting workers from nominal system voltage.
- Inspect temporary protective grounds before using them. ASTM F2499, *Standard Specification for In-Service Test Methods for Temporary Grounding Jumper Assemblies Used on De-Energized Electric Power Lines and Equipment* provides guidance on the proper inspection and testing of ground clusters.
- If the temporary protective grounds are found to be defective, tag and remove them from service.
- To prevent shock hazards, never stand near ground conductors that may be lying on the earth.
- Never coil ground conductors. If they are on carriers, remove all the conductors from the carrier and stretch them out on the earth or floor. When short-circuit current tries to pass through a coil of wire, the coil acts as an inductor or choke. This causes the coil to generate excessive heat, which may cause the conductors to vaporize.
- Inspect the ground conductors after using them and verify that they remain usable. Use ASTM F2499 as a guideline for inspection.
- Regularly test temporary protective grounds. ASTM F2499 states that grounds are to be tested at a "time interval established by the user to ensure that defective grounding jumper assemblies are detected and removed from service in a timely manner." Temporary protective grounds are safety-related devices and should be tested at least annually.

SUMMARY

- Electrical lockout/tagout, referred to as "establishing an electrically safe work condition" in NFPA 70E, is a necessity for protecting electrical workers from the three recognized hazards of electricity. These hazards include electric shock, arc flashes, and arc blasts.
- OSHA 1910.333(a)(1) requires that energized circuits or equipment be deenergized before work is performed on or near them. The safest practice is to turn off the equipment and deenergize it.
- The proper PPE can only reduce the effects that electrical hazards would have on a worker if an accident were to occur. PPE cannot eliminate risks completely.
- Circuits must be grounded to discharge induced voltages to make equipment safe for work.
- Use single-line diagrams, schematic diagrams, or other appropriate means to determine all sources of voltage.
- Disconnecting means such as a circuit breaker, switch, fuse, or disconnect switch must be used to lockout and tag devices.
- A copy of the written lockout/tagout procedures must be kept by the employer and must also be available to all employees.
- Temporary protective grounding equipment must meet ASTM F855 requirements.

Digital Learner Resources
ATPeResources.com/QuickLinks
Access Code: 705798

6

ESTABLISHING AN ELECTRICALLY SAFE WORK CONDITION

Review Questions

Name _____   Date _____

_____ 1. The primary method required by OSHA for protecting qualified electrical workers from electrical hazards is ___.
   A. equipment shutdown
   B. lockout/tagout
   C. posting signs
   D. using appropriate PPE

_____ 2. The process of removing the source of electrical power and installing a lock to prevent power from being turned on is called ___.
   A. lockout
   B. tagout
   C. equipment isolation
   D. forced shutdown

_____ 3. Tags used for lockout/tagout may include ___.
   A. the name of the worker who placed the tag
   B. the phone number of the worker who placed the tag
   C. the length of time the equipment is expected to be out of service
   D. all of the above

T   F   4. The process of placing a danger tag on the source of electrical power, which indicates that the equipment may not be operated until the danger tag is removed, is called tagout.

T   F   5. It is not required to verify that a circuit or equipment has been deenergized as long as all steps of the lockout/tagout procedure have been performed.

_____ 6. An absence-of-voltage test is performed to verify that a conductor or circuit part ___.
   A. contains voltage
   B. contains no voltage
   C. is properly installed
   D. is within tolerance

_____ 7. A ___ is a test instrument that indicates the presence of voltage by illuminating when the test tip is near an energized conductor or circuit part.
   A. proximity voltage tester
   B. solenoid voltage tester
   C. digital multimeter
   D. continuity tester

139

_____ **8.** A complex lockout/tagout procedure is required when ___.
   A. there are multiple energy sources
   B. there are multiple crews on the job site
   C. work is taking place at multiple locations
   D. all of the above

_____ **9.** NFPA 70E ___ identifies a list of requirements employees must fulfill when performing a complex lockout/tagout procedure.
   A. 120.2(B)
   B. 120.2(D)(1)
   C. 120.2(E)(2)
   D. 120.2(D)(2)

_____ **10.** An induced voltage is measured with a(n) ___.
   A. high-input impedance voltmeter
   B. low-input impedance voltmeter
   C. analog multimeter only
   D. digital multimeter only

_____ **11.** An area in which a person or the devices being used are at the same potential as the electrical system ground is a(n) ___ zone.
   A. safe work
   B. grounded
   C. equipotential
   D. high-hazard

_____ **12.** OSHA requires that all live parts to which an employee may be exposed be ___ before the employee works on or near them.
   A. removed
   B. deenergized
   C. isolated
   D. tagged

T   F   **13.** A backfed voltage, or backfeed, is an unexpected voltage that is created by placing a deenergized conductor near a conductor that is energized and carrying a load.

T   F   **14.** Low-voltage circuits routed through cable trays can induce a voltage into deenergized cables that are in the same cable tray.

_____ **15.** When isolating electrical equipment, OSHA allows the use of a tag without a lock when ___.
   A. the equipment has no means of accepting a lock
   B. tagging procedures provide a level of safety equivalent to a lock
   C. there is a single power source and the equipment is within line-of-sight
   D. Both A and B

_____ **16.** ___ cannot be used for isolating electrical equipment for lockout/tagout.
   A. Switches
   B. STOP/START stations
   C. Fused disconnects
   D. Circuit breakers

Chapter 6 – Establishing an Electrically Safe Work Condition

_____ **17.** A(n) ___ is used to reenergize electrical circuits.
　　　　A. automatic transfer switch
　　　　B. protective relay
　　　　C. across-the-line starter
　　　　D. regenerative drive

_____ **18.** Which of the following devices is not recommended for performing absence-of-voltage tests?
　　　　A. noncontact voltage tester
　　　　B. solenoid voltage tester
　　　　C. digital multimeter
　　　　D. all of the above

_____ **19.** OSHA ___ includes electrical lockout/tagout procedures.
　　　　A. 1910.332
　　　　B. 1910.333
　　　　C. 1910.334
　　　　D. 1910.335

T　　F　　**20.** A simple lockout/tagout procedure is not required to be in writing.

T　　F　　**21.** It is acceptable for employees to use their own personal locks for lockout/tagout.

WORKING ON ENERGIZED CONDUCTORS AND CIRCUIT PARTS

Electrical work should be performed on conductors and circuit parts that are placed in an electrically safe work condition. OSHA 1910.333(a) and NFPA 70E 130 both require safe work practices to be employed when employees are performing energized electrical work. Energized conductors and circuit parts will always present hazards such as shock, arc flash, and arc blast. The only way to eliminate these hazards is to place the energized conductors and circuit parts in an electrically safe work condition as required by OSHA and NFPA 70E.

OBJECTIVES

- Explain the importance of identifying when a task is considered energized work.
- List the conditions that may make energized electrical work appropriate.
- Describe the significance of an energized electrical work permit (EEWP).
- Explain the requirements for unqualified personnel working within or near the Limited Approach Boundary.
- Explain the requirements of an arc flash risk assessment.
- List precautions that are important for personal safety.
- List protective equipment that is not considered PPE.
- Determine the minimum approach distance between unqualified personnel and energized overhead lines.
- Explain the difference between touch potential and step potential.
- Describe the purpose of an equipotential zone.
- Explain employee training and job briefs.
- Explain how to properly service live-line tools.
- Describe how to safely apply temporary protective grounds.
- List the safety checks required by OSHA 1910.269(o) for high-voltage and high-power testing.
- Describe the hazard of open-circuiting a secondary winding of a current transformer (CT) when it is under load.

DETERMINING ENERGIZED WORK

It is important to identify when a task is considered energized work in order to follow the appropriate safety procedures for working on energized conductors and circuit parts. *Energized work* is work on or near exposed energized conductors or circuit parts where an employee is exposed to electrical hazards. NFPA 70E Table 130.7(C)(15)(A)(a) provides examples of tasks that are considered energized work, but it is not all-inclusive. As a general rule, any work that exposes an employee to electrical hazards is considered energized work. According to NFPA 70E 130.2(B)(3), it is not required that an energized electrical work permit (EEWP) is completed for troubleshooting, voltage testing, and other diagnostic activities. Even though an EEWP may not be required, the worker is required to comply with all the provisions of NFPA 70E 130.

Many electrical workers may not realize that they are performing energized work as specified by OSHA and NFPA 70E. Michael McCann, who is the Director of Safety Research for the Center to Protect Workers' Rights (CPWR), collaborated on an energized work practices study with Jeffery Potts, who analyzed the results of the study and used it for the basis of his graduate thesis related to occupational safety. In this study, 5000 IBEW electrical workers were sent a survey to examine energized work practices among electricians. Of the 5000 surveys sent out, 1329 were returned. **See Figure 7-1.**

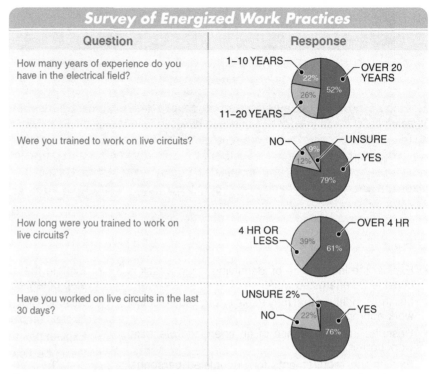

Figure 7-1. Many electrical workers may not realize that they are performing energized work as specified by OSHA and NFPA 70E.

According to the survey, 22% of electrical workers answered "No" to performing work on energized circuits in the last 30 days. However, 85% of that 22% of electrical workers also answered "Yes" to performing one of the energized tasks mentioned in the 2000 edition of NFPA 70E. **See Figure 7-2.**

Unrecognized Work on Energized Conductors and Circuit Parts

Energized Task	Electricians Performing Task
Voltage testing	63%
Operating a circuit breaker or fused switch with the cover open	47%
Opening hinged covers to expose bare, energized parts	47%
Removing/installing circuit breakers or fused switches	38%
Working on energized parts	29%
Repairing energized circuits	16%

Figure 7-2. Many electrical workers who do not realize they are performing energized work are actually performing tasks listed in the 2000 edition of NFPA 70E that are considered to be energized work.

Of the electrical workers responding to the survey, 39% of those who had performed work on energized conductors and circuit parts had less than four hours of training. While practical experience is important for performing electrical work, it is not a safe way to learn and is not supported by OSHA or NFPA 70E. When asked if their employers had a written safety program, 43% of the electrical workers answered "No." Studies have shown that compliance is much higher when there is a written safety program.

Shermco Industries
NFPA 70E Article 130 contains safe work practices for performing tasks such as racking medium-voltage circuit breakers.

WORK INVOLVING ELECTRICAL HAZARDS—NFPA 70E ARTICLE 130

NFPA 70E Article 130 covers when an EEWP is needed and what safety-related work practices are required when work must be performed energized. NFPA 70E 130.1 states that all requirements of Article 130 apply, regardless of whether the table method or an incident energy analysis is used to determine arc-rated clothing and PPE and the Arc Flash Boundary. Selection of arc-rated clothing and PPE is a small portion of NFPA 70E Article 130. Article 130 contains safe work practices concerning inspection of arc-rated clothing and PPE, approach by unqualified persons, illumination, alertness, look-alike equipment, and much more. Regardless of which method is used to determine arc-rated clothing and PPE and the Arc Flash Boundary, the user of NFPA 70E should be familiar with Article 130 in its entirety

ELECTRICALLY SAFE WORKING CONDITIONS—NFPA 70E 130.2

According to OSHA 1910.333(a)(1), *"Live parts to which an employee may be exposed shall be deenergized before the employee works on or near them, unless the employer can demonstrate that deenergizing introduces additional or increased hazards or is infeasible due to equipment design or operational limitations. Live parts that operate at less than 50 volts to ground need not be deenergized if there will be no increased exposure to electrical burns or to explosion due to electric arcs."*

If an employee is exposed to energized conductors and circuit parts, an electrically safe work condition must be created before work is allowed within the Limited Approach Boundary. The Limited Approach Boundary is the closest distance an unqualified person can approach exposed energized conductors or circuit parts that pose a shock hazard. NFPA 70E defines it as the distance at which a shock hazard exists from exposed energized conductors or circuit parts. The three conditions that allow work to be performed on energized conductors and circuit parts are the following:
- if a greater hazard is created or increased by deenergizing the conductors and circuit parts
- if it is infeasible to deenergize the conductors and devices
- if the energized conductors and devices are operating at less than 50 V and there would be no increased exposure to electrical burns or arc flash

Greater Hazards

According to OSHA 1910.333(a)(1), *"Note 1: Examples of increased or additional hazards include interruption of life support equipment, deactivation of emergency alarm systems, shutdown of hazardous location ventilation equipment, or removal of illumination for an area."*

Energized electrical work is permitted on energized conductors and circuit parts if deenergizing them causes additional or increased hazards. For example, the shutdown of ventilation equipment in a hazardous location or the deactivation of life support equipment in a hospital may cause additional or increased hazards. These examples do not suggest that working on energized conductors and devices is always acceptable in either situation. Examples are not exceptions. The decision to work on energized conductors and devices should be made after all other options have been considered.

Infeasibility

According to OSHA 1910.333(a)(1), *"Note 2: Examples of work that may be performed on or near energized circuit parts because of infeasibility due to equipment design or operational limitations include testing of electric circuits that can only be performed with the circuit energized and work on circuits that form an integral part of a continuous industrial process in a chemical plant that would otherwise need to be completely shut down in order to permit work on one circuit or piece of equipment."*

Energized electrical work is permitted if it is infeasible to deenergize the conductors and devices due to operational or design limitations. For example, diagnostics, startup testing, and troubleshooting can only be performed on energized circuits.

Energized Work at Less Than 50 Volts

Energized work is permitted if conductors and devices are energized at less than 50 V to ground. This applies only if the overcurrent protective devices (OCPDs) and the capacity of the power system can prevent an arc flash or explosion hazard. For example, energized work is permitted on a 48 V battery bank with an OCPD that can prevent a possible arc flash. **See Figure 7-3.**

Figure 7-3. An arc flash may occur when working around battery banks.

FIELD NOTES: INFEASIBILITY

OSHA issued a letter of interpretation (LOI) addressing infeasibility and the term "continuous industrial process." This is an area that most managers and other people have difficulty with, because to them, every manufacturing line or process would fall under this exception. In reality, OSHA takes a very narrow view on what constitutes a continuous industrial process. The LOI dated December 19, 2006, states the following:

*"**Scenario:** The manufacturing of our products involves many discrete pieces of equipment whose individual processes are part of the overall manufacture of integrated circuit components. For example, we have ten pieces of manufacturing equipment fed out of a 480-volt three-phase panel. A new project requires that additional feeders and a 225-ampere circuit breaker be added to the panel to supply a new piece of equipment. To perform the work in a de-energized state, it requires the power to the panel must be disconnected and appropriate LOTO devices applied. This activity would result in the shutdown of the ten pieces of equipment, causing a significant interruption to our ability to manufacture integrated circuits.*

*"**Question:** Is the panel considered part of a 'continuous industrial process,' thus allowing the work to be performed while the panel was energized using electrical safe work practices, as per Note 2 in §1910.333(a)(1)?*

*"**Response:** It appears that your panel is not part of a 'continuous industrial process.' The term 'continuous industrial process' was derived from its use in the National Electrical Code (NEC). In the NEC 'continuous industrial process' is used in the context of situations where the orderly shut down of integrated processes and equipment would introduce additional or increased hazards. Therefore, to qualify for the exception found in Note 2 of §1910.333(a)(1), the employer must, on a case-by-case basis, determine if the orderly shutdown of the related equipment (including the panel) and processes would introduce additional or increased hazards. If so, then the employer may perform the work using the electrical safe work practices found in §§1910.331-1910.335, including, but not limited to, insulated tools, shields, barrier, and personal protective equipment. If the orderly shutdown of the related equipment and processes would not introduce additional or increased hazards, but merely alter or interrupt production, then the de-energization of the equipment would be considered feasible, and the exception found in Note 2 of §1910.333(a)(1) would not apply. Based on the limited information you provided, it does not appear that de-energization of the panel in question would introduce additional or increased hazards."*

What most people consider infeasible is really only inconvenient. OSHA is not concerned with people being idle or the cost of shutting down circuits or equipment. OSHA is concerned with preserving human life. There are several definitions of infeasible, but the intended meaning is that there is no other way to perform the task.

An example of a continuous process in an industrial plant is replacing a door-up safety switch on a carbon furnace. Shutting down the whole furnace to replace this switch increases the explosion hazard, which endangers many workers involved in the shutdown and in the restart. Leaving the safety switch off also increases the hazard, which endangers workers in the furnace area. The switch is 120 V, which is a shock hazard. However, replacing a door-up safety switch while the furnace is energized is justifiable under the continuous process and the increased hazard clauses. If a shutdown is imminent, it is safer to perform this task during the shutdown. However, the process of shutting down is inherently more hazardous than the energized work.

NFPA 70E Table 130.7(C)(15)(A)(a) provides some insight as to what is considered to be an arc flash hazard. Working on energized electrical conductors and circuit parts of series-connected cells, including voltage testing, requires arc flash PPE. Voltage testing on individual battery cells or individual multicell units does not require arc flash PPE as long as the equipment is properly installed and maintained, covers are securely in place, and there is no evidence of impending failure. There are also differences in how batteries on open racks are handled versus batteries in enclosures for other tasks listed in this table.

There is no way to deenergize a battery. Batteries are always energized. If the cells are shorted by dropping a wrench or another conductive object, there is not only an arc flash hazard but also an arc blast and acid/electrolyte hazard.

SAFETY FACT

Arcing inside an electrical box creates hazards such as a sudden pressure increase (arc blast) and localized overheating (arc flash). A number of protective strategies are available to reduce these hazards, such as an arc-resistant switchgear, high-resistance grounding, remote racking devices, current-limiting fuses, circuit breakers, and a covered bus.

FIELD NOTES:
EXAMPLES ARE NOT EXCEPTIONS

A service technician walked into the maintenance office of a hospital and requested the automatic transfer switch (ATS) be deenergized so he could work on it. The hospital maintenance staff informed the technician that the ATS in question fed an emergency room and that it was a life-critical power supply. The technician proceeded to work on the switch in an energized state and it blew up, killing the technician.

The OSHA compliance officer who investigated the accident wrote the hospital several willful citations. One citation was that the hospital did not allow the technician to deenergize the ATS before working on it. The attorneys for the hospital challenged the citation by pointing to OSHA 1910.333(a)(1), which states, *"Live parts to which an employee may be exposed shall be deenergized before the employee works on or near them, unless the employer can demonstrate that deenergizing introduces additional or increased hazards or is infeasible due to equipment design or operational limitations. Live parts that operate at less than 50 volts to ground need not be deenergized if there will be no increased exposure to electrical burns or to explosion due to electric arcs. Note 1: Examples of increased or additional hazards include interruption of life support equipment, deactivation of emergency alarm systems, shutdown of hazardous location ventilation equipment, or removal of illumination for an area."*

The compliance officer responded by saying, "The Emergency Room was not in use 24 hours a day, seven days a week. When it is used, it has to be sterilized and at that time a portable generator, such as the one being used to power the Emergency Room now, could have been connected and this man's life could have been spared." He added, "The examples given in the regulation are not exceptions. They are examples of possible situations that might occur."

Do not confuse an example with an exception. Examples given by OSHA and NFPA 70E must be carefully interpreted. Misinterpretation of examples from OSHA and NFPA 70E can have fatal consequences.

Normal Operation

NFPA 70E 130.2(A)(4) is new to the 2015 edition. Over the last few NFPA 70E cycles, some of the users of the standard have voiced the opinion that it is unsafe to walk through a machine room or other area that contains operating electrical equipment and it is unsafe to open or close circuit breakers and switches. In their opinion, arc-rated clothing and PPE would be required in both of these situations.

Two Informational Notes are attached to the definition of an arc flash hazard in NFPA 70E Article 100. Informational Note No. 1 states that an arc flash may occur when energized conductors and circuit parts are within enclosed equipment as long as an individual is interacting with the equipment in a way that can cause an electric arc. Some users argued that opening and closing a switch or circuit breaker was considered interacting with the equipment in a way that can cause an electric arc. This was not the intent of the NFPA 70E Committee.

Informational Note No. 2 directs readers to Table 130.7(C)(15)(A)(a) to illustrate the types of activities that could cause an arc flash hazard. The activities these users are referring to are not listed on Table 130.7(C)(15)(A)(a).

To further clarify the intent of the NFPA 70E Committee, 130.2(A)(4) was added to explain the conditions required to constitute normal operation of electrical equipment. These conditions are the same as the conditions in Table 130.7(C)(15)(A)(a) when arc-rated clothing and PPE are not mandated. These conditions are that the equipment must be properly installed and maintained, doors and covers must be properly installed and secured, and no evidence of impending failure may be present. If these conditions are met, unless a qualified person crosses the Restricted Approach Boundary or the Arc Flash Boundary when the equipment is operating, there is no perceived hazard and arc-rated clothing and PPE are not mandated.

Arc-rated clothing and PPE are not mandated when operating equipment in the manner specified by the manufacturer. This means if using a START/STOP station is specified, arc-rated PPE is not mandated when a START/STOP station is used. If, however, the equipment is operated in a manner that the manufacturer does not specify, such as using a circuit breaker to start the machine when a START/STOP station is specified, then arc-rated PPE is mandated.

One other aspect of this section is that even though arc-rated clothing and PPE may not be mandated, that does not mean it is not required under any and all conditions. The qualified person must assess the risks at the time of operation and determine if the equipment is safe to operate. If he or she cannot say with certainty that the equipment is properly installed and maintained, if any of the doors, covers, or door fasteners are missing, or if there is any sign of impending failure, the qualified person must determine what PPE is needed or if the equipment needs to be shut down and repaired. Companies are free to exceed the minimum requirements given in NFPA 70E or the OSHA regulations. If a company mandates PPE for certain tasks, even though it may not be required by NFPA 70E, then the worker must comply.

The note attached to Table 130.7(C)(15)(A)(a) indicates that the table does not cover all possible conditions or situations, and where it is indicated that arc flash PPE is not mandated, an arc flash is not likely to occur. There are two key parts to this note. First, the NFPA 70E Committee cannot foresee every possible situation and circumstance. The worker has to assess the situation at the time the task is being performed to determine if the table method is suitable. Second, "not likely" does not mean never, nor does it mean it will not happen. "Not likely" means there is a reduced risk. It is important that the recommendations of the table method are not followed blindly.

Test Instruments and Equipment

NFPA 70E 110.4(A)(1) states that only a qualified person is allowed to use test equipment on electrical conductors or circuit parts energized at more than 50 V or where any electrical hazard exists. This may seem obvious, because if a person is not qualified he or she should not cross the Limited Approach Boundary. However, individuals still do so and often are injured or killed.

Section 110.4(A)(2) requires a test instrument and its accessories to be rated for the circuit about to be tested. This includes the test leads and other accessories. NFPA 70E 110.4(A)(3) requires the test instrument to be designed for the purpose it is about to be used for and the environment it will be used in. If it is not designed to be used in the rain, it might short out. Using a Category I test instrument where a Category III test instrument is specified could lead to serious injury or death if it is subjected to a transient. NFPA 70E 110.4(A)(4) requires a visual inspection of the test instrument before each use. Damaged or defective test instruments constitute a major hazard to the user. NFPA 70E 110.4(A)(5) requires verification of the test instrument before and after an absence of voltage test.

ENERGIZED ELECTRICAL WORK PERMITS

An EEWP is required whenever work is performed on conductors or devices that are not in an electrically safe work condition. **See Appendix.** It includes a risk assessment and specifies PPE to be used and other measures that may be required to protect both the employee and others who may be in the area. Signatures of supervisors and technicians involved in the work being performed are also required as a control measure to prevent unauthorized energized work from being performed. **See Figure 7-4.** An EEWP must include, but is not limited to, the following elements:

- description of the equipment or circuits that are to be worked on and where they are located
- justification for the energized work
- safe work practices to be used while performing the task
- results of the electrical risk assessment including shock and arc flash
- required Limited and Restricted Approach Boundaries, as well as the nominal voltage of the circuit, the incident energy available at working distance, and the Arc Flash Boundary
- personal protective equipment (PPE) required to do tasks safely
- steps used to restrict unqualified or unauthorized personnel from entering work zone
- evidence of a completed job briefing by completing a job briefing and planning form
- approval to work on the energized circuit

An EEWP is not required if the Limited Approach Boundary is crossed only for visual inspections and the person performing the inspection will not cross the Restricted Approach Boundary. Other activities that do not require EEWPs include the following:

- testing, troubleshooting, and voltage measuring, including calibration
- thermography and visual inspections if the Restricted Approach Boundary is not crossed

SAFETY FACT
Safe work practices and appropriate PPE are always required even when an EEWP is not.

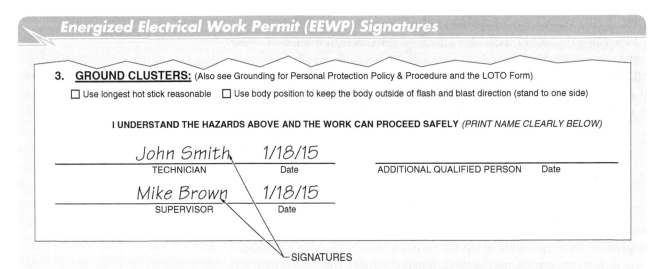

Figure 7-4. Energized electrical work permits (EEWPs) require the signatures of supervisors and technicians involved in the work being performed to prevent unauthorized energized work from being performed.

- access to an area with energized electrical equipment where no electrical work is performed and the Restricted Approach Boundary is not crossed
- housekeeping and other general nonelectrical tasks if the Restricted Approach Boundary is not crossed

The last two activities could pertain to entry into an operating substation. The substation yard is almost always within the Arc Flash Boundary, but groundskeepers and other nonelectrical personnel must have access. These exceptions should provide an appropriate level of safety and still allow employees to perform tasks without completing an EEWP.

Repair Work and Troubleshooting

OSHA and NFPA 70E do not specifically explain the difference between repair work and troubleshooting. If an employee has a tool in their hand, they are doing repair work and must complete an EEWP. If the employee has a test instrument in their hand, they are troubleshooting. Some companies develop specific tasks that fall into one category or another to further clarify the difference between repair work that requires an EEWP and troubleshooting.

APPROACH BOUNDARIES TO ENERGIZED ELECTRICAL CONDUCTORS OR CIRCUIT PARTS FOR SHOCK PROTECTION — NFPA 70E 130.4

Shock protection boundaries include the Limited and Restricted Approach Boundaries. Limited and Restricted Approach Boundaries are implemented whenever personnel may be exposed to energized conductors or devices. NFPA 70E Table 130.4(D)(a) is used to determine the approach boundaries for AC electrical systems and Table 130.4(D)(b) is used for DC electrical systems. The Limited and Restricted Approach Boundaries are independent of the arc flash boundary.

Shock Risk Assessment

A shock risk assessment is used to determine the nominal circuit voltage and approach boundaries (Limited and Restricted) for conductors or circuit parts that are energized and exposed. Once the nominal voltage is determined, the proper PPE (rubber insulating gloves and leather protectors) or insulating shielding can be determined to protect the employee from shock hazards.

Energized Conductors and Devices Operating at 50 V or More

A qualified person may not approach or bring any conductive object closer than the Restricted Approach Boundary unless one of the following applies:
1. The qualified person must be insulated or guarded from the energized conductor or circuit part. An acceptable form of protection would be rubber insulating gloves rated for the voltage. If it is necessary to protect the upper arm and shoulder from contact, rubber insulating sleeves should be used as well.
2. The energized conductors or devices are insulated to the rated voltage of the system.
3. The qualified person is insulated from the energized conductor or devices by using an insulated bucket or platform, such as during live-line work.

Approach by Unqualified Persons

Unqualified personnel are prohibited from crossing the Limited Approach Boundary, which is restricted to qualified personnel. However, there may be times when an unqualified person must enter these areas to perform cleaning or maintenance work. An unqualified person may enter these areas if a qualified person places the energized conductors and devices in an electrically safe condition first.

Working at or Close to the Limited Approach Boundary. When unqualified persons are working near the Limited Approach Boundary, they should be warned of the hazards and warned not to cross the boundary. Unqualified persons working near the Limited Approach Boundary should either be trained to recognize and avoid hazards or monitored by a qualified person. The fact that they are unqualified means they are not able to recognize hazards and cannot be expected to avoid them.

FIELD NOTES:
UNQUALIFIED PERSONNEL

A painter erected a three-story scaffold in Petersburg, VA, to paint a commercial building. As he proceeded to paint the building, he used a metal extension on his roller. He made contact with the weather loop feeding the building's electric power service and was electrocuted. The painter apparently thought the conductors were safe to paint around, not realizing they had weathered and the insulation was cracked in places. When he did not come home after work, his wife called the company he worked for. Coworkers went to his job site and found him dead. Painters, plumbers, carpenters, and other tradesworkers do not work with electricity and do not understand the risks involved or the hazards. They are considered unqualified because they do not recognize the hazards and cannot be expected to avoid them.

Entering the Limited Approach Boundary. If an unqualified person must cross the Limited Approach Boundary, such as for auditing or training purposes, they must be continuously escorted by a qualified person while within the Limited Space. If the qualified person leaves the Limited Space for any reason, the unqualified person must leave the Limited Space as well. Under no circumstance can an unqualified person cross the Restricted Approach Boundary, even when escorted by a qualified person or wearing appropriate PPE. **See Figure 7-5.**

Table 130.4(D)(a), previously 130.4(C)(a), was modified in the 2015 edition of NFPA 70E. The voltage ranges for AC used to be 50 V to 300 V and 301 V to 750 V. They are now 50 V to 150 V and 151 V to 750 V. Note "d" was added to explain that this includes circuits where the voltage does not exceed 120 V. This was required because the table is for phase-to-phase voltages. All phase-to-ground voltages previously had to be multiplied by 1.73 to determine PPE usage. This change allows the worker to change 120 V-class devices without wearing voltage-rated gloves and leather protectors if he or she cannot make contact. Note "a" was also modified to state that when the voltage is above 250 V, the single-phase voltage must be multiplied by 1.73 to determine PPE usage. Previously, all voltages were multiplied by 1.73.

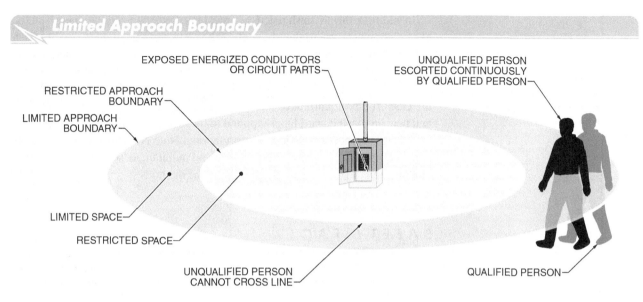

Figure 7-5. Unqualified personnel may cross the Limited Approach Boundary provided a qualified person escorts them continuously.

ARC FLASH RISK ASSESSMENT—NFPA 70E 130.5

NFPA 70E 130.5 was modified greatly for the 2015 edition. The requirements for when an arc flash risk assessment must be performed and what it should contain are now specified. The first step is to determine if an arc flash hazard exists. This can be accomplished by using the table method, if the limits of Table 130.7(C)(15)(A)(b) are not exceeded, or by conducting an incident energy analysis and a JHA/JSA. If there is an arc flash hazard, the appropriate safe work practices, the Arc Flash Boundary, and the arc-rated clothing and PPE must be determined.

An arc flash risk assessment must be completed before approaching energized conductors and circuit parts. An arc flash risk assessment is used to prevent injury by determining potential electrical arc flash hazards, the Arc Flash Boundary, the incident energy exposure at working distance, and the required PPE and arc-rated protective clothing. The arc flash risk assessment should also account for the type, maximum total clearing time, and maintenance of OCPDs.

The condition of maintenance of OCPDs is critical to worker safety. OCPDs almost never operate faster than the manufacturer's specifications. They operate slower. Incident energy is proportional to time, so if the OCPD clears the fault in more time than specified, which is what the table method and an incident energy analysis are based on, the worker could be severely underprotected even though he or she may be wearing the correct arc-rated clothing and PPE. Since the time of operation for faults is a few cycles per second, any increase in operating time causes a large increase in incident energy received by the worker.

An existing arc flash risk assessment must be updated whenever there is a major modification or renovation to an electrical power system. A major modification includes anything that would change the available short-circuit current or the operating time of an OCPD. For example, modifications include the installation of a larger transformer, a transformer with a lower percentage of impedance, larger cables or buses from the transformer into other parts of a system, new and larger motors (over 100 HP), or a large reactor.

In addition, the arc flash risk assessment must be reviewed for changes in the electrical power system that may affect the results of the analysis. An arc flash risk assessment review must take place every five years or less. If there are no changes that affect short-circuit current or incident energy calculations, no further action is required, except for documenting the results of the review. Changes that affect short-circuit current or incident energy calculations must be documented and new arc flash hazard labels applied to the affected equipment or circuits.

A new sentence was added to Informational Note No. 1 in 130.5. It cautions that if equipment is not properly installed and maintained, selection of arc-rated clothing and PPE may be inadequate using the methods specified in Article 130.

SAFETY FACT

One of the concerns of the NFPA 70E Committee was that required maintenance was not being performs on OCPDs, either to save money or due to ignorance of the effects of not performing maintenance. Since incident energy is proportional to time, lack of or failure to perform maintenance will increase the operating time of an OCPD. Doubling the operating time of an OCPD doubles the incident energy a worker receives.

Exceptions to Arc Flash Risk Assessment

The only alternative to performing an arc flash risk assessment is to use NFPA 70E Tables 130.7(C)(15)(A)(a), (b), and 130.7(C)(16). However, these tables have restrictive limitations and cannot be used for many high-energy systems. If the short-circuit current or the maximum total clearing time exceeds the limits specified in the notes of the table, then the tables cannot be used and an arc flash risk assessment must be conducted.

Incident Energy Analysis

An incident energy analysis is part of an arc flash hazard engineering study and is used to determine incident energy. Incident energy is a measurement of thermal energy (in cal/cm^2) at a working distance from an arc fault. Incident energy to a person's face or chest area at working distance is calculated using IEEE 1584, *Guide for Performing Arc Flash Hazard Calculations*. The arc rating of flame-resistant clothing and PPE must equal or exceed the estimated incident energy exposure at the specified working distance. Parts of the body that come even closer to the arc source, such as hands, will receive more incident energy and must be protected accordingly.

A maximum clearing time of 2 sec is often used when calculating incident energy. If a person is exposed to an arc flash, 2 sec is a reasonable amount of time for that person to move out of the way if possible. There are situations when a person may not be able to move from the area, such as when he or she is working from a bucket truck. According to Informative Annex D.2.4(2), if a worker is in an area where egress may be difficult, the arc flash risk assessment must calculate the incident energy at the actual exposure time. This requirement provides the necessary information that workers need if they are in an area with poor egress.

Arc Flash Boundary

The Arc Flash Boundary is the distance from exposed energized conductors or circuit parts where a person would receive the onset of second-degree burns on bare, unprotected skin. An Arc Flash Boundary may be necessary even when the equipment is in a guarded state, such as when inserting or removing a circuit breaker or inserting or removing motor control center (MCC) buckets or bus plugs. To calculate the Arc Flash Boundary, factors that must be known are the voltage of the energized conductors or circuit parts, the maximum total clearing time of the OCPD, and the available short-circuit current. See NFPA 70E Informative Annex D for Arc Flash Boundary calculations and additional information. The Arc Flash Boundary can also be determined using the table method.

Protective Clothing and Other Personal Protective Equipment (PPE) for Application with an Arc Flash Risk Assessment

Two different methods can be used for determining the required arc rating of protective clothing and PPE when working within the Arc Flash Boundary. The first method involves performing an incident energy analysis to determine the potential incident energy exposure to a person's chest and face at the specified working distance. Parts of the body that are closer than the chest or face will receive more incident energy and must be protected accordingly. The second method, the table method, uses NFPA 70E Tables 130.7(C)(15)(A)(a), (b), and 130.7(C)(16). NFPA 70E 130.7(C) can be referenced for additional information on these tables.

One issue that has repeatedly surfaced during the last few NFPA 70E cycles is that of companies that perform arc flash studies or their customers mixing the table method and an incident energy analysis together on arc flash warning labels. This has always been discouraged by the NFPA 70E Committee, and in the 2015 edition of NFPA 70E two statements were added. The first statement, in 130.5(C), states that either an incident energy analysis or the table method, but not both, can be used to determine appropriate arc-rated clothing and PPE for the same piece of equipment. The second statement, in 130.5(C), states that the result of an incident energy analysis cannot be used to specify arc-rated clothing and PPE when using the table method.

SAFETY FACT

All Arc Flash Boundaries are calculated at an incident energy exposure of 1.2 cal/cm² for all voltages.

The type of personal protective equipment (PPE) required when working on energized conductors and circuit parts is determined by performing an incident energy analysis or using the table method.

It is acceptable to use an incident energy analysis to determine PPE down to a certain level in the electric power system, then use the table method. For example, a company could use an incident energy analysis down to their 480 V MCCs. Below that (probably for lighting panels and other low-energy equipment), the company could use the table method. It is not acceptable, however, to use an incident energy analysis and to then specify a PPE category based on the table method on the same label. If using an incident energy analysis, Table H.3(b) in Informative Annex H should be used to specify the arc-rated clothing and PPE required. Tables and labels do not mix.

Equipment Labeling

According to NFPA 70E 130.5(D), electrical equipment such as switchboards, panelboards, industrial control panels, meter socket enclosures, and MCCs that are likely to require examination, adjustments, maintenance, or servicing while they are energized must have a warning label. **See Figure 7-6.** The phrase "such as" means this sentence is not intended to be an all-inclusive list and only illustrates examples of the type of equipment that will need to be field marked. The electrical equipment specified in NFPA 70E 130.5(D) must have a warning label that includes the nominal system voltage, arc flash boundary, and at least one of the following:
- available incident energy and the corresponding working distance
- minimum arc rating of clothing
- required level of PPE
- the PPE category using the table method

In order to prevent wholesale changeout of existing labels, NFPA 70E included a grandfather clause that allows existing labels applied prior to September 30, 2011, to remain in service if they contain the available incident energy or the required level of PPE.

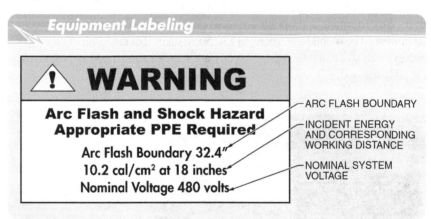

Figure 7-6. Equipment that falls under NFPA 70E 130.5(D) must be field marked with a label.

The basis for the information on the labels has to be documented. If the labels are based on an incident energy analysis, then the documentation from the incident energy analysis must be available. It does not have to be stated on the label, but it must be available for inspection. If the labels are based on the tables in NFPA 70E, those tables must be available and the limits of Table 130.7(C)(15)(A)(b) must not be exceeded. In either case, some engineering judgment must be used to support the method used to choose the arc-rated clothing and PPE.

One of the concerns of the NFPA 70E Committee is that many employees cannot determine the arc flash hazard exposure. This may be due to a lack of information, such as not knowing where to get the available short-circuit current or how to determine the maximum total clearing time of an OCPD, or due to a lack of training or expertise. Employees that lack training and experience may not be able to determine the appropriate arc flash protective equipment and clothing. The required label informs an employee of the appropriate PPE and clothing required for that specific device.

The labels required by NFPA 70E 130.5(D) are intended to warn qualified personnel that once the covers are removed from the equipment there is a hazard of arc flash and shock. These required labels pertain to panelboards, switchboards, industrial control panels, meter socket enclosures, and MCCs that are likely to need servicing, adjustments, or maintenance while energized. However, this requirement does not extend to all electrical equipment such as motors, light switches, or other similar devices.

Two new requirements were added to 130.5(D) of the 2015 edition of NFPA 70E. The first requirement is that the arc flash warning labels are required to be replaced if the review of the incident energy analysis determines the information on them is incorrect or out-of-date. The second requirement is that the equipment owner is responsible for the documentation, installation, and maintenance of the labels. At large facilities it is easy to determine who is responsible for the labeling requirements, but it may not be as clear at small stores or commercial sites. This new requirement ensures the owner of the equipment is the entity responsible for labels.

FIELD NOTES:
INCIDENT ENERGY AND DISTANCE

Incident energy decreases by the inverse square of the distance as a person moves away from an arc source. This describes the relationship between incident energy and distance and also indicates the importance of working distance. The reverse is also true: incident energy increases by the square of the distance as a person gets closer to an arc source.

If an arc flash hazard warning label specifies a level of incident energy at a working distance of 18", changes in distance will affect the incident energy a person receives. This is important to keep in mind if a worker moves closer to an exposed energized conductor in order to see through his or her bifocal glasses. Without realizing it, the worker will have reduced the effectiveness of his or her arc-rated clothing and PPE.

Below are a few examples of how the incident energy increases as distance decreases. The following calculations were performed using IEEE 1584-based software:
- At a working distance of 18", the incident energy is listed on the label as 8.0 cal/cm^2.
- At a distance of 12", the incident energy is 15.7 cal/cm^2.
- At a distance of 6", the incident energy is 49.2 cal/cm^2.

This is an example of an instance when a worker would probably wear arc-rated PPE and equipment rated at 10 cal/cm^2 because it is commonly available and would properly protected the worker for the specified working distance of 18". However, simply by decreasing the working distance, the worker might receive arc-related burns when he or she should not have received any.

OTHER PRECAUTIONS FOR PERSONNEL ACTIVITIES—NFPA 70E 130.6

NFPA 70E provides additional requirements that are important for personnel safety. The categories of precautions for personnel activities include alertness; blind reaching; illumination; conductive articles being worn; conductive materials, tools, and equipment being handled; confined or enclosed work spaces; housekeeping duties; occasional use of flammable materials; anticipating failure; routine opening and closing of circuits; and reclosing circuits after protective device operation.

Alertness

When employees are within the Limited Approach Boundary, they must be instructed to stay alert at all times where electrical hazards may exist. This includes both present hazards and additional or increased hazards that may develop due to a change in the energized electrical work being performed.

If employees show signs of illness, fatigue, or any other type of impairment, they must not be allowed to work where electrical hazards exist. Impairments may be caused by lack of sleep, emotional trauma, or drug use. Employees using alcohol and illegal drugs may show obvious signs of impairment. When a person is taking prescription or over-the-counter drugs, the signs of impairment may not be so obvious. The same drug can affect different people in different ways. When a doctor first prescribes a drug, it may take some time for a person to determine how the drug is affecting him or her and to acclimate to any side effects. It may also take some time for the doctor to determine the proper dosage. Employees taking prescription drugs should avoid working on energized conductors and devices until the proper dosage has been determined and they become acclimated to their medication.

Employees must also be instructed to be alert for any changes that may affect the safety of the job. Additional or increased hazards may be created when the original plan for a job or the scope of work is changed due to unforeseen circumstances. For example, if a guard or barrier is removed in order to gain access to energized equipment, employees must be made aware of the hazards that may result from this action.

Blind Reaching, Illumination, and Obstructed View of Work Area

According to OSHA 1910.333(c)(4)(ii), *"Where lack of illumination or an obstruction precludes observation of the work to be performed, employees may not perform tasks near exposed energized parts. Employees may not reach blindly into areas which may contain energized parts."*

Employees must not reach into equipment or areas that may contain exposed energized conductors or circuit parts, unless adequate illumination is provided. Adequate illumination means that there is enough lighting to clearly see all potential electrical hazards. If the work to be performed cannot be clearly seen, either due to an obstruction or lack of lighting, employees may not perform tasks where electrical hazards exist.

Some PPE, such as arc-rated face shields, have a dark or colored tint that reduces the amount of light transmitted through the face shield and can change the perception of colors. The manufacturers of arc-rated face shields and hoods have made significant improvements over the years to improve light transmission. However, reduced light transmission can still be an issue, especially with older arc-rated hoods. **See Figure 7-7.**

FIELD NOTES: CHANGES IN THE SCOPE OF WORK

On November 24, 1997, my friend Chuck was involved in an electrical accident caused by a change in the scope/conditions of the work. Chuck was a very talented, well-trained technician working for a company in Grapevine, Texas. He was promoted to Service Center Manager in Kansas City. One evening, Chuck was called out to an energy center near Lawrence, Kansas, to troubleshoot a 13.8 kV circuit breaker that was having problems.

Chuck was met by the supervisor and an electrical technician at the site. During the course of the work, it was decided that the barrier covering the energized bus stabs had to be removed. The subsequent OSHA report stated that a grounded piece of the shutter mechanism fell onto the energized stab, creating a fireball 26′ in diameter. The technician was killed instantly. The supervisor died five days later, while Chuck died seven days later. OSHA assessed $455,000 in fines against the energy center and the company that employed Chuck, including six willful violation citations and eight serious violation citations.

One of the citations was for the failure to conduct an additional job briefing when the scope of work changed, exposing the workers to additional or increased hazards. With the shutter in place, the equipment was in a guarded condition. When the decision was made to remove the shutter, a new job briefing should have been held to determine what new hazards might be present and what PPE or other equipment might be necessary to carry out the work safely. In fact, if they had assessed the situation, they probably would have realized that there was no safe way to perform the task. They were within a few dozen yards of a generator, working on a 13.8 kV circuit breaker. The short-circuit current and the incident energy must have been tremendous. The other citations were for the lack of PPE and training and for not maintaining guarding.

Why would three experienced electrical workers attempt such a task energized? There is no way to tell for certain, but they were probably fatigued from working throughout the day and wanted to get the job finished. Workers often want to do a good job at whatever it is that they do. They also tend to ignore risks when they feel that an equipment outage is affecting the operation of a facility, so they plow ahead with the work without taking the time to properly reevaluate the situation.

OSHA provides further incident reports on their website. Every one of these incident reports impacts families and friends, coworkers, and acquaintances.

Figure 7-7. Some PPE, such as arc-rated face shields, have a dark or colored tint that may reduce the amount of light transmitted through the face shield and may change the perception of colors.

Conductive Articles Being Worn

According to OSHA 1910.333(c)(8), *"Conductive articles of jewelry and clothing (such as watch bands, bracelets, rings, key chains, necklaces, metalized aprons, cloth with conductive thread, or metal headgear) may not be worn if they might contact exposed energized parts. However, such articles may be worn if they are rendered nonconductive by covering, wrapping, or other insulating means."*

NFPA 70E considers eyeglasses with conductive frames to be conductive apparel. Eyeglasses can slip or fall off a person's face or make contact with exposed energized conductors or circuit parts. Eyeglasses may be worn if they are restrained from falling or from contact by using goggles or nonconductive restraints.

Metal-Framed Eyeglasses

A majority of the NFPA 70E Committee felt that metal-framed eyeglasses presented a hazard when working within the Limited Approach Boundary. However, according to OSHA's letter of interpretation, wearing eyeglasses with exposed-metal frames or metal parts in the frames normally does not present an electrical contact hazard when they are restrained so that they cannot contact exposed energized electrical conductors or circuit parts. **See Figure 7-8.** The general recommendation for workers wearing metal-framed eyeglasses is to wear safety goggles over them or to purchase prescription safety glasses, which eliminates the entire issue and provides better vision.

Conductive Materials, Tools, and Equipment Being Handled

All conductive objects must be handled in a manner that prevents them from making contact with energized electrical conductors or circuit parts. A *conductive object* is an item that is not rated for the voltage to which it is exposed. Conductive objects or materials are to be restrained from coming closer to exposed energized conductors or circuit parts than the allowable distances in NFPA Table 130.4(D)(a) or 130.4(D)(b), column 2 or 3. However, objects may swing or the load may shift, which can cause contact. According to a study conducted by the National Institute for Occupational Safety and Health (NIOSH), 42% of all electrocutions are caused by contact with overhead lines.

Confined or Enclosed Work Spaces

According to OSHA 1910.333(c)(5), *"When an employee works in a confined or enclosed space (such as a manhole or vault) that contains exposed energized parts, the employer shall provide, and the employee shall use, protective shields, protective barriers, or insulating materials as necessary to avoid inadvertent contact with these parts. Doors, hinged panels, and the like shall be secured to prevent their swinging into an employee and causing the employee to contact exposed energized parts."*

Confined or enclosed work spaces pertain mostly to areas with limited access, such as manholes and vaults. However, confined or enclosed work spaces also apply to areas, such as switchgear and annunciator panels, where movement is limited by an enclosure and exposed energized conductors or circuit parts exist. An employer must provide rubber insulating shielding, barriers, blast blankets, or other types of insulating materials to prevent contact and other hazards that may be present in these areas. Blast blankets may be used in underground manholes or vaults where protection is needed against possible arc flash.

Letter of Interpretation Concerning Metal-Framed Eyeglasses

October 30, 1993

SUBJECT: Eyeglasses with Exposed Metal Parts

This is in response to your memorandum of July 17, 1992, requesting clarification of 1910.333(c)(8) as it may apply to eyeglasses with exposed metal parts. Please accept our apologies for the extensive delay in responding.

Eyeglasses with exposed metal parts are considered "Conductive apparel." As noted in the middle of column 2 of page 32007 of the preamble published in Volume 55, Number 151 of the **Federal Register** on Monday, August 6, 1990, the Electrical Safety Related Work Practice standard at 1910.333(c)(8) prohibits employees from wearing conductive objects in a manner presenting an electrical contact hazard. Normally, the wearing of eyeglasses containing exposed metal frames (or metal parts of frames) is not considered to present an electrical contact hazard. However, when the glasses have a metal type frame and the employee is working with his or her face extremely close to energized parts or when a metallic chain strap is attached to the frame for wearing around the neck, an electrical contact hazard can be present. In such cases, the standard permits the hazard to be removed by eliminating the chain and wearing either a protective face shield or appropriate safety glasses over the metal frame optical glasses.

OSHA Letter of Interpretation dated October 30, 1993

Figure 7-8. According to OSHA's letter of interpretation, wearing eyeglasses with exposed-metal frames or metal parts in the frames normally does not present an electrical contact hazard.

Doors and Hinged Panels

Doors, hinged panels, and other covers must be secured from closing and making contact with employees. If a door or hinged panel could create a hazard, such as bumping into the worker, or could cause injury, either by swinging into the worker or by causing the worker to bump into something, it must be secured.

Clear Spaces

The working space around equipment usually means a three-dimensional area in front of the equipment. NFPA 70E 130.6(H) requires that all sides of equipment that are needed to operate or maintain the equipment must be kept clear. This space cannot be used for storage.

Housekeeping Duties

According to OSHA 1910.333(c)(9), *"Where live parts present an electrical contact hazard, employees may not perform housekeeping duties at such close distances to the parts that there is a possibility of contact, unless adequate safeguards (such as insulating equipment or barriers) are provided. Electrically conductive cleaning materials (including conductive solids such as steel wool, metalized cloth, and silicon carbide, as well as conductive liquid solutions) may not be used in proximity to energized parts unless procedures are followed which will prevent electrical contact."*

Housekeeping duties that present electrical hazards must not be performed within the Limited Approach Boundary unless insulating blankets, shields, or other types of insulating materials are used or barriers are erected to prevent inadvertent contact. Conductive cleaning materials must not be used when working within the Limited Approach Boundary unless procedures to prevent contact are used. *Note:* Only qualified persons are allowed to work within the Limited Approach Boundary. Unqualified personnel must not cross the Limited Approach Boundary unless continuously escorted by a qualified person.

Occasional Use of Flammable Materials

Where flammable liquids or other combustibles such as solvents, fuels, or other types of flammable materials are occasionally present, electric equipment that could ignite them through sparks or arcs must not be used unless precautions are taken to prevent a hazardous condition. Precautions include measures such as ventilating or exhausting vapors and maintaining a clean working environment. Other than solvents and fuels, flammable materials include but are not limited to combustible dust, ignitable fibers or flyings, and flammable gases, vapors, or liquids.

Anticipating Failure

Equipment showing signs of imminent failure must be deenergized before work is performed on or near it. However, it may be necessary to perform work while the equipment is energized if deenergizing the equipment introduces additional or increased hazards or if it is infeasible.

According to NFPA 70E, employees must be protected from any hazard associated with equipment failure until the equipment is either deenergized or repaired. This may involve establishing a clear area around such equipment to prevent injury to personnel, using blast blankets or other types of PPE and equipment to reduce the effects of a failure, or implementing other steps as necessary.

Routine Opening and Closing of Circuits

Only devices specifically designed as a disconnecting means, such as load-rated switches or circuit breakers, can be used to open, close, or reverse circuits that are carrying load current. Only in an emergency may terminal lugs, fuses, cable connectors not of the load-break type, or cable splice connections be used to open, close, or reverse a circuit carrying load current.

SAFETY FACT

The Informational Note after NFPA 70E 130.6(J) refers the reader to the NEC® for areas where flammable materials are continuously present. Refer to NEC® Articles 500 through 516.

Reclosing Circuits after Protective Device Operation

Once an OCPD such as a circuit breaker or fuse has operated, it cannot be closed until it is safe to do so. The repetitive manual reclosing of circuit breakers or reenergizing circuits through replaced fuses is prohibited. Repetitive does not mean a circuit breaker can be reset once. Repetitive refers to any reclosing of a circuit breaker that has tripped. After an OCPD has been opened, inspection of the circuit is required to ensure it is safe. Inspection means the circuit has to be tested to ensure a short circuit does not exist. Reclosing a circuit breaker into a short circuit could cause the circuit breaker to fail violently. OSHA and NFPA 70E state that no further inspection is required if it can be determined from both the design of the OCPD and the circuit involved that an overload rather than a fault condition caused the trip operation.

OTHER PROTECTIVE EQUIPMENT—NFPA 70E 130.7(D)

The protective equipment in NFPA 70E 130.7(D) pertains to non-PPE-related items that are often used on or near exposed energized conductor and circuit parts within the Limited Approach Boundary. These types of protective equipment include but are not limited to insulated tools, fuse and fuse holding equipment, test instruments, ropes and handlines, fiberglass-reinforced plastic (FRP) rods, portable ladders, protective shields, rubber insulating equipment, voltage-rated plastic guard equipment, and physical or mechanical barriers.

Insulated Tools and Equipment

When working in areas where tools or handling equipment could make contact with exposed energized conductors or circuit parts, tools that are insulated from the voltages must be used. All insulated tools used on low-voltage circuits are rated at 1000 V and must have the required ASTM markings. **See Figure 7-9.**

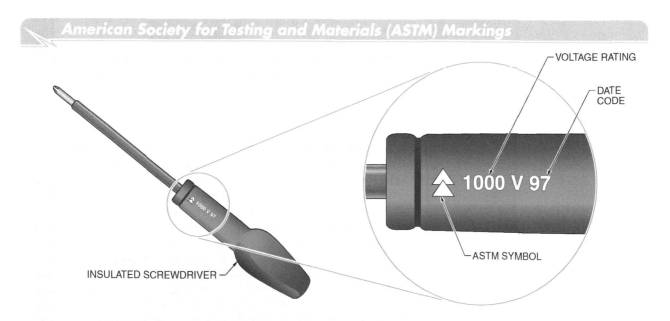

Figure 7-9. Approved insulated hand tools must have the required American Society for Testing and Materials (ASTM) markings on them. This includes the double triangle symbol, the voltage rating, and the date code of when the tool was manufactured.

According to OSHA, insulated hand tools and equipment are required when working on, near, or in areas where tools could make contact with energized conductors or circuit parts. NFPA 70E specifically states that insulated hand tools and equipment are required when working within the Restricted Approach Boundary.

Insulated hand tools and equipment can be used for simple tasks such as removing panel covers. While removing the cover itself does not require insulated hand tools, there is still the risk of creating a hazard if the tool is dropped and makes contact with an energized conductor or circuit part during the removal. For low-voltage systems, an insulated tool should not cause an arc flash if incidental contact is made.

Insulated hand tools and equipment must be protected from damage. The handle of an insulated hand tool is covered with a dielectric material that reduces the chance of electric shock. On two-layer tools the dielectric material is protected by a rubber outer layer that also reduces the chance of electric shock. If the outer layer of an insulated hand tool is damaged to where the inner dielectric layer can be seen, the hand tool must be taken out of service and tagged. **See Figure 7-10.**

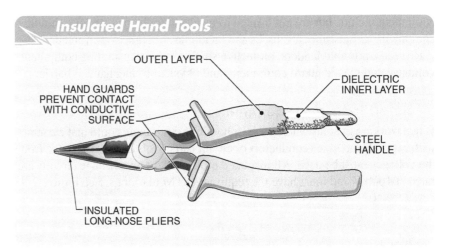

Figure 7-10. If the outer jacket of an insulated hand tool is damaged to where the inner dielectric layer can be seen, the hand tool must be taken out of service and tagged.

Requirements for Insulated Hand Tools. Insulated hand tools must be rated for the voltages of the circuit or equipment into which they may come into contact. They must also be designed for the manner and environment in which they will be used. Before each time an insulated hand tool is used, it should be inspected. If there is any damage found that could affect an employee's safety, the hand tool must be tagged and removed from service.

SAFETY FACT

When an insulated hand tool is damaged, all the insulation should be stripped from the handle so it cannot be mistaken for a properly insulated tool. Since it is not cost-effective to retest insulated or insulating hand tools, two-layer tools cannot be used on energized conductors or circuit parts if they show signs of damage. This includes cuts, rips, or any penetrations to the outer layer that expose the dielectric underlayer. Single-layer tools must be removed from service if the depth of the cut or damage cannot be positively determined.

Fuse and Fuseholder Equipment

Fuse and fuseholder equipment must be rated for the circuit voltage of the energized terminals. However, this does not imply that the fuses may be removed while under load. Unless it is a small control-type circuit, the circuit must be opened with no power feeding it. Trying to remove a fuse under load can cause an arc flash that could seriously injure an employee. Authorized fuse pullers must always be used when removing a fuse. Only newer-style fuse pullers should be used to remove fuses. **See Figure 7-11.** Older-style fuse pullers manufactured from laminates should be retired.

Ropes and Handlines

Ropes and handlines are required to be nonconductive if they are used within the Limited Approach Boundary. Ropes and handlines are often used to raise and lower equipment, tools, and supplies when working from insulated platforms or bucket trucks. Ropes and handlines must be insulated because they may contact energized conductors or circuit parts.

Fiberglass-Reinforced Plastic (FRP) Rods

Live-line tools must meet ASTM Standard F711, *Standard Specification for Fiberglass-Reinforced Plastic (FRP) Rod and Tube Used in Live Line Tools*. They must also be inspected before each use. It is recommended to use IEEE 978 for inspecting and testing live-line tools.

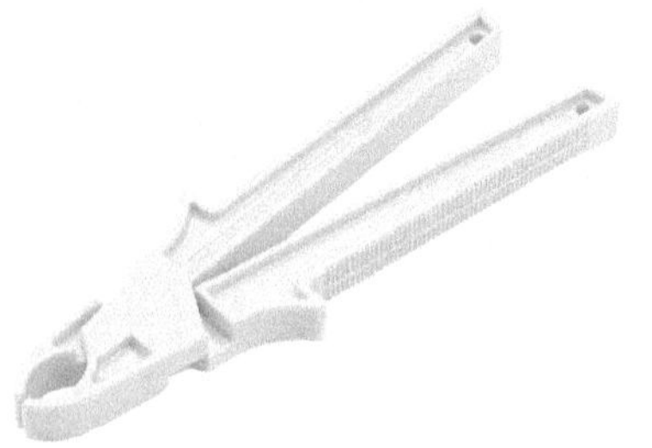

Ideal Industries, Inc.

Figure 7-11. Fuse pullers are used to safely remove cartridge fuses from electrical boxes and cabinets.

Portable Ladders

Ladders must have nonconductive side rails if they are used where they may contact exposed energized conductors or circuit parts. Ladders must also meet ANSI Standard A14.5, *Safety Requirements for Portable Reinforced Plastic Ladders*. Most ladders used today are made of fiberglass. There are also ANSI standards for wooden ladders, but these ladders are not commonly used for performing electrical work. Aluminum ladders should not be used for performing electrical work.

Defective ladders must be tagged and taken out of service. Ladders that are defective and not tagged are in violation of OSHA regulations, even when locked in a room and chained up. Ladder defects include missing nonslip feet, broken or cracked rungs, and cracks in the side rails. OSHA requires at least Type I ladders, which are rated for 250 lb. OSHA calculates that a full-grown man carrying tools will weigh around 250 lb. **See Figure 7-12.**

Protective Shields

When employees are within the Limited Approach Boundary, they must be protected from hazards using protective shields, insulating materials, or protective barriers. Unqualified personnel must be protected whenever covers are removed and exposed energized conductors or circuit parts are present.

Rubber Insulating Equipment

Rubber insulating equipment must meet applicable ASTM standards listed in NFPA 70E Table 130.7(C)(14) and Table 130.7(F). Types of rubber insulating equipment include rubber insulating gloves, blankets, sleeves, and shielding.

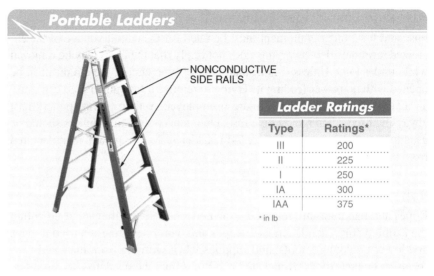

Figure 7-12. To prevent potential hazards, portable ladders must have nonconductive side rails if they are used in areas where they may contact exposed energized conductors or circuit parts.

Voltage-Rated Plastic Guard Equipment

Voltage-rated plastic guard equipment must meet ASTM Standard F712, *Standard Test Methods and Specifications for Electrically Insulating Plastic Guard Equipment for Protection of Workers*. Types of voltage-rated plastic guard equipment include plastic insulating gloves, blankets, sleeves, and shielding.

Physical or Mechanical Barricades

Physical or mechanical barricades may not be installed closer than the Restricted Approach Boundary. If for any reason the Restricted Approach Boundary cannot be maintained, the circuits or equipment must be deenergized. Types of physical or mechanical barricades include approximately waist high safety barricade tape, plastic chains, or plastic barricades.

Alerting Techniques

Alerting techniques are used to warn and protect personnel from potential hazards, such as electric shock, burns, or electric equipment failure, that can result in injury. Alerting techniques include safety signs, barricades, and attendants.

Safety Signs and Tags. Safety signs and other alerting devices are required to warn employees about electrical hazards that may be present. Signs must meet standards from ANSI Standard Z535, *Series of Standards for Safety Signs and Tags*. ANSI Z535 requires a different format than what may be seen on older signage. The new format is intended to provide more information through the use of color and pictograms. **See Figure 7-13.**

Barricades. Barricades are used with safety signs to limit access to areas where there may be exposed energized conductors or devices. Conductive barricades must not be used. Also, barricades may not be placed closer than the Limited Approach Boundary or the Arc Flash Boundary, whichever is the greater distance from the equipment.

Safety Signs and Tags

OLD FORMAT

NEW FORMAT

Figure 7-13. New signage required by ANSI Z535 provides more information in a user-friendly format.

Attendants. Attendants are used if signs or barricades do not prevent people from accessing the work area. The primary purpose of an attendant is to warn unqualified personnel of the hazards that may be present and to prevent them from entering the work area. An attendant must be stationed in hazardous work zone areas when there is reason to believe posted warnings may be violated or if there have been violations in the past. An attendant is required to stay at a work area as long as a hazard exists.

FIELD NOTES: ATTENDANTS

At Shermco we were working on a project in an underground garage. The building manager had made arrangements for employees to park in another area while the work was being performed. We barricaded the entrance to the garage with safety barrier tape and stationed an attendant to inform the employees of the building manager's arrangements.

We were progressing well with our energized work until a driver pulled into the garage. The driver was informed that we were doing energized work and that the building manager had arranged for other parking. The driver looked at the attendant and said, "You're not a cop. You can't tell me what to do."

The driver then drove through the safety barrier tape and into a parking spot. We had to shut down the job until the building manager could locate the driver and enforce the temporary parking arrangements.

This incident highlights a couple of important points. First, ignorance combined with arrogance is extremely dangerous. Second, if an unauthorized or unqualified person enters a safe work zone, the area must be secured and made safe so that no one is injured. As responsible workers, we could not allow the driver to be in harm's way. We had to shut down our work and make it safe until that person left the area.

If you are working in an area that has had a safe work zone violation in the past or if you have reason to think it could be violated, an attendant must be placed.

Figure 7-14. An alerting method must be used to help an employee differentiate between look-alike equipment.

Look-Alike Equipment. One of the three alerting techniques must be used when there is equipment of similar size, shape, and construction in the immediate work area. This helps employees differentiate between look-alike equipment. There are several situations where the problem of look-alike equipment can occur. For example, there may be dozens, even hundreds, of similarly sized and shaped MCCs all placed together. **See Figure 7-14.** If there are several MCCs adjacent to the MCC being worked on, it may be possible for a technician to enter the wrong one if the technician has to leave the area to get parts. Applying a tag, setting up safety barrier tape around the MCC, or even applying a piece of safety barrier tape attached to the enclosure door will provide indication of the correct MCC.

Another situation common in outdoor substations is when there are several circuit breakers in a row. They all look the same, but one of them is deenergized and the others are not. Since circuit breakers do not hum like transformers do, the technician needs to be extra careful to verify he or she is working on the correct piece of equipment. In this case, an absence of voltage test would verify it is deenergized. To prevent any misunderstanding, if the immediate area has to be left for a period of time, one of the three identification means can be employed in this situation.

STANDARDS FOR OTHER PROTECTIVE EQUIPMENT

All protective equipment in NFPA 70E 130.7(D) must conform to the applicable standards in NFPA 70E Table 130.7(F). This is a redundancy designed to cover any piece of PPE or equipment not specifically addressed in NFPA 70E 130.7(D).

WORK WITHIN THE LIMITED APPROACH BOUNDARY OR ARC FLASH BOUNDARY OF UNINSULATED OVERHEAD LINES — NFPA 70E 130.8

According to OSHA 1910.333(c)(3), *"If work is to be performed near overhead lines, the lines shall be deenergized and grounded, or other protective measures shall be provided before work is started. If the lines are to be deenergized, arrangements shall be made with the person or organization that operates or controls the electric circuits involved to deenergize and ground them. If protective measures, such as guarding, isolating, or insulating, are provided, these precautions shall prevent employees from contacting such lines directly with any part of their body or indirectly through conductive materials, tools, or equipment."*

Precautions must be taken so that personnel working near uninsulated overhead lines do not make contact with those lines, neither with their bodies nor with any conductive object they may be carrying. A conductive object does not have an insulation rating for the voltages involved in the work being performed, or its rating is less than the nominal voltage of the circuit with which it may come in contact. This includes objects that are not typically thought of as conductive, such as ropes, 2 × 4s, or vinyl electrical tape.

Deenergizing and Guarding

Overhead lines must be deenergized and grounded if there is a possibility of an employee making contact with them. When lines or equipment are deenergized, the company or utility that operates them must deenergize and ground them at the point where the work is being performed. Other methods used to protect personnel from making contact with their bodies or with any conductive objects include guarding, isolating, or insulating the overhead lines.

Determination of Insulation Rating

According to NFPA 70E 130.8(B), a qualified person must determine whether the insulation rating of overhead lines is adequate for the voltage the lines are operating. The concern is that many workers assume that the overhead line is insulated if it has a jacket over the conductors. In many cases, the jacket is only to provide protection from weathering and does not provide any insulation from electric shock.

Employer and Employee Responsibility

Both the employee at the site and the employer share a responsibility to ensure that lines and equipment are properly secured or that other protection methods are in place. Employees are also responsible for wearing appropriate PPE and implementing the required safe work procedures and practices.

Approach Distances for Unqualified Persons

According to OSHA 1910.333(c)(3)(i)(A), *"When an unqualified person is working in an elevated position near overhead lines, the location shall be such that the person and the longest conductive object he or she may contact cannot come closer to any unguarded, energized overhead line than the following distances: (1) For voltages to ground 50kV or below - 10 feet (305 cm); (2) For voltages to ground over 50kV - 10 feet (305 cm) plus 4 inches (10 cm) for every 10kV over 50kV."*

The minimum safe approach distance near overhead lines for an unqualified person is based on a voltage of 50 V to 50 kV to ground. When an unqualified person is working near overhead lines or equipment, they must comply with the Limited Approach Boundary distance specified in NFPA 70E Table 130.4(D)(a) or (b), column 2. The Limited Approach Boundary is the distance from the energized line to the person or any conductive object the person may be carrying. **See Figure 7-15.**

The Limited Approach Boundary distances given in NFPA 70E Table 130.4(D)(a) or (b), column 2 are for exposed moveable conductors. Overhead power lines fall primarily into this category because they sag and sway in the wind. These approach distances also apply to unqualified persons operating bucket trucks or aerial lift platforms that bounce around and sway in the wind. OSHA requires a spotter when approaching an overhead power line because it can be difficult to judge distances. If the operator of the bucket truck or aerial lift platform is a qualified person and the bucket truck or aerial lift platform is insulated to the circuit voltage, no spotter is required.

SAFETY FACT

Phase-to-ground values are used for determining approach distances in OSHA 1910.333(c). NFPA 70E shows approach distances only in phase-to-phase values. To convert phase-to-ground values to phase-to-phase values, multiply the phase-to-ground value by 1.732. For example, 50 kV phase-to-ground is the same as 72.5 kV phase-to-phase (50 kV × 1.732 = 72.5 kV).

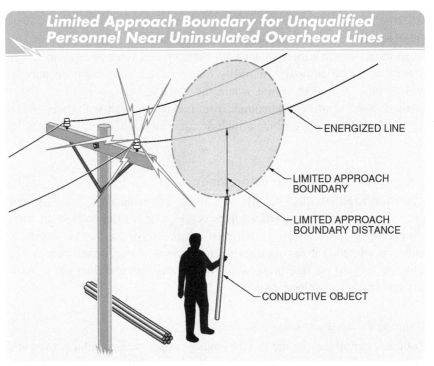

Figure 7-15. The Limited Approach Boundary for an unqualified person working near uninsulated overhead power lines is measured from the energized line to the person or any conductive object the person may be carrying.

Vehicular and Mechanical Equipment

According to OSHA 1910.333(c)(3)(iii)(A), *"Any vehicle or mechanical equipment capable of having parts of its structure elevated near energized overhead lines shall be operated so that a clearance of 10 ft. (305 cm) is maintained. If the voltage is higher than 50kV, the clearance shall be increased 4 in. (10 cm) for every 10kV over that voltage."*

Mechanical equipment that can be elevated and make contact with an overhead power line must stay the minimum distance given in NFPA 70E Table 130.4(D)(a) or (b), column 2. These approach distances may be reduced for the following reasons:

- Mechanical equipment and vehicles that are in transit with their booms retracted so they cannot make contact with overhead power lines or equipment may have the approach distance reduced to as little as 6′ for up to 50 kV to ground (72.5 kV phase-to-phase). For higher voltages, add 4″ for every additional 10 kV to ground.
- Insulated barriers that are not part of an attachment to the vehicle and are rated for the voltage may have the approach distance reduced to the working clearance for the insulating barrier. Line covers and line hoses can be used to provide temporary insulation up to about 34.5 kV. There is no temporary insulating shielding available for systems rated above 34.5 kV. This provision allows the use of rubber barriers and shields on lines and installed equipment, but does not allow the shielding to be used on the crane or other mechanical equipment. The shielding must be placed on the energized lines or equipment.

- An insulated lift that is rated for the voltage and operated by a qualified person may approach the Restricted Approach Boundary specified in NFPA 70E Table 130.4(D)(a) or (b), column 4. If the employee in the lift is wearing proper PPE, they may approach closer to perform live-line work as needed.

If the mechanical equipment is a mobile crane or derrick being used for construction purposes, its use probably would fall under OSHA 1926 Subpart CC. This regulation went into effect November 8, 2010, and contains requirements for approach distances. The work zone for mobile cranes must be established prior to the start of work. The work zone for mobile cranes is an area 360° around the crane extending to its maximum working radius. The mobile crane cannot come any closer than the minimum clearance distance listed in OSHA 1926.1408, Table A. **See Figure 7-16.** Table A provides the minimum acceptable working distances for voltages from 50 V to 1,000,000 V.

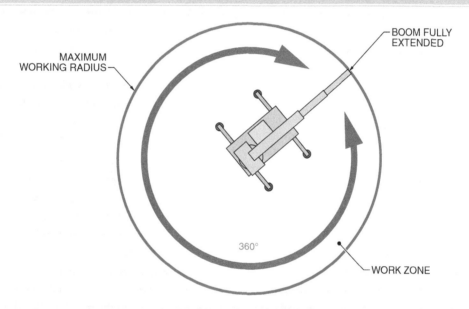

Mobile Crane Clearances

Table A - Minimum Clearances

Voltage (nominal, kV, AC)	Minimum Clearance Distance (ft)
Up to 50	10
Over 50 to 200	15
Over 200 to 350	20
Over 350 to 500	25
Over 500 to 750	35
Over 750 to 1000	45
Over 1000	As established by the power line owner/operator or registered professional engineer who is a qualified person with respect to electrical power transmission and distribution

TABLE A FROM OSHA 1926 SUBPART CC

Table T - Minimum Clearances

Voltage (nominal, kV, AC)	Minimum Clearance Distance (ft)
Up to 0.75	4
Over 0.75 to 50	6
Over 50 to 345	10
Over 345 to 750	16
Over 750 to 1000	20
Over 1000	As established by the power line owner/operator or registered professional engineer who is a qualified person with respect to electrical power transmission and distribution

TABLE T FROM OSHA 1926 SUBPART CC

Figure 7-16. OSHA 1926.1408, Table A, and OSHA 1926.1411, Table T, are used to determine minimum acceptable working distances for mobile cranes.

OSHA 1926.1411, Table T, applies to equipment that has its boom lowered and secured and is in transit. This table also requires that a dedicated spotter who is in constant contact with the equipment operator be used if the mechanical equipment could come closer than 20′. The dedicated spotter must be able to accurately gauge distances. If the mechanical equipment is traveling at night or when visibility is poor, then overhead power lines must be illuminated and the path the equipment will take must be identified and used. There are further requirements if the equipment will be intentionally brought closer than 20′ to an energized overhead power line.

Equipment Contact. Employees on the ground are not permitted to make contact with a vehicle, mechanical equipment, or any attachment due to the possibility of touch potential. *Touch potential* is a difference in voltage between an energized object and a person in contact with that object, which causes current to flow through that person's body. One of the following two methods is used to allow contact with a vehicle or mechanical equipment:

- Protective insulating equipment that is rated for the voltage is used.
- The vehicle or mechanical equipment is positioned so that no uninsulated part, such as a boom, platform, or bucket, can come closer than the distances allowed in NFPA 70E 130.8(F)(1).

Equipment Grounding. When a vehicle or mechanical equipment is intentionally grounded and there is a possibility of contact with overhead power lines, employees may not stand near the grounding point. If the elevated structure makes contact with an energized line, current from the overhead power line will flow through the equipment into the ground, causing ground potentials within several feet of the grounding point. This is referred to as step potential. Precautions to prevent employees from being injured by touch or step potentials must be implemented. Precautions may include using ground mats to create an equipotential zone; requiring employees to wear dielectric overshoes per ASTM F1117, *Standard Specification for Dielectric Footwear;* or insulating or barricading the area beyond where ground potentials might exist.

Touch Potential. Touch potential is a difference in voltage between an energized object and a person in contact with that object, which causes current to flow through that person's body. There is a risk of touch potential whenever a vehicle or mechanical equipment is operating near uninsulated overhead power lines. If the vehicle or mechanical equipment makes contact with an energized line, the current will flow through it to ground. If a person is in contact with the vehicle or mechanical equipment, it can cause a hazardous condition because the current will take all paths to ground. Most of the current will flow through the path of least resistance. However, large amounts of current can flow through alternate paths, such as a person's body, and cause injury. **See Figure 7-17.** If a ground rod with low resistance has not been installed or if the ground rod has a poor connection, a person's body will act as a path for the current, causing shock or electrocution.

Step Potential. *Step potential* is a difference in voltage between each foot of a person when standing near an energized object. For example, step potential may occur when standing next to a ground rod during a fault. There is a risk of step potential whenever a vehicle or mechanical equipment is operated near overhead power lines. If a vehicle or mechanical equipment makes contact with an energized line, current flows through it into the ground. From the ground, current passes through "shells" of earth resistance. These shells are assumed to be equal in thickness and become larger in surface area as they move away from the grounding point.

Chapter 7 – Working on Energized Conductors and Circuit Parts

The grounding point can be a ground rod or the wheels of the vehicle or mechanical equipment making contact with the ground. The shells of earth resistance closest to the grounding point will have the highest resistance. As the current flows through each shell, the resistance decreases because the surface area of each shell becomes larger as it moves away from the ground rod or grounding point. As the resistance decreases, the voltage also decreases. For example, if a person were standing with one foot closer to the ground rod, that foot would be at a 1200 V potential. The foot further away from the ground rod would be at a 700 V potential. The potential difference would be 500 V, which is enough voltage difference to force current through a person's body and cause injury. **See Figure 7-18.**

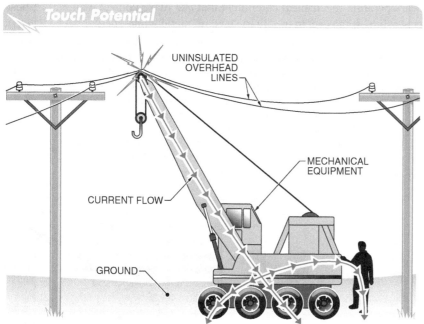

Figure 7-17. When a vehicle or mechanical equipment makes contact with an energized line, the current will flow through all paths to ground.

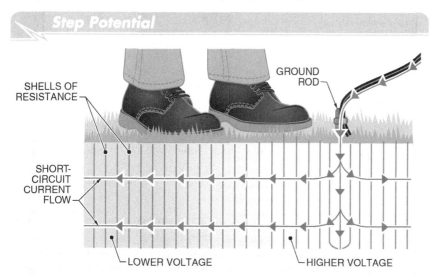

Figure 7-18. As short-circuit current flows through shells of earth resistance, the resistance decreases and causes a difference in voltage from one area to another.

Shermco Industries

Figure 7-19. Ground mats are used to protect employees from step or touch potentials.

Equipotential Zone Created by Ground Mats. An *equipotential zone* is an area in which all conductive elements are bonded or otherwise connected together in a manner that prevents a difference of potential from developing within the area. To avoid the hazards of step and touch potential, an equipotential zone must be created. This is accomplished by using a ground mat made of conductive materials such as metal. **See Figure 7-19.** An equipotential zone is created by placing a ground mat in series with the ground conductor between the switch and the ground rod. This procedure is used any time a ground mat is installed. Ground mats are typically used in outdoor substations or at a gang-operated switch. Ground mats are also used to establish an equipotential zone for employees standing near operating mechanical equipment that can make contact with an overhead power line. It may seem unsafe to stand on metal while operating a switch, but it is actually the safest place to be if the switch were to fail during operation.

When current flows into the ground from a fault (or lightning strike) it travels through shells of earth resistance. As the current flows away from the grounding point and through the shells of earth resistance, a voltage drop occurs due to the resistance of the earth. This can be demonstrated using Ohm's law. **See Figure 7-20.**

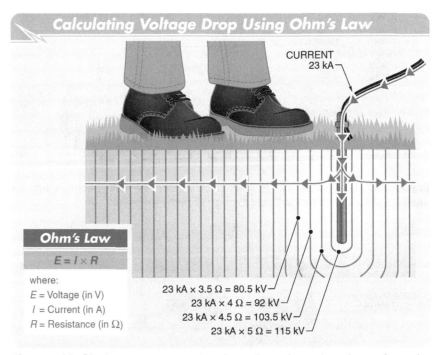

Figure 7-20. Ohm's law can be used to determine voltage drop due to the resistance of the earth.

In the environment of a substation, the actual ground resistance will be well below 1 Ω. Even when resistance is less than 1 Ω, the resulting voltage can be lethal due to the difference in voltage between a person's feet. If a fault occurs due to a failed switch and a person is standing nearby with their feet apart, a step potential can occur. If an employee is holding the switch handle during a fault, a touch potential can occur.

The switch, operator, and ground rod are all at the same potential when an employee is standing on a ground mat. During a fault, the current flows through the switch. But when everything is bonded together, there is no difference in potential. Without a difference in potential, the resistance of a person's body cannot be overcome and there is no current flow through the body. A shock hazard does not exist when there is no difference in potential. **See Figure 7-21.** When using ground mats, it is important to be aware of the following:

- The connection between the switch, ground mat, and ground rod must be checked for tightness. A high-resistance connection at any one of these points can cause a shock hazard to the operator.
- Do not allow remote earth to intrude into the equipotential zone. *Remote earth* is any object that is on or in the ground outside the equipotential zone and extends into the zone. This includes pipes, wire, weeds, or any other similar objects.
- If a fault or failure occurs at the switch or structure, the employee should stay on the ground mat. It is the only safe place to be during a fault. Ground gradients around the equipotential zone can be lethal if a person tries to step off the ground mat or tries to leave the area. This may be especially difficult to remember if there are arcs and sparks flying around. However, as soon as an employee steps off the ground mat, there will be a difference in potential between the employee's feet.

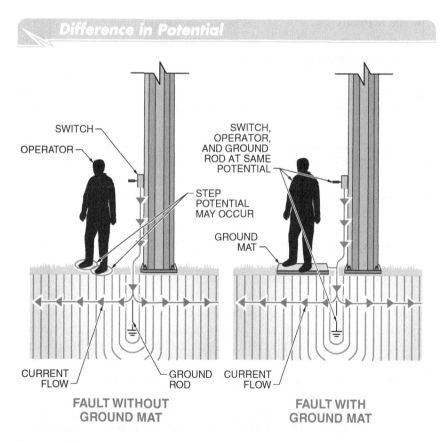

Figure 7-21. Ground mats are used to create an equipotential zone, which protects employees from step or touch potentials when faults occur.

IMPORTANT REQUIREMENTS OF OSHA 1910.269

OSHA 1910.269 covers the operation and maintenance of electric power generation, control, transformation, transmission, and distribution lines and equipment. When OSHA 1910.269 applies to employees, employers must take the appropriate steps to protect them. Many important sections of OSHA 1910.269 are included in NFPA 70E.

Training

According to OSHA 1910.269(a)(2)(i)(A) and (B), *"Employees shall be trained in and familiar with the safety-related work practices, safety procedures, and other safety requirements in this section that pertain to their respective job assignments. Employees shall also be trained in and familiar with any other safety practices, including applicable emergency procedures (such as pole top and manhole rescue), that are not specifically addressed by this section but that are related to their work and are necessary for their safety."*

Tasks that are performed less often than once a year require retraining before the task is performed. Retraining or additional training is also required if one of the following conditions apply:

- The supervision and annual inspections required by OSHA indicate that the employee is noncompliant with the required safety-related work practices.
- New technology, new types of equipment, or changes in procedures necessitate safety-related work practices that are different from those the employee would normally use.
- An employee must apply safety-related work practices that are not normally used during their regular job duties.

The requirement for retraining does not necessarily mean that formal, classroom-style training is the only acceptable method. OSHA and NFPA 70E both find on-the-job training (OJT) acceptable, but it must be performed by a qualified person and it must be documented. NFPA 70E has a requirement for documenting training that includes listing the content of the training, the employee's name, and the dates the training was conducted. What was meant by "content of training" was discussed at length during NFPA 70E Committee deliberations. The intent was to have more detail than just a description, which was one of the words considered by the NFPA 70E Committee, but not to go as far as requiring a copy of the training materials. It was intended to be more like a detailed outline and description that provides enough detail to determine if the training was adequate.

Training requirements were all relocated to Section 110.2 in the 2015 edition of NFPA 70E. NFPA 70E 110.2 requires a qualified person to be trained to understand the specific hazards associated with electrical energy, understand the safety-related work practices and procedural requirements needed to protect the worker from the hazards and risks associated with their job, and identify and understand the relationship between electrical hazards and the injuries they could receive from those hazards.

SAFETY FACT

The use of personal protective equipment (PPE) is required whenever work may occur on or near energized exposed conductors or circuit parts. For maximum safety, PPE must be used as specified in NFPA 70E, OSHA 1910.132 through 1910.138, and other applicable safety mandates.

NFPA 70E 110.2(C) covers emergency response training. There were a lot of revisions to this section for the 2015 edition of NFPA 70E. The revisions include the following:
- Employees who are exposed to the hazard of electric shock are to be trained in methods of safe release of the victim. Making bare-handed contact with someone who is being shocked is almost certain to result in additional injuries or fatalities. Current will flow through the person being shocked to the person who has made contact. In an emergency situation, a natural first reaction is to try to save the person being shocked immediately. This instinct has to be controlled to prevent additional injuries.

  The best method of rescue is to use a rescue hook. A *rescue hook* is a fiberglass pole with a large metal hook on the end, much like a shepherd's crook. The rescue hook provides the distance needed to prevent contact and insulation to prevent shock to the rescuer. Other methods that may or may not be appropriate include using a belt to hook around the person to pull him or her away from the energize circuit, using a 2 × 4 to push him or her away from the circuit, or wearing insulated gloves while pulling him or her away from the circuit. Each of these methods may be successful, or under the wrong circumstances they could lead to an injury for the rescuer. Before work begins, workers should determine where the OCPD that protects the circuit is located and how to deenergize the circuit in case of emergency.

- Employees who are first responders are to be trained in first aid, emergency procedures, and cardiopulmonary resuscitation (CPR). CPR training is to be verified annually. The American Heart Association recommends a two-year interval for this training. However, the NFPA 70E Committee believes that most people do not practice the techniques required to stay proficient in CPR. In an emergency situation, it will be difficult for a person to remember what steps are to be taken and how to perform them unless they receive refresher training every year.

- If the company's emergency response plan includes the use of an AED, training of first responders is also necessary. The 2012 edition of NFPA 70E did not differentiate between companies that have AEDs and those that do not. This made it sound as if everyone required training, whether they had an AED or not. That was not the intent of the NFPA 70E Committee.

- The emergency response training is to be verified annually to make certain that everyone required to have training is up to date.

Additional requirements for qualified persons are listed in 110.2(D)(1).

Existing Conditions

According to OSHA 1910.269(a)(4), *"Existing characteristics and conditions of electric lines and equipment that are related to the safety of the work to be performed shall be determined before work on or near the lines or equipment is started. Such characteristics and conditions include, but are not limited to, the nominal voltages of lines and equipment, the maximum switching-transient voltages, the presence of hazardous induced voltages, the presence of protective grounds and equipment grounding conductors, the locations of circuits and equipment, including electric supply lines, communication lines, and fire-protective signaling circuits, the condition of protective grounds and equipment grounding conductors, the condition of poles, and environmental conditions relating to safety."*

Retraining or additional training is required if new technology, new types of equipment, or procedural changes necessitate safety-related work practices that are different from those the employee would normally use.

OSHA advises employees to perform a risk assessment. A risk assessment is required any time energized lines or equipment are unguarded. They may be unguarded because distance has been reduced or guards have been physically removed to expose energized conductors or circuit parts.

Job Briefing—NFPA 70E 110.1(H)

A *job briefing* is a discussion that takes place between all employees engaged in a task and occurs before work is performed on energized conductors and devices. **See Figure 7-22.** Job briefings are covered in NFPA 70E 110.1(H). The language in NFPA 70E 110.1(H) is almost identical to OSHA 1910.269(c).

Shermco Industries

Figure 7-22. Job briefings should cover hazards associated with the job, work procedures involved, special precautions, energy source controls, and personal protective equipment (PPE) requirements.

According to OSHA 1910.269(c), *"(1)(ii) The employer shall ensure that the employee in charge conducts a job briefing that meets paragraphs (c)(2), (c)(3), and (c)(4) of this section with the employees involved before they start each job. (2) The briefing shall cover at least the following subjects: hazards associated with the job, work procedures involved, special precautions, energy-source controls, and personal protective equipment requirements. (3)(i) If the work or operations to be performed during the work day or shift are repetitive and similar, at least one job briefing shall be conducted before the start of the first job of each day or shift. (3)(ii) Additional job briefings shall be held if significant changes, which might affect the safety of the employees, occur during the course of the work. (4)(i) A brief discussion is satisfactory if the work involved is routine and if the employees, by virtue of training and experience, can reasonably be expected to recognize and avoid the hazards involved in the job. (4)(ii) A more extensive discussion shall be conducted: (4)(ii)(A) If the work is complicated or particularly hazardous, or (4)(ii)(B) if the employee cannot be expected to recognize and avoid the hazards involved in the job. Note to paragraph (c)(4): The briefing must address all the subjects listed in paragraph (c)(2) of this section. (5) An employee working alone need not conduct a job briefing. However, the employer shall ensure that the tasks to be performed are planned as if a briefing were required."*

Live-Line Tools

Live-line tools, also known as hot sticks, switching sticks, or shotguns, are insulated tools used for working on energized lines and equipment with AC voltage ratings above 1000 V. Live-line tools are most commonly used in industrial facilities to place and remove personal protective ground sets, as well as to hold a voltage detector during absence of voltage testing. Live-line tools are used instead of hands for overhead line work when voltages exceed the available ratings of rubber insulating PPE. There are many live-line tool attachments that enable an experienced employee to perform a variety of tasks without making contact with the line or equipment.

Live-Line Tool Design. Live-line tools are typically made of fiberglass-reinforced plastic (FRP). However, older types of live-line tools were made of laminated wood. Most FRP live-line tools are filled with dense foam. Dense foam provides strength and seals the inside of the tool to prevent moisture penetration. Fiberglass is porous and therefore uses a hard wax finish to keep moisture out. **See Figure 7-23.** FRP live-line tools have a high dielectric strength (100 kV/ft for new tools) and require far less maintenance than the older laminated-wood live-line tools. Live-line tools must be inspected before and after each use. Any chip to the outside varnish of laminated-wood live-line tools can expose the wood to air and allow it to absorb moisture. FRP live-line tools must be inspected according to IEEE 978. New live-line tool rods, tubes, and poles must be designed and constructed to withstand the following minimum tests:

- FRP live-line tool rods, tubes, and poles must withstand 100 kV/ft (3281 V/cm) for 5 min. *Note:* Live-line tools using rods and tubes that meet ASTM Standard F711, *Standard Specification for Fiberglass-Reinforced Plastic (FRP) Rod and Tube Used in Live-Line Tools,* must conform to this minimum test as well.
- Fiberglass live-line tools must pass an in-service test at 75 kV/ft for 1 minute under wet conditions. The most common method to accomplish this when using a portable electronic tool tester is to use a spray bottle to wet the stick down.
- Laminated wood live-line tools must pass a 50 kV/ft test for 1 minute. These tools are not common anymore, but some may still be in use.
- Live-line tool rods, tubes, and poles must withstand any other equivalent tests conducted by the employer.

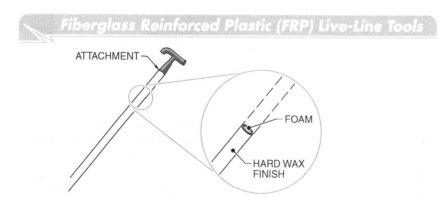

Figure 7-23. Most fiberglass-reinforced plastic (FRP) live-line tools are filled with dense foam and coated on the outside with a hard wax finish to keep moisture out.

Live-Line Tool Maintenance. Each live-line tool must be wiped clean and visually inspected for defects before use each day. If any defect or contamination that adversely affects the insulating qualities or mechanical integrity of the live-line tool is present after wiping, the tool must be removed from service and examined and tested before being returned to service. According to OSHA 1910.269(j)(2)(iii), live-line tools used for primary employee protection must be removed from service every two years for examination, cleaning, repair, and testing. Even though OSHA requires testing every two years, live-line tools are safety devices and should be tested every year. Primary employee protection means that the tool may be used without rubber insulating gloves. Most industrial facilities require the use of rubber insulating gloves when live-line tools are used.

Each live-line tool must be thoroughly examined for defects or contamination. If a defect or contamination that adversely affects the insulating qualities or mechanical integrity of the live-line tool is found, it must be either repaired and refinished or permanently removed from service. If no such defect or contamination is found, the live-line tool must then be cleaned and waxed.

> **SAFETY FACT**
> *Live-line tools should be tested at least every year. OSHA has no specific testing requirement for live-line tools used with rubber insulating gloves where the tool is not used for primary employee protection.*

IEEE Guidelines for Live-Line Tools. Guidelines for the examination, cleaning, repairing, and in-service testing of live-line tools are found in IEEE 978, *Guide for In-Service Maintenance and Electrical Testing of Live-Line Tools*. IEEE 978 does not provide specific testing intervals, but it does provide guidelines. It is recommended that live-line tools be tested at least once per year.

IEEE 978 provides examples that show when a live-line tool should be removed from service. For example, a live-line tool must be taken out of service if an employee feels a tingling sensation when the tool is in contact with an energized conductor or component. A tingling sensation in the hands indicates that a live-line tool is near failure. Other reasons to remove a live-line tool from service according to IEEE 978 include the following:

- The tool's surface has deep cuts, scratches, nicks, gouges, dents, or has been delaminated.
- The glossy surface is lost or deteriorated.
- The tool is improperly stored or improperly exposed to weather.
- The tool shows signs of being mechanically overstressed.
- The tool shows signs of being electrically overstressed, including tracking, burning, or heat blistering.
- The tool fails to pass a test performed with a portable electronic tool tester.
- The tool is inadvertently cleaned with a soap-based cleaner. *Note:* Soap and dishwashing detergents remove all traces of wax from a live-line tool. Live-line tools must only be cleaned with the recommended cleaner.

Electronic Tool Testers. Electronic tool testers perform a dielectric test on live-line tools in one-foot sections along the entire length of the tool. The voltage applied during testing for in-service live-line tools includes the following:

- If the tool is made of FRP, 75 kV/ft (2461 V/cm) is applied for 1 min. Tests made on FRP live-line tools must also be performed under wet conditions.
- If the tool is made of laminated wood, 50 kV/ft (1640 V/cm) is applied for 1 min.
- Other equivalent tests can be applied by the employer.

The tool does not have to be tested if it has an FRP rod or a foam-filled FRP tube and the employer can demonstrate that the tool has no defects that can cause it to fail during use. Testing must verify the integrity of the entire working length of the tool. If the tool is made of FRP, its integrity must also be tested under wet conditions.

Temporary Protective Grounds

Temporary protective grounding is used to ensure a direct path to ground for unwanted fault current and to safeguard equipment and personnel against the hazards of electric shock. Temporary protective grounds are often referred to as ground clusters or ground jumpers and are connected from a deenergized line or bushing to the ground grid or ground bus. **See Figure 7-24.** Temporary protective grounds guard against accidentally reenergized lines or equipment, induced voltages caused by parallel power lines, upstream switch failure, and lightning strikes.

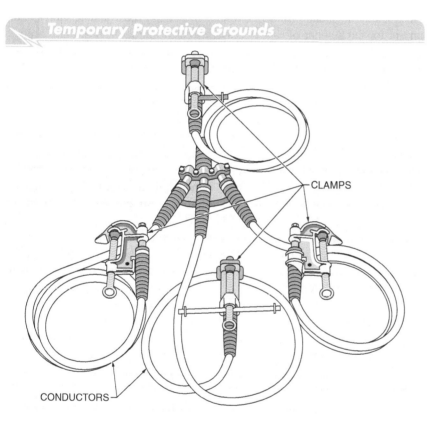

Figure 7-24. Temporary protective grounds are often referred to as ground clusters or ground jumpers and are connected from a deenergized line or bushing to the ground grid.

SAFETY FACT

For temporary personal protective grounds, 20' is the maximum length recommended for conductors. Conductors longer than 20' will cause the impedance to increase significantly, lowering the amount of short circuit current the conductor can handle.

To prevent each employee from being exposed to hazardous differences in electrical potential, it is necessary to place temporary protective grounds in the proper location and arrangement. If there is a difference in potential, there is a shock hazard. Temporary protective grounds must be placed so that all parts of the potential current path are bonded to a grounding point so there can be no difference in potential between components, the person in the equipotential zone, and the grounding point.

According to OSHA 1910.269(n)(4), *"(i) Protective grounding equipment shall be capable of conducting the maximum fault current that could flow at the point of grounding for the time necessary to clear the fault. (ii) Protective grounding equipment shall have an ampacity greater than or equal to that of No. 2 AWG copper. (iii) Protective grounds shall have an impedance low enough so that they do not delay the operation of protective devices in case of accidental energizing of the lines or equipment. Note to paragraph (n)(4): American Society for Testing and Materials* Standard Specifications for Temporary Protective Grounds to Be Used on De-Energized Electric Power Lines and Equipment, *ASTM F855-09, contains guidelines for protective grounding equipment. The Institute of Electrical Engineers* Guide for Protective Grounding of Power Lines, *IEEE Std 1048-2003, contains guidelines for selecting and installing protective grounding equipment."*

OSHA 1910.269(n) applies to the grounding of transmission and distribution lines and equipment for the purpose of protecting employees. In order to be effective, the equipment used for temporary protective grounding must be properly constructed, rated, and installed. Grounds must carry the available short-circuit current for the amount of time necessary to clear the fault. If grounds are not properly sized, they may melt during a short circuit, which can create an arc flash, arc blast, or other hazards. The minimum ground conductor size allowed is No. 2 AWG copper. All parts of the ground cluster, including the cable, ferrules, and clamps, must be sized to carry the amount of current for the time needed to clear the fault. If any part is undersized, the entire assembly will fail.

The type of ground clamp is important. Different types of clamps can be used depending on the type of bus they are being connected to. A flat bus, tubular bus, and angled bus all use a different clamp. The jaws of the clamp can be serrated to cut into the metal and hold firmly. A nonserrated or spring-loaded clamp, such as a welding clamp and automotive jumper cables, may blow off the bus as soon as short-circuit current flows through them.

If lines or equipment are accidentally energized, temporary protective grounds must have low enough impedance to cause the immediate operation of protective devices. Ground sets that are not sized properly will fuse or melt. Loose or damaged connections and defective cable sets or clamps may increase the impedance of the ground set and cause failure when they clear a fault. The ball-and-socket configuration, where it can be used, provides the lowest impedance connection and is the most secure. The grounding ball is mounted onto the equipment or bus, and the socket is part of the cable assembly. These are available from a number of sources. **See Figure 7-25.**

Some companies may have a "one-size-fits-all" ground set that is usually sized at 4/0. This size may or may not be adequate since it may be capable of carrying the available short-circuit current, but it may not carry the short-circuit current for as long as it takes the OCPD to operate.

Testing. Unless a previously installed ground is present, lines and equipment must be tested and found absent of nominal voltage before any ground is installed. A ground should never be connected to a line or device without first testing to ensure the absence of voltage. Power lines may have a voltage induced into them by nearby overhead lines. This voltage is much lower than the nominal voltage and will not cause damage when grounded. However, a voltage being supplied by a generation source may cause an arc flash or arc blast, thereby creating a safety hazard for the employee installing the ground.

SAFETY FACT

Reference ASTM F855-1990, Standard Specifications for Temporary Grounding Systems to be Used on De-Energized Electric Power Lines and Equipment *for sizing ground conductors and other requirements.*

Temporary Protective Grounding Equipment—Ball-and-Socket

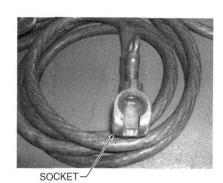

Figure 7-25. The ball-and-socket configuration for temporary protective grounds provides the lowest impedance connection and is the most securely fastened.

According to OSHA 1910.269(n)(6)(i), *"The employer shall ensure that, when an employee attaches a ground to a line or to equipment, the employee attaches the ground-end connection first and then attaches the other end by means of a live-line tool. For lines or equipment operating at 600 volts or less, the employer may permit the employee to use insulating equipment other than a live-line tool if the employer ensures that the line or equipment is not energized at the time the ground is connected or if the employer can demonstrate that each employee is protected from hazards that may develop if the line or equipment is energized."*

If the lines or equipment end is connected first, there may be an induced voltage on a ground cluster. Induced voltages are a significant hazard. Employees must be constantly reminded of the hazards of induced voltages.

According to OSHA 1910.269(n)(6)(ii), *"The employer shall ensure that, when an employee removes a ground, the employee removes the grounding device from the line or equipment using a live-line tool before he or she removes the ground-end connection. For lines or equipment operating at 600 volts or less, the employer may permit the employee to use insulating equipment other than a live-line tool if the employer ensures that the line or equipment is not energized at the time the ground is disconnected or if the employer can demonstrate that each employee is protected from hazards that may develop if the line or equipment is energized."*

FIELD NOTES: TESTING FOR NOMINAL VOLTAGE

I had a student in one of my classes relate the following story of an accident he had been involved in and spent 18 months recovering from. At a steel mill about 23 years prior to the class, the student and another worker were preparing to change the carbon electrode on an arc furnace transformer. These electrodes are 2½' to 3' in diameter and are used to melt the scrap steel in a melting pot. The arc furnace transformer was in one room and its switch was in another room. The student and his associate had performed this task many times and were very familiar with the steel mill and its systems.

The two workers opened the 13.8 kV switch and walked into the room where the transformer was located. Following their procedure, the student took a broomstick handle that had a No. 12 AWG wire attached to it (the other end of the wire was connected to ground) and "bumped" the exposed 13.8 kV bus. The No. 12 AWG wire exploded on contact with the 13.8 kV bus, setting the student's jacket and shirt on fire. He suffered massive burns because they had opened the wrong switch. His injuries were severe enough that he needed 18 months to recover from this accident.

"How reckless," one might say. However, we had the same procedure at the power plant where I worked. Sometimes people used chains; sometimes they used a grounded wire. We never gave it a thought, but we should have in order to avoid similar accidents.

The ground-end connection is removed last to safely ground out induced voltage. Induced voltages can be several thousand volts, depending on the clearance between lines, the length the lines run parallel, voltage, and the loading of the energized lines. Proper safety guidelines must be followed when using personal protective grounding equipment. **See Figure 7-26.**

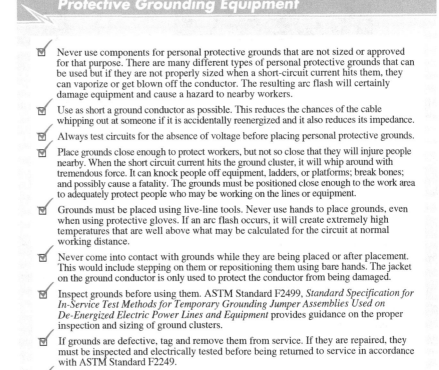

Safety Guidelines for Protective Grounding Equipment

- ☑ Never use components for personal protective grounds that are not sized or approved for that purpose. There are many different types of personal protective grounds that can be used but if they are not properly sized when a short-circuit current hits them, they can vaporize or get blown off the conductor. The resulting arc flash will certainly damage equipment and cause a hazard to nearby workers.
- ☑ Use as short a ground conductor as possible. This reduces the chances of the cable whipping out at someone if it is accidentally reenergized and it also reduces its impedance.
- ☑ Always test circuits for the absence of voltage before placing personal protective grounds.
- ☑ Place grounds close enough to protect workers, but not so close that they will injure people nearby. When the short circuit current hits the ground cluster, it will whip around with tremendous force. It can knock people off equipment, ladders, or platforms; break bones; and possibly cause a fatality. The grounds must be positioned close enough to the work area to adequately protect people who may be working on the lines or equipment.
- ☑ Grounds must be placed using live-line tools. Never use hands to place grounds, even when using protective gloves. If an arc flash occurs, it will create extremely high temperatures that are well above what may be calculated for the circuit at normal working distance.
- ☑ Never come into contact with grounds while they are being placed or after placement. This would include stepping on them or repositioning them using bare hands. The jacket on the ground conductor is only used to protect the conductor from being damaged.
- ☑ Inspect grounds before using them. ASTM Standard F2499, *Standard Specification for In-Service Test Methods for Temporary Grounding Jumper Assemblies Used on De-Energized Electric Power Lines and Equipment* provides guidance on the proper inspection and sizing of ground clusters.
- ☑ If grounds are defective, tag and remove them from service. If they are repaired, they must be inspected and electrically tested before being returned to service in accordance with ASTM Standard F2249.
- ☑ Never stand near grounds that may be lying on the ground. If they are reenergized they will try to straighten out with great force.
- ☑ Never coil grounds. If they are on carriers, remove the entire conductor from the carrier and stretch out on the ground (earth). When short-circuit current tries to pass through a coil of wire, the coil acts as an inductor or choke. This often results in a vaporized conductor.
- ☑ Inspect the grounds after each use. Ensure the grounds are still serviceable. ASTM Standard F2499 can be used as a guide for inspection.
- ☑ Test personal protective grounds on a regular basis. According to ASTM Standard F2499, grounds are to be tested at a time interval established by the user to ensure that defective grounding jumper assemblies are detected and removed from service in a timely manner. Personal protective grounds are safety-related devices and it is recommended that they be tested annually.

Figure 7-26. Proper safety guidelines must be followed when using personal protective grounding equipment.

Testing and Test Facilities

OSHA 1910.269(o) provides regulations for safe work practices when performing high-voltage and high-power tests in laboratories, shops, substations, and in the field, as well as on electric transmission and distribution lines and equipment. This only applies to testing that involves interim measurements utilizing high voltage, high power, or combinations of both. It does not apply to testing that involves continuous measurements, such as routine metering, relaying, and normal line work.

An employer must establish and enforce work practices to protect each employee from the hazards of high-voltage or high-power testing at all test areas, temporary or permanent. At a minimum, these work practices must include test area guarding, grounding, and the safe use of measuring and control circuits. A means providing for periodic safety checks of field test areas must also be included. All test areas must have safe work practices, which must be enforced.

Employees must be trained in safe work practices upon their initial assignment to the test area. They must also be provided with periodic reviews and updates. Initial training is intended to provide the level of training the employee would need to be qualified to perform the tasks safely. Retraining is to be provided in accordance with NFPA 70E 110.2(D)(3) and OSHA 1910.269(a).

According to OSHA 1910.269(o)(3), *"(ii) The employer shall guard permanent test areas with walls, fences, or other barriers designed to keep employees out of the test areas. (iii) In field testing, or at a temporary test site not guarded by permanent fences and gates, the employer shall ensure the use of one of the following means to prevent employees without authorization from entering: (iii) (A) Distinctively colored safety tape supported approximately waist high with safety signs attached to it, (iii)(B) A barrier or barricade that limits access to the test area to a degree equivalent, physically and visually, to the barricade specified in paragraph (o)(3)(iii)(A) of this section, or (iii)(C) one or more test observers stationed so that they can monitor the entire area."*

Due to the potential for unaware or unqualified employees to be injured, test areas must be guarded to prevent unauthorized employees from entering during testing. For example, if high-voltage or high-power testing is performed in the field under field test conditions, the test area must be guarded. This includes field testing inside fenced-in substations. Proper precautions should be taken to guard this area. Waist-high safety barrier tape is most often used for a temporary barrier when field testing, even inside a fenced-in substation. If there are other employees with keys who can access a substation, guarding of the test area is required to make it a controlled area.

The barriers required when field testing or at a temporary test site should be removed when no longer needed to protect unauthorized employees. Safety barrier tape that is no longer needed but is left in position may cause employees to ignore safety barrier tape that is still in service.

If either direct or inductive coupling during testing can energize test equipment or the apparatus under testing, access to the test area must be guarded to prevent employees from accidentally making contact with any energized parts. For example, access to a fenced-in substation must be limited when it is considered a test site. **See Figure 7-27.**

Figure 7-27. Test sites such as substations must be barricaded and may also be fenced in to control access to test equipment.

Shermco Industries
When performing high-voltage tests on electric transmission and distribution lines, refer to the regulations in OSHA 1910.269(o).

According to OSHA 1910.269(o)(4)(i), *"The employer shall establish and implement safe grounding practices for the test facility."*

Employers are required to establish and implement safe grounding practices, which means that the practices must be enforced. The safe grounding practices are listed in OSHA 1910.269(o)(4)(i)(A) through 1910.269(o)(4)(vi).

According to OSHA 1910.269(o)(4)(i)(A), *"The employer shall maintain at ground potential all conductive parts accessible to the test operator while the equipment is operating at high voltage."*

The safety ground on high-voltage test equipment, such as hipot testers, should be connected when employees are performing tests. However, it is often not connected. The safety ground should be connected from the ground lug of the test instrument to the case of the device being tested. This ensures that the test instrument is grounded in case of an internal failure. It also eliminates the possibility of touch potential between the test instrument and the device being tested. **See Figure 7-28.**

According to OSHA 1910.269(o)(4)(i)(B), *"Wherever ungrounded terminals of test equipment or apparatus under test may be present, they shall be treated as energized until tests demonstrate they are deenergized."*

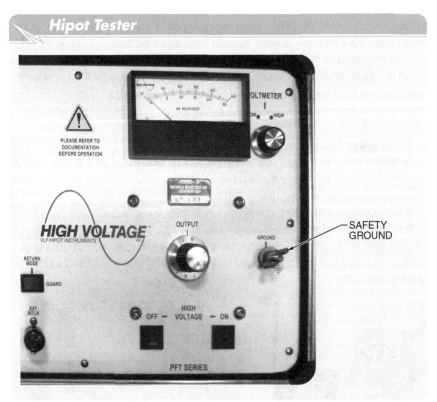

Shermco Industries

Figure 7-28. The safety ground on a hipot tester should be connected from the ground lug of the test instrument to the case of the device being tested to ensure the test instrument is grounded in case of an internal failure or to eliminate the possibility of touch potential.

Applying a DC voltage or current to a transformer can create an inductive charge that is stored in the windings until the test leads are disconnected. Before transformer ohmmeters were developed, line-powered micro-ohmmeters were used to perform DC winding resistance tests on transformers. Removing the line-powered micro-ohmmeter test leads between tests can create a hazard, especially on larger transformers, due to the arc that can occur. Voltage-rated gloves should always be worn when connecting or removing test leads on any device being tested.

SAFETY FACT
When performing DC winding resistance testing on transformers, enough time should always be allowed for the transformer to completely discharge after being tested. Newer winding resistance testers have indicators that signal when it is safe to remove the leads.

According to OSHA 1910.269(o)(4)(ii), *"The employer shall ensure either that visible grounds are applied automatically, or that employees using properly insulated tools manually apply visible grounds, to the high-voltage circuits after they are deenergized and before any employee performs work on the circuit or on the item or apparatus under test. Common ground connections shall be solidly connected to the test equipment and the apparatus under test."*

Visible grounds can be personal protective grounds or, when the device is completely isolated, static grounds. Static grounds are used to prevent induced or static voltages from appearing on the bushings of devices being tested. The use of properly insulated tools or rubber insulating gloves is required when placing static grounds. Static grounds are applied before testing, between tests, and upon completion of tests. Generally, bushings should either be grounded or connected to the test instrument in an outdoor substation environment. However, there are a few exceptions that allow bushings to float or be disconnected. For example, bushings can be disconnected when performing a tank-loss-index test on an oil circuit breaker. A *tank-loss-index (TLI)* is a diagnostic test performed using an insulation power factor test set. One such test set is a Doble® M4100. In this test, bushings must be left unconnected to either ground or the test set, which is commonly referred to as "floating" the bushings.

According to OSHA 1910.269(o)(4)(iii) and (iv), *"(iii) In high-power testing, the employer shall provide an isolated ground-return conductor system designed to prevent the intentional passage of current, with its attendant voltage rise, from occurring in the ground grid or in the earth. However, the employer need not provide an isolated ground-return conductor if the employer can demonstrate that both of the following conditions exist: (A) The employer cannot provide an isolated ground-return conductor due to the distance of the test site from the electric energy source, and (B) The employer protects employees from any hazardous step and touch potentials that may develop during the test. Note to paragraph (o)(4)(iii)(B): See Appendix C to this section for information on measures that employers can take to protect employees from hazardous step and touch potentials. (iv) For tests in which using the equipment grounding conductor in the equipment power cord to ground the test equipment would result in greater hazards to test personnel or prevent the taking of satisfactory measurements, the employer may use a ground clearly indicated in the test set-up if the employer can demonstrate that this ground affords protection for employees equivalent to the protection afforded by an equipment grounding conductor in the power supply cord."*

The need to use a separate ground conductor is uncommon, but it may have to be done during certain tests. A separate ground conductor can be used when the normal ground cannot be used as long as the separate ground conductor is clearly identified.

According to OSHA 1910.269(o)(4)(v), *"The employer shall ensure that, when any employee enters the test area after equipment is deenergized, a ground is placed on the high-voltage terminal and any other exposed terminals."*

This regulation refers to static grounds. Induced voltages can always be present in outdoor substations. The best protection in this environment is grounding. Induced potentials (voltages) can occur in outdoor substations that are energized. Depending on the nominal voltage of the lines, the orientation of the lines, and the clearances between the overhead lines and objects in the substation, the voltages that are induced can be high. Any conductive object, such as conduits, circuit breakers, or transformer frames that are not properly grounded, can be energized if the expanding and contracting magnetic field cuts it.

According to OSHA 1910.269(o)(4)(v)(A) and (B), *"(A) Before any employee applies a direct ground, the employer shall discharge high capacitance equipment through a resistor rated for the available energy. (B) A direct ground shall be applied to the exposed terminals after the stored energy drops to a level at which it is safe to do so."*

Larger capacitor banks have grounding or discharge resistors built into their cases, but they are not available for testing and can fail. A grounding stick is recommended when grounding capacitors or lines that are connected to capacitors. A 1 MΩ resistor inside the grounding stick allows the stored voltage to discharge slowly. **See Figure 7-29.** After waiting at least 5 min, the direct ground connection can be applied. The 5 min wait refers to requirements for capacitors found in OSHA 1910.269(w)(1)(i).

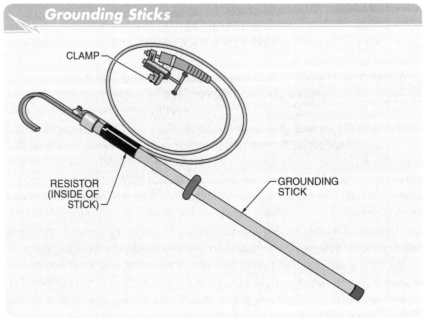

Figure 7-29. Grounding sticks contain a 1 MΩ resistor, which allows the stored voltage in a capacitor or capacitor lines to slowly discharge.

According to OSHA 1910.269(o)(5)(i) and (ii), *"(i) The employer may not run control wiring, meter connections, test leads, or cables from a test area unless contained in a grounded metallic sheath and terminated in a grounded metallic enclosure or unless the employer takes other precautions that it can demonstrate will provide employees with equivalent safety. (ii) The employer shall isolate meters and other instruments with accessible terminals or parts from test personnel to protect against hazards that could arise should such terminals and parts become energized during testing. If the employer provides this isolation by locating test equipment in metal compartments with viewing windows, the employer shall provide interlocks to interrupt the power supply when someone opens the compartment cover."*

In the past, many utilities would manufacture their own test instruments, and some still do. OSHA 1910.269(o)(5) is meant to apply to homemade or shop-built test instruments. Test instruments purchased from reputable manufacturers will already meet these requirements.

According to OSHA 1910.269(o)(5)(iii), *"The employer shall protect temporary wiring and its connections against damage, accidental interruptions, and other hazards. To the maximum extent possible, the employer shall keep signal, control, ground, and power cables separate from each other."*

The intent of this regulation is to prevent voltages from being induced from the high-voltage conductors into the low-voltage signal and control cables. However, this is not always a concern because test equipment manufacturers often use single or double-shielded cable to prevent damage, accidental interruptions, and other hazards.

According to OSHA 1910.269(o)(5)(iv), *"If any employee will be present in the test area during testing, a test observer shall be present. The test observer shall be capable of implementing the immediate deenergizing of test circuits for safety purposes."*

This regulation refers to employees other than the ones performing the testing. For example, there may be multiple work crews at a substation during a preventative maintenance (PM) shutdown. Test observers may be part of a work crew. Their task is to prevent the employees not performing testing from contacting energized components.

According to OSHA 1910.269(o)(6), *"(i) Safety practices governing employee work at temporary or field test areas shall provide, at the beginning of each series of tests, for a routine safety check of such test areas. (ii) The test operator in charge shall conduct these routine safety checks before each series of tests and shall verify at least the following conditions: (A) Barriers and safeguards are in workable condition and placed properly to isolate hazardous areas; (B) System test status signals, if used, are in operable condition; (C) Clearly marked test-power disconnects are readily available in an emergency; (D) Ground connections are clearly identifiable; (E) Personal protective equipment is provided and used as required by Subpart I of this part and by this section; and (F) Proper separation between signal, ground, and power cables."*

Capacitors

A *capacitor* is an electric device that stores electrical energy and is used for power factor correction in electrical power systems. Large motors are often the cause of poor electrical system power factor. Utility companies will often charge a penalty if their customer's sites have poor electrical system power factor.

Power Systems Capacitors

Figure 7-30. Capacitors can be installed to reduce the inductance caused by large motors, but they must be disconnected from an energized source, shorted together, and grounded for 5 min before they are worked on.

This penalty can be avoided by installing power factor correction capacitors to reduce the inductance caused by the large motors. **See Figure 7-30.** The isolation of capacitors must be in compliance with OSHA 1910.269(m) and 1910.269(n), which pertain to the deenergizing and grounding of capacitor installations.

According to OSHA 1910.269(w)(1), *"(i) Before employees work on capacitors, the employer shall disconnect the capacitors from energized sources and short circuit the capacitors. The employer shall ensure that the employee short circuiting the capacitors waits at least 5 minutes from the time of disconnection before applying the short circuit, (ii) before employees handle the units, the employer shall short circuit each unit in series-parallel capacitor banks between all terminals and the capacitor case or its rack. If the cases of capacitors are on ungrounded substation racks, the employer shall bond the racks to ground. (iii) The employer shall short circuit any line connected to capacitors before the line is treated as deenergized."*

A grounding stick should be used for the short-circuiting operations described in OSHA 1910.269(w)(1)(iii). When capacitors are connected to a power line, that line must be short-circuited in order to be deenergized. Capacitor banks and the lines connected to them are considered energized until they are grounded.

Capacitors must be handled carefully. Large capacitors, such as those used for power factor correction in banks or as individual units on large induction motors, must be handled carefully because they can store an electrical charge that can cause serious injury or death if touched. Uninterruptible power supply (UPS) systems and variable-speed drives can have large sets of capacitors, which can be hazardous for unqualified persons.

Current Transformer Secondary Windings

According to OSHA 1910.269(w)(2), *"The employer shall ensure that employees do not open the secondary of a current transformer while the transformer is energized. If the employer cannot deenergize the primary of the current transformer before employees perform work on an instrument, a relay, or other section of a current transformer secondary circuit, the employer shall bridge the circuit so that the current transformer secondary does not experience an open-circuit condition."*

A *current transformer (CT)* is a transformer used to step down line current to make it easier and safer to measure. **See Figure 7-31.** The secondary winding on a CT is a universal hazard. Open-circuiting the secondary winding of a CT under load can create a voltage that is several thousand volts between terminals.

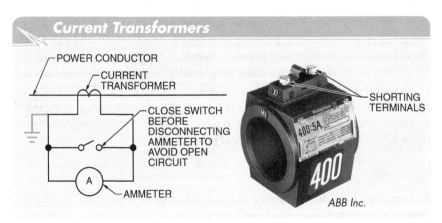

Figure 7-31. A current transformer (CT) is used to step down line current to make it easier to measure.

Current Transformer Operation. The primary winding of a CT is typically a conductor or busbar running through the center of the transformer. The secondary winding of a CT consists of many turns of wire coiled around a core. **See Figure 7-32.** The number of turns is determined by the desired turns ratio of the CT. The *turns ratio* is the number of turns on the primary winding (usually one turn for a CT) of a transformer to the number of turns on the secondary winding. The current ratio of a CT and the turns ratio are the same. For example, a 100:5 CT has 100 A in the primary and 5 A in the secondary. The turns ratio and the current ratio can be simplified to 20:1. This means that for every one turn on the primary there will be 20 turns on the secondary.

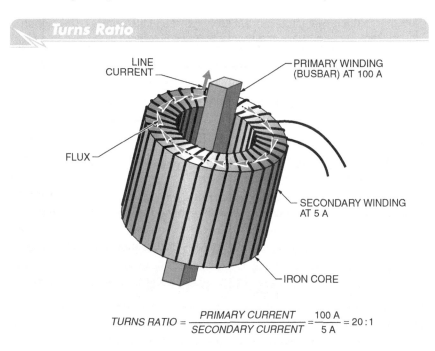

$$\text{TURNS RATIO} = \frac{\text{PRIMARY CURRENT}}{\text{SECONDARY CURRENT}} = \frac{100 \text{ A}}{5 \text{ A}} = 20:1$$

Figure 7-32. The secondary winding of a CT consists of many turns of wire coiled around a core. The number of turns is determined by the desired turns ratio.

By design, CTs try to achieve a balance between the primary ampere-turns and the secondary ampere-turns. An *ampere-turn* is the number of amperes in a circuit multiplied by the number of turns of the primary or secondary winding. For the example 100:5 CT at full load, 100 ampere-turns will be on the primary [1 turn (primary) × 100 A = 100 ampere-turns] and 100 ampere-turns will be on the secondary [20 turns (secondary) × 5 A = 100 ampere-turns].

When the load current in the primary changes, the secondary current will change to match the primary. As long as the secondary circuit stays closed, there should always be a balance in current between the primary and secondary. If the secondary is open-circuited, 100 A will still be on the primary and 0 A will be on the secondary. When this happens, the CT tries to achieve a balance between the ampere-turns of the primary and the ampere-turns of the secondary. The CT cannot achieve this balance when the secondary is open. This causes the core to fill with magnetic flux. A CT is not an efficient voltage transformer, but the magnetic flux can produce high voltages up to several thousand volts. At those levels, the voltage can jump a considerable gap to the employee that open-circuited the CT secondary.

SAFETY FACT

An example of an EEWP and a job briefing and planning form are both located in the Informative Annexes of NFPA 70E. These are not mandatory forms, which means they can be modified to better fit the particular requirements of a company.

FIELD NOTES:
OPEN-CIRCUITING A CURRENT TRANSFORMER

I've seen the various results of open-circuited CTs in the past. For example, a 2000:5 CT opened and it glowed red-hot and started to melt before the circuit was deenergized. A 4000:5 CT blew out a section of switchgear when its secondary open-circuited. A small metering CT open-circuited and did nothing at all.

The effects of open-circuiting a CT depend on a number of factors. One factor is the mass of the iron in the core. For example, a large core will hold more flux and develop higher voltages. Another factor is the turns ratio. The higher the turns ratio of the CT, the higher the voltage it will produce. Another factor is the actual current flowing through the primary. The higher the current in the primary, the more magnetic flux there is in the core.

SUMMARY

- Only a qualified person is allowed to work on exposed energized conductors or circuit parts.
- NFPA 70E requires the use of an energized electrical work permit (EEWP) anytime employees are performing work on exposed, energized conductors and circuit parts.
- EEWPs are not needed when an employee is troubleshooting, performing diagnostics, or voltage testing.
- An arc flash risk assessment is used to prevent injury by determining potential electrical arc flash hazards, the Arc Flash Boundary, the incident energy exposure at working distance, and the required PPE and arc-rated protective clothing.
- Overhead lines must be deenergized and grounded if there is a possibility of making contact with them.
- There is a risk of touch potential whenever a vehicle or mechanical equipment is operating near uninsulated overhead power lines.
- Ground mats are typically used to establish an equipotential zone in outdoor substations or at a gang-operated switch.
- Guidelines for the examination, cleaning, repairing, and in-service testing of live-line tools are listed in IEEE 978.
- Personal protective grounds guard against accidentally reenergized lines or equipment, induced voltages caused by parallel power lines, upstream switch failure, and lightning strikes.
- Requirements for the application of personal protective grounds for circuits and equipment are listed in OSHA 1910.269(n).
- If the secondary winding of a current transformer is opened while the transformer is energized, the voltage can rise to a dangerous level.
- Insulated hand tools and equipment are required when working on or near energized conductors or circuit parts.
- Alerting techniques, such as signs, barricades, and attendants, are used to warn and protect personnel from potential hazards, such as electric shock, burns, or electric equipment failure, that can result in injury.

Digital Learner Resources
ATPeResources.com/QuickLinks
Access Code: 705798

WORKING ON ENERGIZED CONDUCTORS AND CIRCUIT PARTS

Review Questions

Name _____   Date _____

_____ **1.** Energized conductors and devices must be ___ before any employee works on or near them.
- A. inspected
- B. insulated
- C. isolated
- D. deenergized

_____ **2.** Energized work is permitted when ___.
- A. deenergizing will slow down production
- B. deenergizing will cause long delays
- C. large numbers of people will be idle for long periods of time
- D. a greater hazard is created by deenergizing

_____ **3.** An unqualified person must stay a minimum distance of ___′ from energized overhead power lines 50 kV or less.
- A. 4
- B. 6
- C. 8
- D. 10

T   F   **4.** When working in a confined space, doors, hinged panels, and other covers must be secured so they do not close or make contact with an employee working inside the space.

T   F   **5.** A fuse may be pulled while under load as long as the fuse puller is rated for the voltage.

_____ **6.** The minimum safe approach distance near overhead lines for an unqualified person is based on a voltage of ___.
- A. 50 V to 50 kV to ground
- B. 480 V to 750 V phase-to-phase
- C. 2.3 kV to 69 kV phase-to-phase
- D. 15 kV to 50 kV to ground

_____ **7.** After an OCPD has been opened due to an automatic operation, it is required to ___ before closing the OCPD.
- A. disassemble the OCPD and inspect it for damage
- B. inspect the circuit to ensure it is safe
- C. perform an insulation resistance test on the OCPD to verify the case or insulation has not been damaged
- D. replace the OCPD with one that has a higher rating

8. Only a qualified person is allowed to perform testing inside the Limited Approach Boundary when voltages are above ___ V.
   A. 25
   B. 50
   C. 75
   D. 100

9. A(n) ___ is an area in which all conductive elements are bonded or otherwise connected together in a manner that prevents a difference of potential from developing within the area.
   A. touch potential
   B. step potential
   C. equipotential zone
   D. safe zone

10. The ___ Boundary is the distance from exposed energized conductors or circuit parts where a person would receive the onset of second-degree burns on bare, unprotected skin.
    A. Prohibited Approach
    B. Restricted Approach
    C. Limited Approach
    D. Arc Flash

11. A(n) ___ object is any object that is not insulated for the voltages to which it may be exposed.
    A. conductive
    B. nonconductive
    C. electrical
    D. unprotected

12. A(n) ___ mat may be used in an industrial substation to create an equipotential zone.
    A. step potential
    B. electrical
    C. conductive
    D. ground

13. ___ is a difference in voltage between each foot of a person when standing near an energized object.
    A. Step potential
    B. Touch potential
    C. Bonding
    D. Arcing

14. Which of the following is not an approved alerting technique used to warn and protect personnel from potential hazards, such as electric shock, burns, or electric equipment failure, that can result in injury?
    A. safety signs
    B. barricades
    C. attendants
    D. locks and tags

Chapter 7 – Working on Energized Conductors and Circuit Parts

T F 15. Ground mats are typically used in outdoor substations or at a gang-operated switch.

T F 16. Insulated hand tools that show signs of damage may be repaired with electrical tape.

T F 17. Under no circumstance can an unqualified person cross the Restricted Approach Boundary.

_____ 18. An energized electrical work permit is not required when ___.
    A. performing work on low-voltage circuits
    B. using proper PPE
    C. troubleshooting
    D. applying personal protective grounds

_____ 19. An unqualified person may cross the Limited Approach Boundary ___.
    A. when approved by their company
    B. when wearing the appropriate PPE
    C. when continuously escorted by a qualified person
    D. temporarily (less than five minutes)

_____ 20. ___ hand tools are required when working on or near energized conductors or circuit parts.
    A. Insulated
    B. Uninsulated
    C. Electrical
    D. Nonsparking

_____ 21. An energized electrical work permit is required when ___.
    A. performing work on conductors or devices that are not in an electrically safe work condition
    B. troubleshooting
    C. voltage testing
    D. performing diagnostics

_____ 22. There is a risk of touch potential when ___.
    A. vehicular or mechanical equipment is operating near uninsulated overhead lines
    B. working near energized metal-enclosed electrical equipment
    C. electrical equipment is operating
    D. electrical equipment fails

_____ 23. It is not required to update an arc flash risk assessment when ___.
    A. there is a major modification or renovation to an electrical power system
    B. an OCPD setting is modified
    C. changing the size of a transformer
    D. replacing an identical OCPD

PORTABLE ELECTRIC TOOLS AND FLEXIBLE CORDS

Portable electric tools and flexible cords are used on a daily basis. They are common enough that many people do not think of them as hazardous, but anything that uses electricity may create electrical hazards. Electric shock from contact with machines, tools, and appliances results in 16% of all electrocutions. In general, OSHA and the NFPA 70E require the inspection of any tool that is about to be used to ensure it is safe. It is especially important to inspect portable electric tools and flexible cords because these items are in direct contact with an employee's hand and can shock the employee holding the tool.

OBJECTIVES

- Explain flexible cord installation and use.
- List the guidelines for flexible cords that OSHA lists in 1910.334.
- Explain the requirements for headlamps, receptacles, cord connectors, attachment plugs, and portable and vehicle-mounted generators.
- Identify the NFPA 70E standards for handling and inspecting portable electric equipment.
- Identify the various types of GFCIs.
- Explain the regulations concerning overcurrent protection modification.

FLEXIBLE CORDS

A *flexible cord* is an assembly of two or more insulated conductors contained within a jacket and is used for the connection of equipment to a power source. Common types of flexible cords include portable power cables (extension cords) and power cords that supply power to portable electrical equipment. Flexible cords are equipped with an attachment plug. The attachment plug connects to a receptacle outlet or a cord connector. Double-insulated and battery-powered portable electric hand tools are common. Many of the larger portable electric tools, such as table saws and core dills, will have a three-prong grounded cord that attaches to a standard three-prong grounded receptacle. **See Figure 8-1.**

TEMPORARY WIRING

According to OSHA 1910.305(a)(2)(i), *"Temporary electrical power and lighting installations of 600 volts, nominal, or less may be used only as follows: (A) During and for remodeling, maintenance, or repair of buildings, structures, or equipment, and similar activities; (B) For a period not to exceed 90 days for Christmas decorative lighting, carnivals, and similar purposes; or (C) For experimental or development work, and during emergencies."*

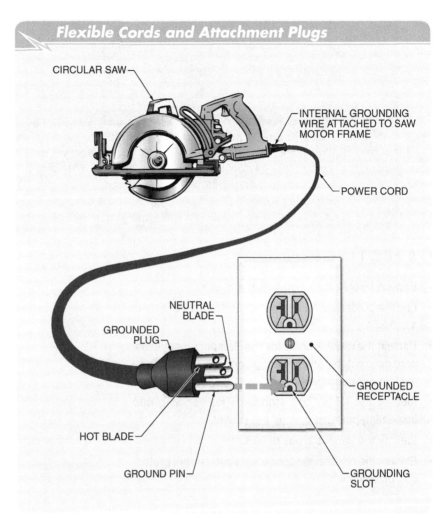

Figure 8-1. Flexible cords are assemblies of two or more insulated conductors contained within an overall outer covering and are used for the connection of equipment to a power source.

When flexible cords are attached to a building surface or secured so as to be out of the way, such as over a walkway, they are considered temporary wiring. During an audit, OSHA inspects for evidence that flexible cords have been in place longer than the 90 days allowed for temporary wiring. Evidence may include a buildup of dirt and a faded or weathered jacket around the cord.

Flexible cords may not pass through pinch points such as windows or doorways, be secured to a building surface such as a wall, or be installed above or behind walls, ceilings, or floors. Under these conditions, the flexible cords may be exposed to damage. Furthermore, they would act as a substitute for fixed or permanent wiring, which is not allowed by OSHA.

According to OSHA 1910.305(a)(2)(ii), *"Temporary wiring shall be removed immediately upon completion of the project or purpose for which the wiring was installed."*

There is a strict limit on the amount of time temporary wiring may be installed. When a project is completed, temporary wiring must be removed because it is not designed or intended to be permanent and will deteriorate over time.

Uses for Flexible Cords and Cables

According to OSHA 1910.305(g)(1)(i), *"Flexible cords and cables shall be approved for conditions of use and location."*

There are a variety of flexible cords with a range of approved uses. For example, a flexible cord that is approved for outdoor use should always be used for outdoor applications. An indoor flexible cord should never be used in an outdoor setting. The cord connectors for indoor flexible cords are not approved for wet or damp outdoor locations, which will cause the jacket and insulation to deteriorate faster. According to OSHA 1910.305(g)(1)(ii), acceptable uses for flexible cords include the following:

- pendants
- wiring of fixtures
- connection of portable lamps or appliances
- portable and mobile signs
- elevator cables
- wiring of cranes and hoists
- connection of stationary equipment to facilitate their frequent interchange
- prevention of the transmission of noise or vibration
- appliances where the fastening means and mechanical connections are designed to permit removal for maintenance and repair
- data processing cables approved as a part of the data processing system
- connection of moving parts
- temporary wiring as permitted by the regulations in OSHA 1910.305(a)(2)

Flexible cords that are approved for use per OSHA 1910.305(g)(1)(ii) must be provided with an attachment plug. The power source energizing the flexible cord must be supplied from an approved receptacle outlet.

OSHA 1910.305(g)(1)(iv) describes instances when flexible cords are prohibited. The instances when flexible cords and cables may not be used include the following:

- as a substitute for the fixed wiring of a structure
- where run through holes in walls, ceilings, or floors
- where run through doorways, windows, or similar openings
- where attached to building surfaces
- where concealed behind building walls, ceilings, or floors
- where installed in raceways, except as otherwise permitted according to OSHA

When determining an acceptable use for a flexible cord, a person should first refer to OSHA 1910.305(g)(1)(ii). If the specific use cannot be found there, the person should then check the prohibited uses from 1910.305(g)(1)(iv). If it cannot be found on either list, the person may use the closest related example.

The tags on a flexible cord provide information on the proper use of the flexible cord.

FIELD NOTES:
POWER STRIPS

Power strips are used anywhere multiple receptacles are needed or to protect electronic equipment from power surges. OSHA issued a letter of interpretation (LOI) written November 18, 2002, dealing with relocatable power taps (RPTs). The LOI is as follows:

"*Question:* What is the current compliance status on the use of 'power strips'?

"*Reply:* 'Power strips' (as they are most commonly referred to) 'Surge/Spike Protectors' or 'Portable Outlets,' typically consist of several components, such as multiple electrical receptacles, on/off power switch, circuit breaker, and a grounded flexible power cord. One nationally recognized testing laboratory, Underwriters Laboratories (UL), refers to power strips as Relocatable Power Taps (RPTs) and, in its 'General Information for Electrical Equipment Directory' (sometimes called the UL white book or UL Directory), describes RPTs as 'relocatable multiple outlet extensions of a branch circuit to supply laboratory equipment, home workshops, home movie lighting controls, musical instrumentation, and to provide outlet receptacles for computers, audio and video equipment and other equipment.' Power strips may contain other electronic components intended to provide electrical noise filtering or surge protection. UL defines and lists such devices in UL 1283, Standard for Electromagnetic Interference Filters and UL 1449, Transient Voltage Surge Suppressors (TVSS); TVSSs are dual-listed by UL and meet the requirements of UL 1363, Relocatable Power Taps.

"OSHA's standard at 29 CFR §1910.303(b)(2), Installation and use, requires that 'Listed or labeled equipment shall be installed and used in accordance with any instructions included in the listing or labeling.' Manufacturers and nationally recognized testing laboratories determine the proper uses for power strips. For example, the UL Directory contains instructions that require UL-listed RPTs to be directly connected to a permanently installed branch circuit receptacle; they are not to be series-connected to other RPTs or connected to extension cords. UL also specifies that RPTs are not intended for use at construction sites and similar locations.

"Power strips are designed for use with a number of low-powered loads, such as computers, peripherals, or audio/video components. Power loads are addressed by 29 CFR §1910.304(b)(2), Outlet devices: 'Outlet devices shall have an ampere rating not less than the load to be served.' Power strips are not designed for high power loads such as space heaters, refrigerators and microwave ovens, which can easily exceed the recommended ampere ratings on many power strips. They must also meet the requirements of §1910.305(g)(1), Use of flexible cords and cables. For example, the flexible power cord is not to be routed through walls, windows, ceilings, floors, or similar openings."

Flexible Cord Types

According to OSHA 1910.305(g)(1)(v), *"Flexible cords used in show windows and showcases shall be Type S, SE, SEO, SEOO, SJ, SJE, SJEO, SJEOO, SJO, SJOO, SJT, SJTO, SJTOO, SO, SOO, ST, STO, or STOO, except for the wiring of chain-supported lighting fixtures and supply cords for portable lamps and other merchandise being displayed or exhibited."*

The designation of a flexible cord is marked along the outside jacket. Each letter represents a characteristic of the cord. There are a variety of cords with different designations. **See Figure 8-2.**

Flexible Cords

Common Flexible Cord Designations

Designation	Meaning
S	Stranded wire, hard service flexible cord rated for 600 V
J	Junior hard service rated for 300 V
T	Insulation and jacket material is PVC thermoplastic
E	Insulation and jacket material is thermoplastic elastomer (TPE) rubber
O	Cord is oil resistant

Note: Designations without a "T" or "E" are assumed to have rubber insulation and jackets.

Common Flexible Cords

Designation	Meaning
SEO	Hard service rated for 600 V; thermoplastic insulation and jacket; oil-resistant
SJE	Junior hard service; rated for 300 V; thermoplastic insulation and jacket
SJEO	Junior hard service; rated for 300 V; thermoplastic insulation and jacket; oil-resistant
SO	Hard service rated for 600 V; rubber insulated and jacketed; oil-resistant
ST	Hard service rated for 600 V; PVC thermoplastic insulation and jacket
STO	Hard service rated for 600 V; PVC thermoplastic insulation and jacket; oil-resistant

Figure 8-2. The designation of a cord is marked along the outside jacket, with each letter representing a characteristic of the cord.

For example, an "SJ" designation indicates that the cord is "stranded wire" (S) and "junior hard service" (J) and is suitable for outdoor use or where it may receive light damage. The "J" after the "S" represents a rating of 300 V. Without the "J," the cable is rated for 600 V. The jacket for an SJ cable is a water-resistant commercial grade. This type of cord is typically used for portable lamps and chain-supported light fixtures.

Identifying Conductors in a Flexible Cord

According to OSHA 1910.305(g)(2)(i), *"A conductor of a flexible cord or cable that is used as a grounded conductor or an equipment grounding conductor shall be distinguishable from other conductors. Types S, SC, SCE, SCT, SE, SEO, SEOO, SJ, SJE, SJEO, SJEOO, SJO, SJT, SJTO, SJTOO, SO, SOO, ST, STO, and STOO flexible cords and Types G, G-GC, PPE, and W flexible cables shall be durably marked on the surface at intervals not exceeding 610 mm (24 in.) with the type designation, size, and number of conductors."*

The National Electrical Code® (NEC®) requires that a grounded conductor (neutral) be color-coded white and an equipment grounding conductor (ground) be color-coded green. However, people working with equipment and electrical devices manufactured in Europe should be aware that green may not be the grounding conductor. Green with a yellow tracer is often used as the energized conductor for equipment manufactured in Europe. The type designation, size, and number of conductors must be marked on the outside of the cable. **See Figure 8-3.**

SAFETY FACT

A flexible cord with a "J" designation is rated for 300 V and is designed to withstand moderate rough use. Flexible cords without the "J" designation are rated for 600 V and have thicker insulation and a thicker jacket to withstand rougher use.

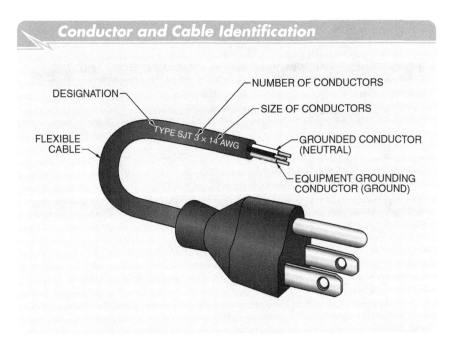

Figure 8-3. The NEC® requires that the grounded conductor (neutral) be color-coded white and the equipment grounding conductor (ground) be color-coded green. The type designation, size, and number of conductors must be marked on the outside of the cable.

Repairing Flexible Cords

According to OSHA 1910.305(g)(2)(ii), *"Flexible cords may be used only in continuous lengths without splice or tap. Hard-service cord and junior hard-service cord No. 14 AWG and larger may be repaired if spliced so that the splice retains the insulation, outer sheath properties, and usage characteristics of the cord being spliced."*

Flexible cords may not be repaired if they are smaller than No. 14 AWG. However, if they are No. 14 AWG or larger, the splice must be mechanically and electrically continuous, and the insulation and jacket must have the same characteristics as the original. Electrical tape, heat-shrink tubing, and other types of splicing are not appropriate for the repair of flexible cords. Many companies do not allow flexible cord sets to be repaired and opt for replacement instead.

Connecting Flexible Cords and Cables

According to OSHA 1910.305(g)(2)(iii), *"Flexible cords and cables shall be connected to devices and fittings so that strain relief is provided that will prevent pull from being directly transmitted to joints or terminal screws."*

In the past, cords had open-faced cord connectors and an electrician's knot was used to prevent strain from transmitting to the joints or terminal screw. Closed cord caps are now used to prevent strain on joints and terminal screws. **See Figure 8-4.** Closed cord caps are designed to compress the conductors between the two halves of the cord cap to secure the conductors. Flexible cords and cables can only be repaired by a qualified person that understands polarity, grounding, and proper repair techniques.

Closed Cord Caps

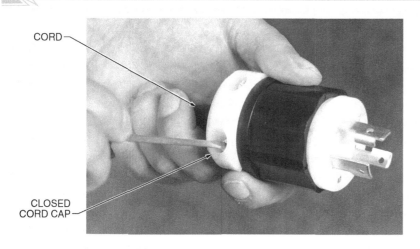

Figure 8-4. Closed cord caps are used to prevent strain from being directly transmitted to joints or terminal screws.

PORTABLE ELECTRIC EQUIPMENT PER OSHA

According to OSHA 1910.334(a)(1), *"Portable equipment shall be handled in a manner which will not cause damage. Flexible electric cords connected to equipment may not be used for raising or lowering the equipment. Flexible cords may not be fastened with staples or otherwise hung in such a fashion as could damage the outer jacket or insulation."*

It is common sense not to use equipment in such a manner that can cause damage, but people often do not realize when they are damaging the equipment. For example, an employee on a ladder can cause damage to an electric tool by using the cord to raise and lower the tool. The cord has a strain-relief device where it enters the tool. **See Figure 8-5.** The strain-relief device is not designed to carry the weight of the tool. Too much strain on the cord can cause the conductors inside to pull loose from their connections, which could short against the tool housing. Any strain on the cord is transmitted directly to the internal connections of the tool.

Flexible cords are prone to damage and must not be installed in a manner that could cause damage to the insulation or outer jacket. For example, cords must not be stapled in place. The maximum 90 day period OSHA allows for the use of flexible cords as temporary wiring is due to concerns of damage over a period of time.

Appropriate measures should be taken when transporting portable electric equipment.

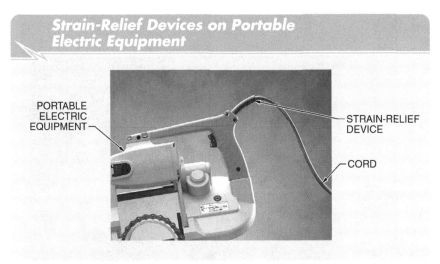

Milwaukee Electric Tool Corporation

Figure 8-5. Flexible electric cords connected to equipment must have a strain-relief device that protects the connections from small pulls and tugs. However, the device is not designed to carry the weight of the tool.

Visual Inspection of Portable Electric Tools and Flexible Cords

According to OSHA 1910.334(a)(2)(i), *"Portable cord and plug connected equipment and flexible cord sets (extension cords) shall be visually inspected before use on any shift for external defects (such as loose parts, deformed and missing pins, or damage to outer jacket or insulation) and for evidence of possible internal damage (such as pinched or crushed outer jacket). Cord and plug connected equipment and flexible cord sets (extension cords) which remain connected once they are put in place and are not exposed to damage need not be visually inspected until they are relocated."*

Portable electric tools and flexible cord sets must be visually inspected before each use. The tool or cord set should be taken out of service if any defect that may create a safety hazard is noticed. While there are many defects to look for when inspecting portable electric tools and flexible cords, common defects include loose parts, deformed or missing pins, and damage to the outer jacket, such as pinching or crushing. Portable electric equipment and flexible cords are allowed to stay in place indefinitely as long as they are not subject to damage.

According to OSHA 1910.334(a)(2)(ii), *"If there is a defect or evidence of damage that might expose an employee to injury, the defective or damaged item shall be removed from service, and no employee may use it until repairs and tests necessary to render the equipment safe have been made."*

Damaged equipment must be tagged to signify it is out of service. Repairs may be made to portable electric tools and flexible cords, but this would typically require replacing the cords and connector. Once repairs have been made, the tool or cord set must be tested to verify it is safe to use. Such tests may include performing a polarity check, testing insulation resistance, testing the resistance of the ground conductor to the frame if it is a grounded tool, and performing an operational check.

According to OSHA 1910.334(a)(2)(iii), *"When an attachment plug is to be connected to a receptacle (including on a cord set), the relationship of the plug and receptacle contacts shall first be checked to ensure that they are of proper mating configurations."*

OSHA does not allow personnel to field modify cord connector pins to match receptacles that are not designed to fit together. This includes twisting pins to fit into a receptacle for which the plug was not designed to be used.

FIELD NOTES:
TESTING ELECTRICAL TOOLS BEFORE THEY ARE PLACED INTO SERVICE

At Shermco we had a large shutdown scheduled. We purchased twelve extension cords from our regular supplier in preparation. We routinely test all electrical tools, devices, and equipment before they are put into service. When the twelve extension cords were tested, six of the twelve were found to have no continuity between the ground pins. We wondered: Were they defective or counterfeit? However, there was no way to tell without cutting them open. Instead, we returned the six defective cord sets and received new ones that tested good.

Portable electrical tools and flexible cords must be inspected visually before each use. It is also good practice to routinely test electrical tools and cords before each use. Even new equipment can be faulty.

Grounding-Type Equipment

According to OSHA 1910.334(a)(3)(i), *"A flexible cord used with grounding type equipment shall contain an equipment grounding conductor."*

A continuous ground path must be maintained from the tool to ground if the tool is grounded. This does not apply to double-insulated tools. However, a ground fault circuit interrupter (GFCI) is required if a grounded flexible cord is used with a double-insulated tool. A GFCI protects the grounded flexible cord, not the tool.

According to OSHA 1910.334(a)(3)(ii), *"Attachment plugs and receptacles may not be connected or altered in a manner which would prevent proper continuity of the equipment grounding conductor at the point where plugs are attached to receptacles. Additionally, these devices may not be altered to allow the grounding pole of a plug to be inserted into slots intended for connection to the current-carrying conductors."*

Many homes built in the 1960s or earlier do not have grounded receptacles. Attachment plugs typically include a ground pin and will not fit into a receptacle without a ground. Some individuals do not follow OSHA regulations and cut off the ground pin from the attachment plug for flexible cords and portable electric tools. However, they do not realize that the ground pin is used to protect them if the energized conductor shorts to a conductive part of the tool. Instead of altering a device, a grounding adapter should be used to accommodate an attachment plug with a ground pin.

Grounding adapters are commonly used when the receptacle does not have a ground but the electrical equipment does. Grounding adapters have a metal tab that is attached to the receptacle by the wall plate screw. **See Figure 8-6.** The problem with grounding adapters is that the wiring in an old home may be asbestos/rubber and may not have a metal sheath. This means that the equipment is not connected to the grounding system of the house and that a grounding

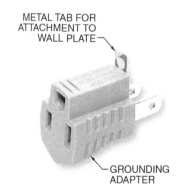

Shermco Industries

Figure 8-6. Grounding adapters are used when a receptacle does not have a ground but the electrical equipment does.

adapter will not provide protection if there is a short circuit. Without a ground, overcurrent protective devices (OCPDs) cannot detect a phase-to-ground fault. Old homes often cost more to insure because they have ungrounded electrical systems, which can increase the chance of fire or injury.

SAFETY FACT

According to OSHA 1910.334(a)(3)(iii), "Adapters which interrupt the continuity of the equipment grounding connection may not be used." This regulation includes grounding adapters used on a two-wire electrical system, unless the box is metallic and connected to ground.

Conductive Work Locations

According to OSHA 1910.334(a)(4), *"Portable electric equipment and flexible cords used in highly conductive work locations (such as those inundated with water or other conductive liquids), or in job locations where employees are likely to contact water or conductive liquids, shall be approved for those locations."*

Working in wet locations requires special tools and training. Often, battery-powered tools are used in wet locations. Tools using low-voltage isolation transformers can also be used. Each tool must be approved for the specific use and environment. Also, tools should be inspected carefully before they are used.

FIELD NOTES:
WORKING IN WET LOCATIONS

There was a report in a Dallas newspaper that a local homeowner called a plumber to repair a leaky water line under his home. Most homes in the Dallas-Fort Worth area are built on concrete slabs due to the shifting soils, but the house that the plumber was visiting was constructed on a pier-and-beam foundation, which allowed it to have a crawl space. The house was an older construction, probably built in the 1940s or 1950s, and used a metallic water system.

The homeowner stated that he was talking to the plumber through the floor while the plumber was under the house. The plumber remarked to the homeowner that when he touched the water pipe with his electric drill, sparks came off the end of it. The plumber apparently thought this was amusing. The homeowner went to the store and when he returned, he noticed the plumber's truck was still parked in front of his house. When the homeowner called out to the plumber, he did not answer. The homeowner then dialed 911.

The emergency responders found the plumber electrocuted under the house. The water pipe was used as the principal ground for the home's electrical system, and it interrupted the ground circuit when it broke. The earth under the home was damp from the broken water pipe and when the plumber made the connection between the water pipe and his electric tool, he became the pathway to ground. This goes to show that it really pays for most tradesworkers to receive some electrical safety training.

Connecting Attachment Plugs

According to OSHA 1910.334(a)(5)(i), *"Employees' hands may not be wet when plugging and unplugging flexible cords and cord and plug connected equipment, if energized equipment is involved."*

Most people know not to connect or disconnect cord-connected equipment with wet hands. However, they may not realize this also includes perspiration. Perspiration contains water and salt, both of which can be dangerous around electricity. A person should ensure their hands are dry when they are using flexible cords and cord-and-plug-connected equipment.

According to OSHA 1910.334(a)(5)(ii), *"Energized plug and receptacle connections may be handled only with insulating protective equipment if the condition of the connection could provide a conducting path to the employee's hand (if, for example, a cord connector is wet from being immersed in water)."*

This OSHA regulation demonstrates the importance of always carrying rubber insulating gloves. Employees are more likely to use rubber insulating gloves if they are carrying gloves or gloves are readily available.

According to OSHA 1910.334(a)(5)(iii), *"Locking type connectors shall be properly secured after connection."*

Locking plugs fit into locking receptacles by firmly inserting and twisting to lock in place. Once locked in place, locking plugs cannot be pulled out without first twisting them. OSHA requires this type of locking connector to be completely locked into the receptacle to prevent it from coming loose during usage. **See Figure 8-7.**

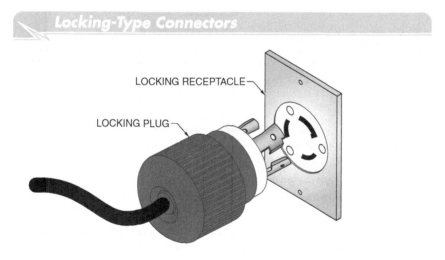

Figure 8-7. Locking connectors must be completely locked in to prevent them from coming loose during usage.

Handlamps

According to OSHA 1910.305(j)(1)(ii), *"Handlamps of the portable type supplied through flexible cords shall be equipped with a handle of molded composition or other material identified for the purpose, and a substantial guard shall be attached to the lampholder or the handle. Metal shell, paper-lined lampholders may not be used."*

Portable handlamps must have a handle to prevent shocks from occurring while a person is holding the handlamp and a guard that will hold up during use. **See Figure 8-8.** The word "substantial" is subjective, but it means that the guard should not collapse or break during normal use. Older handlamps that use metal guards or metal-shell lampholders are not considered to be safe and should not be used in the workplace or at home.

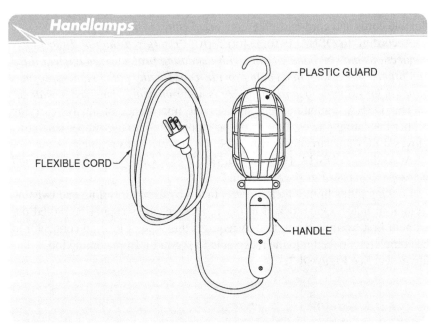

Figure 8-8. Portable handlamps must have a handle to prevent shocks from occurring while a person is holding the handlamp and a guard that will hold up during use.

Receptacles, Cord Connectors, and Attachment Plugs

According to OSHA 1910.305(j)(2)(i), *"All 15- and 20-ampere attachment plugs and connectors shall be constructed so that there are no exposed current-carrying parts except the prongs, blades, or pins. The cover for wire terminations shall be a part that is essential for the operation of an attachment plug or connector (dead-front construction). Attachment plugs shall be installed so that their prongs, blades, or pins are not energized unless inserted into an energized receptacle. No receptacles may be installed so as to require an energized attachment plug as its source of supply."*

Commercially available portable electric equipment meets the design requirements stated in this paragraph, but the tool or flexible cord must be inspected to ensure none of these conditions exist. It should be noted that a dead-front construction is required for cord connectors. Older style open-faced connectors with fiber covers may not be used.

According to OSHA 1910.305(j)(2)(ii), *"Receptacles, cord connectors, and attachment plugs shall be constructed so that no receptacle or cord connector will accept an attachment plug with a different voltage or current rating than that for which the device is intended. However, a 20 A T-slot receptacle or cord connector may accept a 15 A attachment plug of the same voltage rating."*

Receptacles are available in a variety of shapes and sizes depending on the application. For example, receptacles rated for 30 A have a different blade configuration than receptacles rated for 15 A. Receptacles rated for 110 V have a different blade configuration than receptacles rated for 220 V. It is important not to modify the cord connector blades. Any modifications may nullify the protection offered by the specific configurations. **See Figure 8-9.**

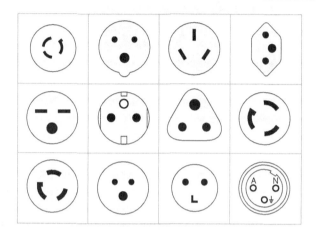

Figure 8-9. Receptacles are available in a variety of shapes and sizes depending on the application.

According to OSHA 1910.305(j)(2)(iii), *"Nongrounding-type receptacles and connectors may not be used for grounding-type attachment plugs."*

According to this OSHA regulation, a portable electric tool that has a ground must be connected to a receptacle that has a ground. The ground must be continuous from the tool to the earth in order to be effective.

According to OSHA 1910.305(j)(2)(iv), *"A receptacle installed in a wet or damp location shall be suitable for the location."*

Receptacles installed in wet or damp locations present additional hazards to a person using portable electrical equipment. Wet locations require specially designed receptacles to prevent water intrusion.

Portable and Vehicle-Mounted Generators

According to OSHA 1910.269(i)(3)(i), *"The generator may only supply equipment located on the generator or the vehicle and cord- and plug-connected equipment through receptacles mounted on the generator or the vehicle."*

The word "equipment" usually refers to a receptacle mounted onto the frame of a generator. It does not refer to cord-connected portable electric tools or equipment. **See Figure 8-10.**

According to OSHA 1910.269(i)(3), *"(ii) The non-current-carrying metal parts of equipment and the equipment grounding conductor terminals of the receptacles shall be bonded to the generator frame. (iii) For vehicle-mounted generators, the frame of the generator shall be bonded to the vehicle frame. (iv) Any neutral conductor shall be bonded to the generator frame."*

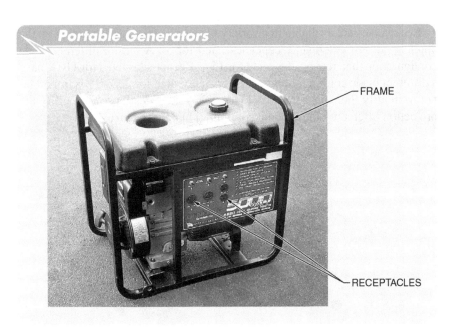

Figure 8-10. The frame of a portable generator is used as a grounding electrode for cord-connected hand tools with grounding-type cords and caps.

OSHA requires that most parts of equipment be bonded to the frame of a generator, and that the frame of the generator be bonded to the frame or chassis of the vehicle, if one is used. The generator frame becomes the return path for the current because current returns to its source. Bonding between the frame of the generator and the vehicle prevents the possibility of touch potential.

PORTABLE ELECTRIC EQUIPMENT PER NFPA 70E 110.4(B)

NFPA 70E standards for handling and inspecting portable electric equipment and flexible cord sets are very similar to OSHA regulations. These requirements closely parallel OSHA requirements, following them word-for-word in some places.

Handling and Storing Equipment

According to NFPA 70E 110.4(B)(1), portable electric tools and flexible cords must be handled in a manner that will not cause damage. This includes raising or lowering tools by their cords. Flexible cords and cords connected to equipment are not to be stapled or hung in a manner that could cause damage.

Storing extension cords and portable electric tools must be done in a manner that will protect them from damage. Wadding up a cord and throwing it into a toolbox is not an appropriate storage technique. Portable electric tools should be stored in their original equipment manufacturer (OEM) storage cases, if possible. If no case is available, they should be stored with their cord either wound beside them and secured with a reusable retainer or lightly wound around the tool. Lightly wound means there is no strain on the cord or strain relief component. Winding the cord tightly could cause damage to the jacket or strain relief component.

Grounding-Type Equipment

According to NFPA 70E 110.4(B)(2), flexible cords used with grounding-type equipment are to contain a ground conductor as well. Cord connectors (attachment plugs) and receptacles cannot be modified to be used in a manner not intended by the manufacturer. Neither can an adapter be used that interrupts the continuity of the ground, nor can the cord connector or receptacle be modified in a way that would interrupt the ground conductor.

Inspection of Portable Cord-and-Plug-Connected Equipment and Flexible Cords

According to NFPA 70E 110.4(B)(3)(a), portable electric equipment and flexible cords must be inspected for external defects before each use. Common defects listed for inspection, which are also stated in OSHA 1910.334(a)(2)(i), include loose parts, deformed or missing pins, and evidence of internal damage such as crushed or pinched outer jacket. Inspections are not required on cord-and-plug-connected equipment that will not be subjected to damage and will not be relocated once they are in place. If they are relocated, they will then require an inspection. Visual and mechanical inspections are required along with operational and electrical tests to ensure a portable electric power tool or flexible cord is safe to use.

Visual Inspection. Both OSHA 1910.334 and NFPA 70E 110.4(B)(3)(a) require that portable electric tools and flexible cords be inspected prior to use. Visual and mechanical inspections are performed first. To complete a visual inspection of portable electric tools and flexible cords, the following tasks should be performed:

- Inspect for indications of overheating. Areas showing discoloration and soot or smoke damage are signs that tools may be overheating. Signs that cords are overheating may be areas of discoloration or bubbling of the jacket.
- Check for a secure connection of the flexible cord to the tool housing or connector. Look for signs of loose or detached flexible cords. If the flexible cord or jacket is pulled from the connector or tool, take the tool out of service immediately.
- Inspect for cuts, breaks, and crushing damage to the jacket of flexible cords. Cuts and breaks to the jacket may indicate damage to the insulation as well. Also look for evidence of repairs to the flexible cord. Electrical tape, heat shrink, and other quick-fix remedies are unacceptable because they do not have the same characteristics as the original jacket. If the conductor is smaller than No. 14 AWG, it should not be repaired. Crushing or pinching of the jacket may indicate damage to the insulation beneath it.
- Inspect for missing or altered pins on the cord connector (attachment plug). Do not use a tool or flexible cord with a missing ground pin. If there is no path through the tool to ground, the current will use a person's body as the path to ground. Never twist or alter pins on a cord connector to mate with a receptacle of a different configuration.
- Check for any open-faced cord connectors. These older types of cord connectors use a fiber disk over the energized terminals. They require an electrician's knot to prevent strain from being transmitted directly to the terminals, but it is often not done correctly. Flexible cords with open-faced cord connectors must be replaced immediately with an approved closed-style cord connector or tagged and removed from service.

Figure 8-11. When visually inspecting portable electric tools and flexible cords, it is necessary to look for labels showing that the tool or cord is approved and listed by a nationally recognized testing laboratory (NRTL).

- Inspect for labeling that shows that the tool or cord is approved and listed by a nationally recognized testing laboratory (NRTL), such as Underwriters Laboratories (UL). **See Figure 8-11.** NRTLs test and certify that equipment and devices meet the standards that they state they meet. There are over a dozen NRTLs recognized by OSHA. UL is the most well-known of the NRTLs. The labeling should be inspected for both grounded and double-insulated tools. Double-insulated tools must also have either the international symbol for double-insulation or have wording indicating it is double-insulated or both.
- Inspect for loose or missing parts that may affect the safety of the tool or flexible cord.
- Check for loose guards or guards that are not functioning or positioned properly.
- If a lock button is used, make sure it is properly designed. Either a constant pressure switch or a momentary contact on-off control is required. These regulations are covered in OSHA 1910.243(a)(2)(ii) and 1926.300(d)(1) through 1926.300(d)(3). A constant pressure switch delivers power to the tool only when the control switch is depressed. When the switch is released, the tool stops operating. Momentary contact on-off control means that when the switch is depressed, the tool turns on, and when the switch is depressed again, the tool turns off. Many companies prohibit tools that have lock buttons, unless they are disabled.
- Ensure that a properly sized flexible cord with the required markings is used. Flexible cords should be a type approved for commercial use and must be durably marked every 24″ or less. OSHA 1910.305(g)(2)(i) covers the required markings for flexible cords. Also look for the NRTL marking. Commercial flexible cords should be No. 12 AWG. Use care when connecting flexible cords together to ensure the voltage drop does not damage equipment.

Mechanical Inspection. Once the visual inspection is finished, a tool should be mechanically inspected. The tool should be inspected for indications of malfunctions while it is operating. Running the tool for a few seconds will provide insight as to whether it is going to perform as required. Indications of a possible malfunction during an operational test include the following:

- The switch is sticking.
- The tool does not immediately turn off or slowly returns to the OFF position when power to the power switch is turned off.
- There is excessive sparking from the motor while it is operating. There will always be some small amount of sparking from the brushes, but excessive sparking may indicate a problem with the tool.
- The tool is operating sluggishly or vibrating oddly.
- There are strange noises such as grinding, buzzing, and squealing.
- There is excessive heating, which may cause smoke to exit the tool.
- The tool does not hold the bit or blade properly.

Electrical Testing. A well-developed power tool maintenance program should include electrically testing the tool at least once a year. Power tools are handled roughly in the field. They are dropped, thrown around, and bounced around in the back of pickup trucks. Electrically testing power tools ensures that they are safe to use and will not cause injury. Tests that are commonly performed on grounded portable electric power tools include the following:

- Testing the insulation resistance of the motor—The stationary windings of a motor can be tested by taking a megohm measurement between the neautral and case. The insulation resistance should be as high as several hundred megohms, but can be as low as 1 MΩ. According to ANSI/NETA MTS-2011, *Standard for Maintenance Testing Specifications for Electrical Power Equipment and Systems,* the minimum insulation resistance for insulation on all low-voltage motors is 5 MΩ.
- Testing the continuity between the grounding conductor (green wire and ground pin) and the frame of the tool—These tests can only be performed on grounded tools using a multimeter. A reading of less than 0.2 Ω is considered good continuity.
- Testing for excessive current leakage—A portable electric tool tester can be used to check for running current and excessive leakage current, as well as insulation resistance and continuity. **See Figure 8-12.**

Conductor Sizing for Flexible Cord Sets

The conductor sizing of a flexible cord is important when operating portable electric tools. Portable electric power tools run on power, which is measured in watts (W). To determine power, the following formula is applied:

$P = E \times I$

where

P = power (in W)

E = voltage (in V)

I = current (in A)

For example, what is the power of the motor for a 110 V, 4.5 A circular saw?

$P = E \times I$

$P = 110 \times 4.5$

$P = 495$ W

The motor of a 110 V, 4.5 A circular saw uses 495 W of power. If voltage to the motor drops, the motor still requires 495 W of power. It will also increase the amount of current in the circuit to make up for the voltage drop caused by the resistance of the conductor. This increase in current is not a problem for a 100′ flexible cord. At a length of 100′, the No. 14 AWG copper conductor in the flexible cord can carry 5 A to 7 A. **See Figure 8-13.** However, if the length is increased to 200′ by a daisy chain of two 100′ flexible cords, then that same No. 14 AWG copper conductor can only carry 2 A to 5 A. It will only carry up to 2 A continuously and up to as much as 5 A as the duty cycle decreases. *Duty cycle* is the amount of time an electrical tool is carrying a specific amount of current.

The increase in current flowing through the flexible cord creates additional heat. This creates additional resistance, which further increases the voltage drop. Increasing the voltage drop causes the motor to draw more current, which creates additional heat. As the conductor heats, its resistance and voltage drop increase. Because the motor must have the wattage required for it to run, each time the voltage drops it will balance out the equation by drawing more current. This cycle of voltage drop and current increase may cause the tool to overheat. It is important to refer to the owner's manual to determine the appropriate length and AWG of flexible cord for portable electrical equipment.

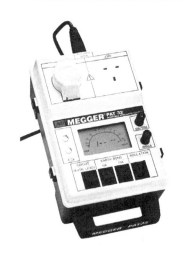

Megger Group Limited

Figure 8-12. Portable electric tool testers are used to test grounded portable electric power tools.

Cord Length[†]	Multiconductor Cable Size*					
	Current Rating[‡]					
	0–2	2–5	5–7	7–10	10–12	12–15
25	16	16	16	16	14	14
50	16	16	16	14	14	12
100	16	16	14	12	12	—
150	16	14	12	12	—	—
200	14	14	12	10	—	—

* in AWG
[†] in ft
[‡] in A

Figure 8-13. Circuit resistance increases as the length of a flexible cord increases. The increase in resistance causes a decrease in current flow through the cord.

Defective Equipment

Defective equipment must not be used. It must be tagged to identify it as defective and removed from service. Defective equipment cannot be used again until it has been repaired and tested to ensure it is safe to use.

Proper Mating

When connecting a cord to a receptacle, it is necessary to ensure that the blades on the cord connector have the same configuration as the receptacle. OSHA 1910.334 and NFPA 110.4(B)(2) prohibit field modifying the configuration of the cord connector or receptacle.

Conductive Work Locations

Portable electric equipment and flexible cords must be approved for use in areas where they may become wet or inundated with water. GFCIs are required whenever employees may encounter damp or wet areas. Good work practices involve using GFCIs any time portable electric tools or flexible cords are used.

Connecting Attachment Plugs

According to NFPA 70E 110.4(B)(5), an employee's hands must be dry when plugging and unplugging flexible cords and portable electric equipment if the equipment is energized. If the condition of the connection can cause a shock hazard, rubber insulating equipment must be used while plugging in or unplugging equipment that is energized.

For example, an employee must use rubber insulating gloves if a cord connector is wet and can provide a conductive path to the employee. Rubber insulating shielding or insulated tools can be used as well, depending on the size and configuration of the connector. Also, a cord connector does not need to be visibly wet to provide a conductive path to an employee. A path can be created if an employee has sweaty hands because sweat contains water and salts that make it conductive.

Locking-type connectors should be securely fastened. Locking-type connectors must also be inspected to ensure they are safe to use.

GFCI Protection Devices

A *ground fault circuit interrupter (GFCI)* is a device that protects against electric shock by detecting an imbalance of current in the energized conductor and the neutral conductor. An imbalance of 5 mA or more will cause the GFCI to open the circuit. GFCIs are common, everyday electrical devices found in homes and on job sites. GFCIs are not ground fault protection devices. Ground fault protection is used to protect electrical equipment and power systems.

GFCIs are available as receptacle types, portable types, or circuit breaker types. GFCI receptacles provide ground fault protection at the point of installation. **See Figure 8-14.** Portable GFCIs are designed to be easily moved from one location to another. GFCI circuit breakers that are installed in a panelboard provide GFCI protection and conventional circuit overcurrent protection for all branch circuit components connected to the circuit breaker.

The NEC® first required GFCIs in 1968. A GFCI manufactured prior to 2003 stays energized and operates like a normal receptacle even if the GFCI portion is defective. A newer GFCI manufactured after 2003 will not function at all if the GFCI portion of the receptacle is defective. GFCIs from before 2003 should be replaced with the newer style to prevent workers from having a false sense of security.

NFPA 70E 110.4(C)(1) requires that employees be provided GFCI devices where required by local, state, or federal law. It also states that portable GFCI devices listed for portable use shall be permitted. When the words "shall be permitted" are used, it means that the devices are not mandatory and that they cannot be excluded from possible use. In this instance, a worker may use a portable GFCI in place of a wall-mounted, permanent GFCI.

NFPA 70E 110.4(C)(2) notes that some cord-and-plug-connected equipment may not be of the standard 125 V, 15 A, 20 A, or 30 A ratings. In cases where the portable electric tool or device is being used outdoors (or in another wet area) and exceeds the standard ratings, an assured equipment grounding conductor program must be implemented because there are no commercially available GFCIs for that type of equipment. An *assured equipment grounding conductor program* is the process of testing and verifying that a ground path is unbroken and continuous.

GFCI Testing. According to NFPA 70E 110.4(D), GFCI devices must be tested in accordance with the manufacturer's instructions. According to the manufacturers, GFCIs should be tested once per month by pressing the PUSH TO TEST button on the receptacle. Pressing the button would allow current to flow through an internal resistor and simulate an imbalance condition. As long as the GFCI trips when the button is pushed, the electronic circuitry is working. With older GFCIs, a defective GFCI circuit may not disable the receptacle and a person would have a false sense of security, thinking that they are protected when, in fact, they are not.

Figure 8-14. Like every other electrical device, GFCIs require maintenance and must be tested in accordance with the manufacturer's recommendations.

GFCI Receptacles

SAFETY FACT

Since all manufacturers require the operational testing of a GFCI every 30 days, NFPA 70E 110.4(D) could make it difficult for companies with massive numbers of GFCIs installed in their buildings and facilities to be compliant with NFPA 70E.

GFCI testers are available to test GFCI receptacles. These testers are used to verify whether GFCIs are functioning in accordance with the manufacturer's specifications. **See Figure 8-15.** If there is no ground connection, the tester will not function. The actual current required to cause the GFCI to operate is displayed. If the GFCI fails to trip the tester will display INVALID. In addition, GFCI testers can detect the following:
- normal wiring
- open ground
- reversed polarity
- reversed hot conductor and ground conductor
- open hot conductor or neutral conductor

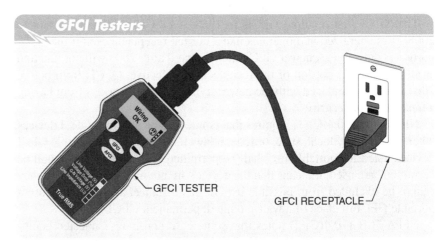

Figure 8-15. GFCI testers are used to verify GFCI receptacles are functioning properly.

It is worth the time and money to replace all the older GFCIs with new two-core GFCIs. Under normal (balanced) conditions with an electrical device connected to a GFCI, the current flow into the receptacle is approximately equal to the current flow out of the receptacle. There may be some minor losses due to heat loss in the connected electrical equipment, but GFCIs typically must detect at least a 5 mA (±1 mA) difference between what goes into the GFCI and what leaves the GFCI to trip it. **See Figure 8-16.**

SAFETY FACT

A potentially dangerous ground fault is any amount of current that may deliver a dangerous shock. Any current over 8 mA is considered potentially dangerous, depending on the path the current takes, the physical condition of the person receiving the shock, and the amount of time the person is exposed to the shock.

Overcurrent Protection Modification

Overcurrent protective devices (OCPDs) such as fuses, circuit breakers, and protective relays cannot be modified, even temporarily, except for what is allowed in codes and standards for overcurrent protection. Changing fuse ratings and/or circuit breaker or protective relay settings can interfere with the selective coordination that was engineered into the electrical power system when it was initially installed.

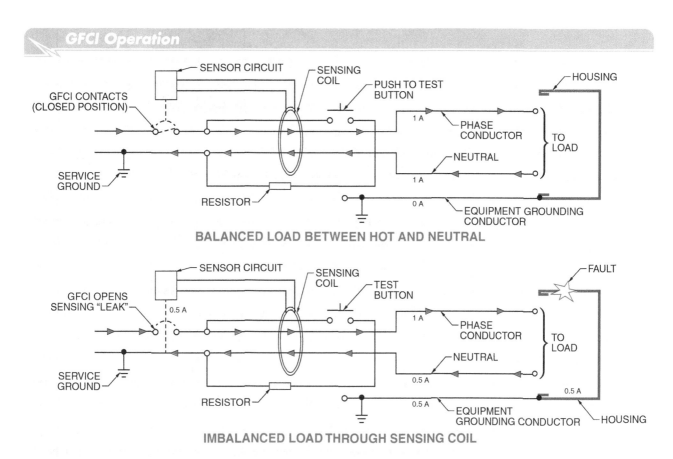

Figure 8-16. Under normal (balanced) conditions with an electrical device connected to a GFCI, the current flow into the receptacle is approximately equal to the current flow out of the receptacle.

Selective coordination is the act of isolating a faulted circuit from the remainder of the electrical system, thereby eliminating unnecessary power outages. The faulted circuit is isolated by the selective operation of only the OCPD closest to the overcurrent condition. However, if the settings are changed on circuit breakers or fuses are replaced with fuses of a different rating, selective coordination can be nullified. This typically results in nuisance tripping or in some cases a no-trip situation. In an effort to stop the nuisance tripping, employees sometimes change the settings on other OCPDs to balance the system. This can lead to a dangerous environment for electrical workers.

One of the basic principles of selective coordination is that instantaneous devices in series cannot be coordinated. For example, OCPDs are typically connected in a series-parallel circuit. If all the OCPDs have an instantaneous function only, any of the circuit breakers may operate if there is a high enough fault current. **See Figure 8-17.** Instantaneous means that there is no intentional time delay, but there will always be some mechanical delay, which is typically 2 to 4 cycles (0.3 to 0.7 sec). If a 25 kA fault is created by the load failing, it could cause the main breaker to operate, while the load breaker may not operate when it should. Since short-circuit current flows through the main breaker first, it will operate if the current is above its setpoint. The current also may reach the secondary main circuit breaker. If it does, that breaker may also operate if the current exceeds its minimum pickup point (instantaneous pickup setting). The circuit breaker most likely not to operate is the load breaker.

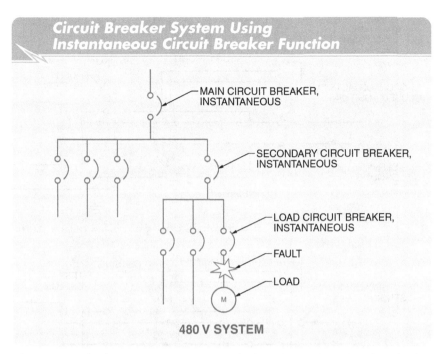

Figure 8-17. If a fault occurs in a system designed with circuit breakers having only the instantaneous function, the main circuit breaker is most likely to trip first, causing the rest of the power system to shut down.

In a properly coordinated electrical system, OCPDs are set so the OCPD closest to the load will operate in case of a fault. If it does not operate, then the next device upstream will operate. If that device does not operate, then the next device even farther upstream will operate, and so on. This usually will continue up to a transformer secondary winding. **See Figure 8-18.** In this example, two of the three circuit breakers have a short-time delay (STD) function while the load breaker does not. The STD function provides an intentional delay in the operation of the circuit breaker during short-circuit conditions. The delays given are 0.18 sec and 0.33 sec. These delay times are nominal times, meaning that the circuit breaker will not operate at that specific time but rather within a time band that has maximum and minimum operating times. The STD function can provide a time delay of up to 0.5 sec.

As short-circuit current flows through the main circuit breaker, it exceeds the STD setting, but it cannot trip until the time delay has passed. The current also flows through the secondary circuit breaker, but it also has a time delay and cannot trip. The current reaches the load breaker and, having its instantaneous function enabled, trips instantaneously. The current flow through the system ends, and the main and secondary circuit breakers reset.

However, if the load circuit breaker does not operate, the STD function on the secondary breaker will time out and trip. If for some reason the secondary breaker does not operate, then the STD function of the main circuit breaker will time out and trip. More of the power system is deenergized at each level, but the fault condition still must be cleared. Incident energy is proportional to time. Increasing the time increases the heat received by a worker during an arc flash event. Electrical workers must be aware of the incident energy that may be available when they are working on power systems using short-time delays. They must also be aware of the hazardous effects of high levels of incident energy.

Circuit Breaker System Using Short-Time Delay (STD) Circuit Breaker Function

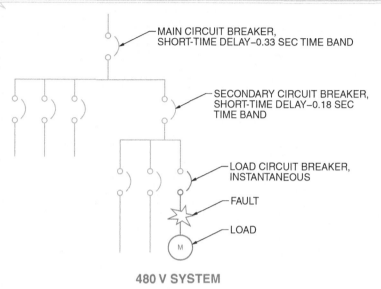

Figure 8-18. If a fault occurs in a system designed with circuit breakers having an STD function, the load circuit breaker trips, while the rest of the power system only operates if needed.

SUMMARY

- During an audit, OSHA looks for evidence that flexible cords have been in place longer than 90 days.
- Flexible cords and cables must be approved for conditions of use and location.
- Portable electric tools and flexible cord sets must be visually inspected before each use.
- Flexible cords that are No. 14 AWG or larger may be repaired only if the splice is mechanically and electrically continuous and the insulation and jacket has the same characteristics as the original.
- Portable electric tools and flexible cords must be handled in a manner that will not cause damage.
- Receptacles, cord connectors, and attachment plugs must be constructed so that no receptacle or cord connector will accept an attachment plug with a different voltage or current rating than that for which the device is intended except for a 20 A T-slot receptacle or connector with a 15 A attachment plug of the same voltage rating.
- GFCIs must be tested per the manufacturer's recommendations.

Digital Learner Resources
ATPeResources.com/QuickLinks
Access Code: 705798

PORTABLE ELECTRIC TOOLS AND FLEXIBLE CORDS

Review Questions

Name _____   Date _____

_____ 1. Flexible cords may be used ___.
  A. as a replacement for permanent wiring
  B. to install portable signs
  C. through walls or other openings
  D. when attached to building surfaces

_____ 2. A flexible cord with an "SJ" designation is ___.
  A. insulated with natural rubber
  B. solid wire and hard service
  C. stranded wire and junior hard service
  D. oil-resistant

_____ 3. Portable electric tools and flexible cords must be visually inspected ___.
  A. before each use
  B. every day
  C. once per month
  D. once a year

_____ 4. Which of the following is not a common defect to look for when inspecting portable electric tools and flexible cords?
  A. loose parts
  B. deformed or missing pins
  C. damage to the outer jacket of the cord
  D. sparking

T   F   5. Flexible cord connector blades and pins may be modified to match particular receptacles.

T   F   6. Flexible cords can be repaired using heat-shrink tubing.

_____ 7. The required markings on flexible cords should be displayed every ___" or less.
  A. 12
  B. 24
  C. 36
  D. 48

_____ 8. A well-developed power tool maintenance program should include electrically testing the tool every ___.
  A. year
  B. two years
  C. three years
  D. four years

_____ 9. A ground fault circuit interrupter (GFCI) is a device that protects against electric shock by detecting an imbalance of ___ in the energized conductor and the neutral conductor.
   A. voltage
   B. current
   C. resistance
   D. capacitance

_____ 10. GFCIs should be tested ___ by pressing the PUSH TO TEST button on the receptacle.
   A. before each use
   B. every day
   C. once per month
   D. once a year

_____ 11. A good continuity reading between the grounding conductor and the frame of an electric tool is ___ Ω.
   A. less than 0.2
   B. 1
   C. 5
   D. 10

_____ 12. Extension cords may only be repaired by ___.
   A. a person authorized by the employer
   B. sending it out to a repair shop
   C. sending it to the manufacturer
   D. a qualified person

_____ 13. What is the minimum extension cord conductor size that can be repaired?
   A. 16 AWG
   B. 14 AWG
   C. 12 AWG
   D. 10 AWG

_____ 14. When are overcurrent protective devices (OCPDs), such as fuses, circuit breakers, and protective relays, allowed to be modified?
   A. when it is required by company policies and procedures
   B. when it is needed to get equipment started
   C. OCPDs cannot be modified at any time except for what is allowed in codes and standards
   D. any time

_____ 15. Duty cycle is defined as the ___.
   A. amount of time an electrical tool is carrying a specific amount of current
   B. amount of time a tool has to be turned off to prevent overheating
   C. total amount of time a tool can be used in a 24 hr period
   D. specific amount of current a tool can carry for a 3 hr period

_____ 16. The correct method used to prevent strain on joints or terminal screws of cord connectors is by ___.
   A. using an electrician's knot
   B. applying heat-shrink tubing to the outside
   C. using a closed-style cord connector
   D. taping the connector using an approved insulating tape

CHOOSING AND INSPECTING PERSONAL PROTECTIVE EQUIPMENT

Using personal protective equipment (PPE) does not eliminate electrical hazards. Rather, PPE only reduces the effect those hazards can have on an individual wearing it. The PPE requirements in NFPA 70E are not intended to prevent all injuries; they are intended to ensure injuries received during an electrical accident are survivable. Even when proper PPE is worn, there is always a chance of injury. However, these injuries will be minor compared to the injuries sustained if PPE is not worn.

OBJECTIVES

- Define arc thermal performance value (ATPV).
- Explain the meaning of the words "use of" and "appropriate" as stated in OSHA 1910.
- Explain the importance of head protection.
- Describe the inspection and storage process for rubber insulating gloves.
- Define leather protectors.
- Explain rubber insulating blanket and sleeve inspection and testing.
- List the forms of eye protection.
- Define arc flash protective clothing.
- Explain the methods used to determine PPE per NFPA 70E.
- Explain how the tables from NFPA 70E 130.7(C) are permitted to be used to determine personal protective equipment (PPE).
- List factors to consider when selecting protective clothing.
- List the types of arc flash protective equipment according to NFPA 70E.
- Describe PPE material and clothing characteristics.

PERSONAL PROTECTIVE EQUIPMENT

Personal protective equipment (PPE) is clothing and/or equipment worn by an employee to reduce the possibility of injury in the work area. The use of PPE is required whenever work may occur on or near exposed energized conductors or circuit parts. PPE is designed to protect specific parts of the body, such as eyes, face, ears, and torso.

The arc rating is listed on arc flash PPE. One type of rating is arc thermal performance value. The *arc thermal performance value (ATPV)* is the incident energy that results in sufficient heat transfer through PPE to cause the crossing of the Stoll curve burn injury model, which is designed to prevent second-degree burns. The onset of a second-degree burn is a severe first-degree burn with possible blistering of the skin, but it is not a life-threatening injury. The ATPV listed on PPE does not suggest that the user is completely protected from burn injuries, but the injuries would be survivable.

Another type of rating is the energy breakopen threshold. The *energy of breakopen threshold (E_{BT})* is the point at which a fabric allows a 1″ crack or a ½″ hole, but no burn is registered. Knit fabrics typically are more insulative than they are strong, so they receive an E_{BT} value. Woven materials typically receive an ATPV. Research shows these are functionally equivalent so neither is better than the other. An arc rating and ASTM F1506 value should be identified for fabrics. An ASTM F1891 value should be listed for rainwear. Typically, wearing PPE to the level of the tables or incident energy calculations has prevented most workers from receiving burns. Extreme circumstances found in utilities and a few industrial locations can be more hazardous than expected, but this is not the norm. Arc-rated PPE has been found to be very protective, but PPE should not be used as an excuse for doing energized work.

SAFETY FACT

ATPV is based on the Stoll curve, not a specific time period. The time may vary during the actual testing. ASTM F1959, Standard Test Method for Determining the Arc Rating of Materials for Clothing, *provides an equation for predicting a second-degree burn under arc-rated clothing and PPE. The Stoll response is $cal/cm^2 = 1.1991 \times t_i^{0.2901}$ where t is the elapsed time from the initiation of the arc.*

Individual components of PPE are worn together to provide a complete system to protect the body. All required components of PPE must be worn in order to provide adequate protection from the hazards that may be present. **See Figure 9-1.** An arc flash presents a number of hazards: heat (convective and radiated), a pressure wave (arc blast), an acoustic wave (sound), light radiation (UV and visible), toxic gases, and shrapnel. If just one component of the PPE system is missing, injury could result. For example, the sound created by an arc has been measured in excess of 165 dB. A level of 140 dB can cause a loss of hearing if proper hearing protection is not used.

Personal Protective Equipment

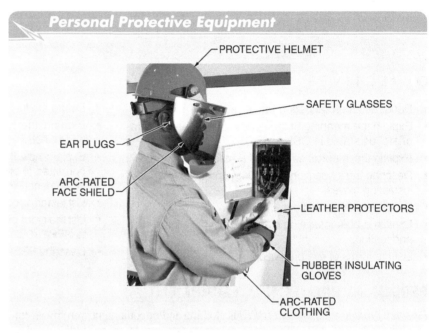

Figure 9-1. PPE is a combination of separate items that are designed to protect specific parts of the body and provide overall protection when they are worn together.

THE USE OF PPE PER OSHA

One of the requirements for a qualified electrical worker is given in OSHA 1910.333(c)(2), *"Such persons shall be capable of working safely on energized circuits and shall be familiar with the proper use of special precautionary techniques, personal protective equipment, insulating and shielding materials, and insulated tools."*

The phrase "use of" may not be completely clear when it is first read. In this case, the phrase "use of" refers to the selection, use, limitations of use, storage, required testing (if any), and maintenance of PPE. This is further illustrated in OSHA 1910.132(f)(1), *"The employer shall provide training to each employee who is required by this section to use PPE. Each such employee shall be trained to know at least the following: (i) When PPE is necessary; (ii) What PPE is necessary; (iii) How to properly don, doff, adjust, and wear PPE; (iv) The limitations of the PPE; and, (v) The proper care, maintenance, useful life and disposal of the PPE."*

OSHA 1910.132(f)(2) states, *"Each affected employee shall demonstrate an understanding of the training specified in paragraph (f)(1) of this section, and the ability to use PPE properly, before being allowed to perform work requiring the use of PPE."*

Assessing a Workplace for Hazards

According to OSHA 1910.132(d)(1), *"The employer shall assess the workplace to determine if hazards are present, or are likely to be present, which necessitate the use of personal protective equipment (PPE). If such hazards are present, or likely to be present, the employer shall: (i) Select, and have each affected employee use, the types of PPE that will protect the affected employee from the hazards identified in the hazard assessment."*

If it is determined that hazards will be present in a workplace, a risk assessment is needed to assess the hazards, evaluate the risks associated with the task, and select the appropriate level of PPE. This regulation applies to all worksites, tasks, and occupations and provides justification for employers to perform a risk assessment. Even if the only energized task being performed is placing the equipment in an electrically safe work condition, a risk assessment is required.

When to Use PPE

According to OSHA 1910.335(a)(1)(i), *"Employees working in areas where there are potential electrical hazards shall be provided with, and shall use, electrical protective equipment that is appropriate for the specific parts of the body to be protected and for the work to be performed."*

Even though the language in this regulation is not specific, the intent is to protect employees from electrical hazards. The intent and meaning of this regulation can be broken down and clarified into the following:

- *"Employees working in areas where there are potential electrical hazards..."* The three recognized electrical hazards are shock, arc flash, and arc blast. Proof of a hazard in the area where work is about to begin is not required; there only has to be the potential for a hazard. If removing a cover exposes energized conductors or circuit parts, there is a potential hazard.

- *"Employees... shall be provided with, and shall use, personal protective equipment..."* "Shall" is legal terminology for "must." This means that it is required and there are no other options. The employer must provide PPE and the employee must use that PPE. If the employee refuses to wear the appropriate PPE as provided by the employer, OSHA expects the employer to escalate disciplinary action to encourage employees to comply. If the employer does not enforce the OSHA regulations, OSHA will cite the employer with a willful citation. Safety policies that are not enforced are not safety policies according to OSHA. OSHA will cite the employer for not following OSHA regulations or the employer's written safety policies, whichever offers the highest level of protection.

- *"... electrical protective equipment that is appropriate..."* The key word in this section is "appropriate." The word "appropriate" means that enough equipment is used to protect the employee, but it does not interfere with the performance of the task. One simplified method of choosing appropriate arc-rated clothing and PPE is found in Table H.2 in NFPA 70E Informative Annex H, "Guidance on Selection of Protective Clothing and Other Personal Protective Equipment," where arc-rated daily workwear (PPE Category 2) is supplemented by PPE Category 4 arc-rated clothing and PPE as needed. "Appropriate" also means that the protection must be appropriate for the task. For example, Class 2 rubber insulating gloves could be worn to work on a 115 V receptacle, but they would not be appropriate because wearing this level of protection while performing the task could lead to additional or increased hazards. Therefore, Class 2 gloves are not allowed for this type of work.

- *"... for the specific parts of the body to be protected and for the work to be performed."* PPE is designed for the specific part of the body it is protecting. For example, rubber foam ear inserts provide hearing protection, but they do not provide protection for other parts of the body. To protect the face or eyes, other types of PPE must be worn. Also, if the PPE interferes with the work to be performed to the extent that the PPE itself creates a hazard, then it is no longer protecting the employee and is not considered appropriate.

PPE Maintenance

According to OSHA 1910.335(a)(1)(ii), *"Protective equipment shall be maintained in a safe, reliable condition and shall be periodically inspected or tested, as required by 1910.137."*

PPE and clothing must be maintained in a safe and reliable condition to function as designed. All PPE and arc flash protective clothing has storage and care requirements.

According to OSHA 1910.137(b)(2)(ii), *"Insulating equipment shall be inspected for damage before each day's use and immediately following any incident that can reasonably be suspected of having caused damage. Insulating gloves shall be given an air test, along with the inspection."*

After the appropriate PPE has been selected, it should not be used until it has been inspected. Prior damage or defects can make the PPE unsafe for work. Inspecting the equipment before each use is the most important part of PPE maintenance.

According to OSHA 1910.132(e), *"Defective and damaged equipment. Defective or damaged personal protective equipment shall not be used."*

PPE must be inspected for defects. If any defects are found during inspection, the PPE must not be used. Defective PPE and other electrical protective equipment must be taken out of service and tagged as defective. The tag is needed to ensure that the equipment is not used later by someone who is unaware of its defects.

TYPES OF PPE

PPE is selected on the basis of the nominal or design voltage of the equipment or circuits to be worked on or near. PPE for electrical workers can include head protection, hand protection, rubber insulating blankets and sleeves, eye protection, and arc flash protective clothing. Nominal voltages can be found on electrical equipment nameplates, dataplates, schematics, and single-line diagrams.

Head Protection

Head protection requires the use of a hard hat. A *hard hat* is a protective helmet that is used in the workplace to prevent injury from the impact of falling or flying objects and electric shock. The shell of a hard hat is composed of durable, lightweight materials such as high-density polyethylene or fiberglass-impregnated polycarbonate. *Hard hat suspension* is a shock-absorbing lining that keeps the shell away from the head to provide protection from falling objects striking the hard hat. Each hard hat has a manufactured date and is identified by class.

According to OSHA 1910.335(a)(1)(iv), *"Employees shall wear nonconductive head protection wherever there is a danger of head injury from electric shock or burns due to contact with exposed energized parts."*

Besides providing protection from falling objects, hard hats may also prevent contact with exposed energized conductors and circuit parts, as well as provide some protection from arc flashes. This is why there can be no metal on a hard hat. Stickers and labels must be nonmetallic and must be at least ½″ away from the brim.

Hard Hat Inspection. Hard hats must be inspected to ensure that they will function according to their design. Some inspection points include deformation and deterioration of the shell, loose suspension, unauthorized alterations, manufactured date, and class.

Deformation of the hard hat shell includes cracks or dents. Cracks can often develop at the connection points or mounting lugs for accessories such as hearing protectors or face shields. **See Figure 9-2.** This type of damage is often caused by mounting a face shield or other accessory to the hard hat and then trying to remove it later. Hard hat accessories are not intended to be changed out, and since the mounting lug is a weak point, cracks will often occur. Small cracks may look insignificant, but they weaken the hard hat structurally and will invoke an OSHA citation.

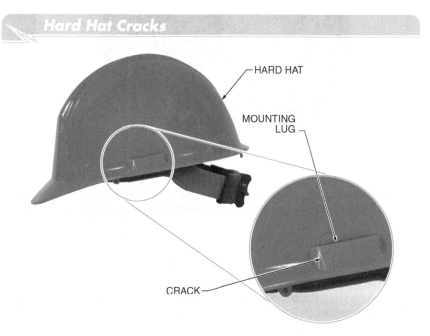

Figure 9-2. Hard hats must be removed from service if their mounting lugs are damaged.

The hard hat shell should also be checked for deterioration. Over time, the shell of a hard hat deteriorates due to aging or exposure to sunlight. Because ultraviolet rays from the sun will damage plastic and fiberglass, hard hats should not be stored outdoors. Also, fiberglass or plastic hard hats should never be painted. Solvents should not be used to clean a hard hat.

The shell of a plastic hard hat can be tested by performing a squeeze test. A squeeze test is performed by grasping a hard hat by its sides, which are then firmly pressed in and released. The hard hat should immediately spring back to its original size and shape. If it does not return to its original shape, it should be replaced. Squeeze tests should not be performed on fiberglass hard hats or hard hats that have a face shield mounted to them. Fiberglass will crack and the face shield or its mounting may become damaged.

The suspension inside a hard hat should be inspected. The suspension inside a hard hat must be secure to the shell. There should not be any movement in or out of the connection points. Otherwise, the suspension may slip out of the shell. **See Figure 9-3.** The suspension and connection points should be checked for cracks or other damage. Cracks can sometimes develop in the thin portions of the hard hat suspension.

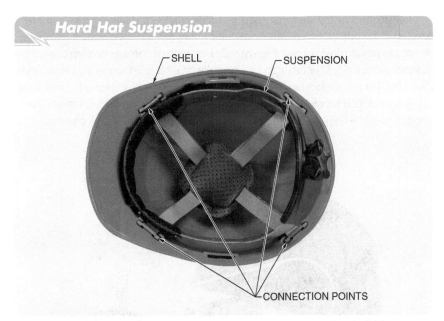

Figure 9-3. The suspension inside a hard hat must be secure to the shell. There must not be any cracks or other damage to the connection points or the suspension itself.

SAFETY FACT

Employees commonly have stickers on their hard hats. Stickers are acceptable as long as either the inside or the outside shell is visible for inspection. For example, if a sticker provided by the manufacturer is located on the inside of a hard hat, another sticker cannot be placed on the outside of the hard hat in the same location.

A hard hat should be inspected for alterations. It is common to see workers drill holes in their hard hats to make them cooler to wear in the summer months. Holes weaken the hard hat and may reduce protection from falling objects. Other alterations include drilling attachment holes for ear protectors (ear muffs). Alterations to any PPE must be approved by the manufacturer.

Manufacture Date. Every hard hat displays the date on which it was manufactured. This date is usually molded into the brim of the hard hat. **See Figure 9-4.** Most manufacturers recommend replacing hard hats every five years. Some OSHA compliance officers check the date codes on hard hats when conducting an audit of a company's in-house safety program. A company may get cited for not following the manufacturer recommendations if the hard hat is more than five years old.

Date Coding

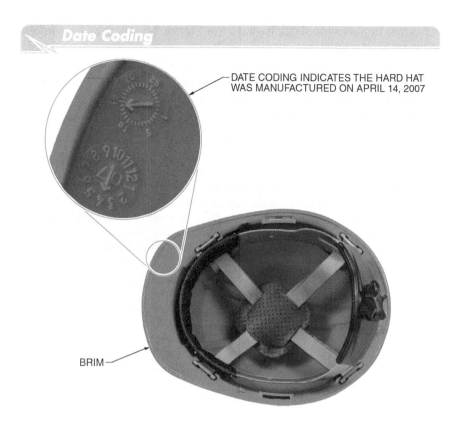

Figure 9-4. The date on which a hard hat was manufactured is usually molded into the brim of the hard hat.

Class. Hard hats are identified by class. Class G, E, and C hard hats are used for construction and industrial applications. Class G (general) hard hats, which are commonly used in construction and manufacturing facilities, protect against impacts and voltages up to 2200 V. Class E (electrical) hard hats protect against impacts and voltages up to 20,000 V. Class C (conductive) hard hats are manufactured with lighter materials and provide adequate impact protection, but they do not provide electrical protection. Employees working on electrical equipment must wear hard hats that are either Class E or Class G. Hard hats must also meet ANSI Standard Z89.1 for impact resistance.

SAFETY FACT

Do not use flammable materials, such as terry cloth headbands, paper towels, or rags, inside a hard hat. Anything that is flammable inside a hard hat can ignite during an arc flash event.

Rubber Insulating Gloves

Rubber insulating gloves are gloves made of natural latex rubber (Type I) or a synthetic ozone-resistant material (Type II) and are used to provide protection from electric shock hazards. **See Figure 9-5.** The primary purpose of rubber insulating gloves is to insulate hands and lower arms from possible contact with live conductors or circuit parts. In addition to electric shock protection, rubber insulating gloves, when used with leather protectors, provide substantial protection from the heat of an electrical arc.

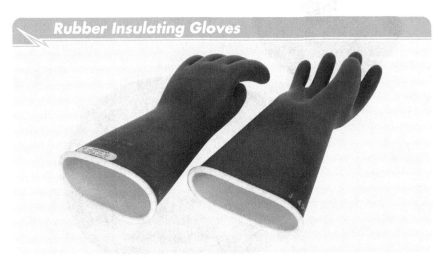

Figure 9-5. Rubber insulating gloves provide protection from electric shock.

ASTM F1236, *Standard Guide for Visual Inspection of Electrical Protective Rubber Products,* offers a detailed guide for the inspection, maintenance, and testing of rubber insulating gloves, as well as other types of rubber insulating protective equipment. Manufacturers also publish in-service inspection and maintenance guides that are used for inspecting and maintaining rubber insulating gloves.

Rubber Insulating Glove Inspection. The entire surface of a pair of rubber insulating gloves must be field tested (visual inspection and air testing) before each use. In addition, they must also be electrically tested every six months. Visual inspection of rubber insulating gloves is performed by stretching a small area (particularly the fingertips) and checking for defects such as ozone cutting, ultraviolet (UV) radiation damage, punctures or pinholes, embedded or foreign material, deep scratches or cracks, cuts or snags, or deterioration caused by oil, heat, grease, insulating compounds, or any other substance that can harm rubber.

In order to properly inspect rubber insulating gloves, any dirt or contaminants should be cleaned off the rubber. Then the rubber should be expanded. This can be accomplished either by rolling the cuff tightly toward the palm in such a manner that air is trapped inside the glove or by using a mechanical inflation device. **See Figure 9-6.** When using a mechanical inflation device, care must be taken to avoid over inflation. Type I gloves made of natural rubber should be inflated to only 1.5 times their original size. Type II gloves should be inflated to only 1.25 times their original size. After being inflated, the gloves can be inspected under a bright light for damage or deterioration.

Inspecting Rubber Insulating Gloves

Figure 9-6. Mechanical inflation devices can be used to inflate rubber insulating gloves to inspect them for punctures or other defects.

Any kind of hole or opening, no matter how small, renders rubber insulating gloves unsafe. Pinholes can be detected by holding an inflated glove up to the face and feeling or listening for air leaks. It is also important to check for embedded foreign objects. Objects such as wire, splinters of wood, metal, or thorns may become embedded in rubber insulating gloves and render them unsafe.

Pinholes in a rubber insulating glove can be detected by trapping air inside the glove and holding it up to the face to hear or feel for air leaks.

An employee must also inspect rubber insulating gloves for tears and snags. Snags are punctures that may not penetrate all the way through the glove. A glove with a snag is unsafe and must be removed from service. **See Figure 9-7.**

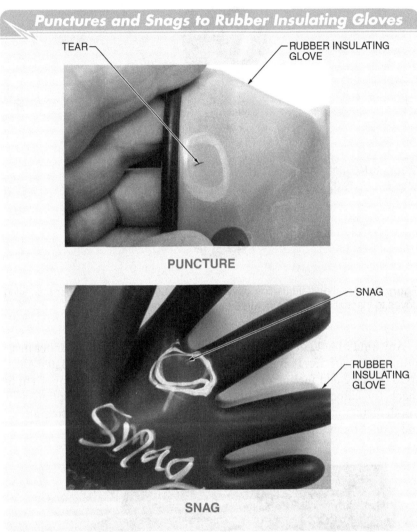

Figure 9-7. The entire surface of rubber insulating gloves must be visually inspected before each use. A glove with a snag is unsafe and must be removed from service.

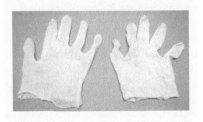

Figure 9-8. Cotton glove liners may be used underneath rubber insulating gloves.

Chemical Damage. Contact from certain chemicals, such as petroleum-based products, can damage rubber insulating gloves. Although it is common for employees to use glove powder on their hands to make it easier to put on and take off rubber insulating gloves, baby powder should never be used. Some baby powders use petroleum-based products as a binder for their fragrance ingredients. Cotton glove liners may be used instead of powders. **See Figure 9-8.** They are inexpensive and keep hands warmer in the winter and soak up sweat in the summer. Other products that may contain petroleum-based products include hand lotions and the adhesive on electrical tape. Substances such as grease, oils, and fuels can mechanically weaken the gloves and cause oxidation at a faster rate. **See Figure 9-9.**

Damage Caused by Grease

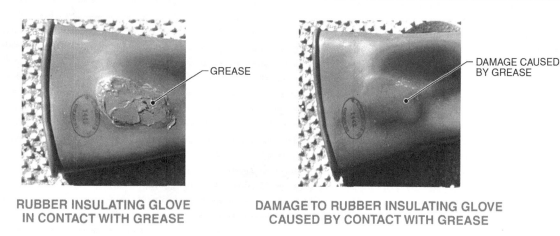

Figure 9-9. Grease can mechanically weaken gloves and cause oxidation at a faster rate.

Salisbury

It is important to inspect for changes in the texture of the rubber insulating glove such as softening or hardening. Gloves should also be inspected for areas that are sticky or inelastic. Changes in texture can be caused by aging or exposure to caustic chemicals.

Ozone Cutting. Rubber insulating gloves must be inspected for ozone cutting. *Ozone cutting* is the damage caused to natural rubber through exposure to ozone. Ozone is created by exposing air to the dielectric stress of high voltages. Ozone breaks down the long-chain molecules in natural rubber, which causes cracking in the rubber.

Corona is the ionization of air caused by exposure to the dielectric stress of a high-voltage electrical field. Corona causes natural rubber to deteriorate at a fast rate. Natural rubber normally begins to age after about ten years as oxygen in the air breaks down the long-chain molecules in the rubber, causing cracking. Corona attacks the natural rubber much more aggressively than oxidation, causing the same type of deterioration in a matter of weeks.

Much of the damage attributed to corona is actually the result of improper storage or exposure to petroleum products. **See Figure 9-10.** When natural rubber is damaged by being folded or creased for extended periods of time, it is mechanically weakened. Mechanical weakening allows oxygen to deteriorate the rubber at a much faster rate than normal, similar to corona.

Another way gloves can be damaged due to improper storage is by a heavy object pressing on the rubber for an extended period of time, causing an imprint. **See Figure 9-11.** Testing laboratories categorize all damage of this type as corona because they cannot determine the difference between oxidation caused by mechanical damage or corona.

Glove manufacturers make a Type II corona-resistant glove, which is a glove composed of synthetic material that does not deteriorate from the effects of corona. Corona-resistant gloves are not required for either low-voltage applications or high-voltage applications, provided that care is taken not to expose the rubber insulating gloves that are used for extended periods of contact with high voltages.

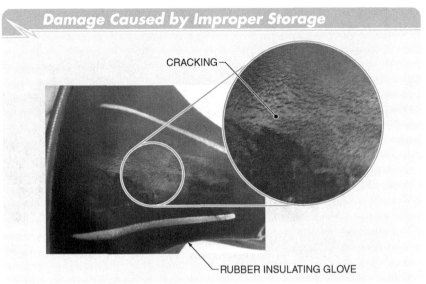

Shermco Industries

Figure 9-10. Improper storage can cause cracking.

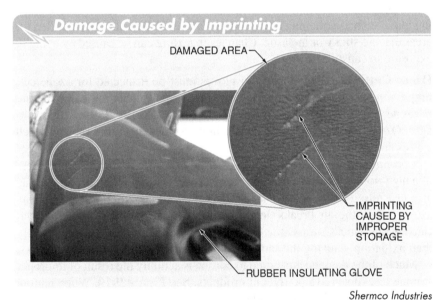

Shermco Industries

Figure 9-11. Imprinting is caused by a heavy object pressing the rubber for an extended period of time.

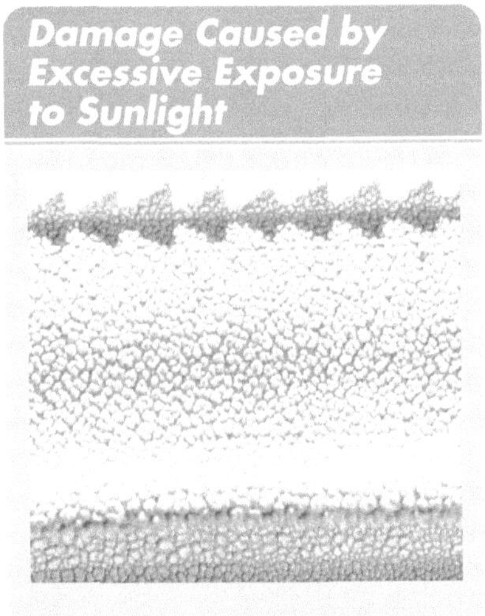

Figure 9-12. A line hose that has been exposed to sunlight for an extended period of time may become etched. Similar damage can affect rubber insulating gloves.

Excessive Exposure to Sunlight. Natural rubber, as well as plastic and fiberglass, deteriorates when exposed to sunlight over extended periods of time. Rubber insulating gloves develop a white, powdery substance called "bloom" during the early stage of sunlight exposure. At this stage the gloves probably have not been damaged, but should be retested to verify they are still suitable for continuing service. Prolonged exposure to sunlight will cause the gloves to become etched. Similar damage can be seen on a line hose that has been exposed to sunlight for an extended period of time. **See Figure 9-12.** Etching of the surface or other damage caused by the UV radiation in sunlight requires the gloves be removed from service.

Testing Rubber Insulating Gloves. Rubber insulating gloves must be electrically tested every six months once they are issued for service. OSHA allows a one-time exception for new gloves. New gloves can go twelve months without being tested, as long as they have not been issued for service. This requirement is found in OSHA 1910.137 Table I-5, "Rubber Insulating Equipment Test Intervals," and footnote (1). Once rubber insulating gloves have been tested, they are no longer new, regardless of whether or not they have been issued for service. Gloves can only be "new" once.

Rubber insulating gloves can be electrically tested by the user if the user follows the requirements of ASTM D120. Most companies send their gloves to an approved testing lab rather than test them in house. Gloves that are electrically tested are usually marked with an ink stamp showing the date they were tested, but a logbook tracking the individual glove is also acceptable. Gloves can have a unique tracking/ID number written on the inside of the cuff, as long as the marker is not conductive. **See Figure 9-13.**

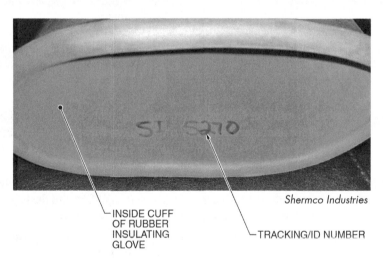

Shermco Industries

INSIDE CUFF OF RUBBER INSULATING GLOVE

TRACKING/ID NUMBER

Figure 9-13. A logbook can be used to track the testing history of individual rubber insulating gloves. Rubber insulating gloves can have a unique tracking/ID number written on the inside of the cuff for identification.

Rubber insulating gloves are broken down into classes. Each class is represented by a colored label that identifies the maximum use voltage. **See Figure 9-14.** The *maximum use voltage* is the maximum amount of voltage a rubber insulating glove can be exposed to without suffering dielectric deterioration over time.

Each glove class also has a proof test voltage, also known as the maintenance test voltage. The *proof test voltage* is the amount of voltage applied to test rubber insulating gloves for weakening or damage. The proof test performed on rubber insulating gloves is similar to proof tests that are performed on new electrical equipment to ensure that they will not fail when placed into service.

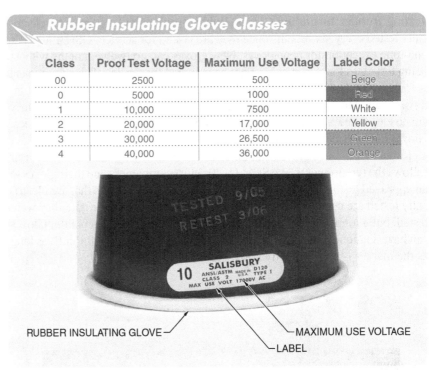

Rubber Insulating Glove Classes

Class	Proof Test Voltage	Maximum Use Voltage	Label Color
00	2500	500	Beige
0	5000	1000	Red
1	10,000	7500	White
2	20,000	17,000	Yellow
3	30,000	26,500	Green
4	40,000	36,000	Orange

Figure 9-14. Rubber insulating gloves are broken down into classes. Each class is represented by a colored label that identifies the maximum use voltage.

Figure 9-15. Rubber insulating gloves should be placed cuff down into a bag to prevent dirt and other debris from getting inside them.

Storing Rubber Insulating Gloves. After each use, rubber insulating gloves must be inspected for damage, cleaned by rinsing them with clear water, dried, and then properly stored. The following guidelines are recommended for storing rubber insulating gloves:

- Insert rubber insulating gloves into their leather protectors, and then place them into the glove bag cuff down. Placing the gloves fingers first into the bag may allow for dirt and other debris to get inside the glove. **See Figure 9-15.**
- Never fold or crease any rubber insulating product for extended periods of time. Creases or folds cause mechanical weakening along the fold, causing the glove to quickly deteriorate.
- Store rubber insulating gloves in a relaxed condition. Never put PPE or other equipment on top of rubber insulating gloves, always place the gloves on top of the PPE and other equipment to prevent creasing.
- Never store rubber insulating gloves inside out. Turning the glove inside out for extended periods of time creates mechanical stress over the entire surface of the glove, causing premature failure. Gloves can be turned inside out for inspection, but they should be turned right-side out afterward.
- Do not store rubber insulating gloves in direct sunlight or hanging from a hook on a truck bed.
- Store rubber insulating gloves in areas with a moderate temperature. Avoid storing gloves in outdoor storage areas that are exposed to extreme temperatures.
- Store only one pair of rubber insulating gloves in a bag that is designed for one pair of gloves. Trying to squeeze two or more pairs of gloves inside a bag designed to hold one pair may crease and damage the gloves.

Leather Protectors

According to OSHA 1910.335(a)(1)(iii), *"If the insulating capability of protective equipment may be subject to damage during use, the insulating material shall be protected. (For example, an outer covering of leather is sometimes used for the protection of rubber insulating material.)"*

Leather protectors are gloves worn over rubber insulating gloves to prevent damage to the rubber insulating glove. Leather protectors have no dielectric qualities or rating. Leather protectors sold with rubber insulating gloves are manufactured to strict ANSI standards. Off-the-shelf leather gloves from hardware stores cannot be used as a substitute. Rubber insulating gloves, especially low-voltage Class 0 and 00 gloves, are especially susceptible to damage because they are very thin.

OSHA allows the use of rubber insulating gloves without leather protectors in limited cases where dexterity is required. However, this means that the rubber insulating glove is at much greater risk of damage. When rubber insulating gloves are used without leather protectors, they must be derated. Class 00 gloves are derated to 250 V, Class 0 gloves are derated to 500 V, and the other glove classes are derated by one glove class. Once rubber insulating gloves have been used without leather protectors they cannot be used at their full voltage rating until they have been electrically tested to verify that they have not been damaged.

Leather Protector Inspection. Leather protectors should be inspected for cuts, rips, or tears that go through the leather. Surface cuts or blemishes are not a problem. The inspection should also include checking for open or loose seams, especially at high-stress points, such as between the fingers. This is performed by pulling the glove and spreading the fingers to inspect the seam between them. However, the glove should not be pulled too hard because that may cause a problem with the stitching. Also, it is necessary to lightly pull seams at other locations on the protector to inspect them.

Leather protectors also need to be inspected for oil or grease that is penetrating through the leather. If it appears that oil or grease could contact the rubber insulating glove, the leather protector should be disposed of. Protectors have a shiny finish to keep grease and oil from penetrating through the leather. A clean rag with a small amount of solvent can be used to clean off surface grease or oil.

Leather protector inspection also includes checking for embedded foreign objects. An employee should put a hand inside the protector and rub firmly over the surface. A foreign object that pokes the employee's hand will also poke the insulating rubber glove. Leather protectors that have embedded foreign objects should never be used.

The distance between the rubber insulating glove bead and leather protector cuff should also be checked. A minimum distance must be maintained to prevent shock from voltage creep or flash over. Voltage creep occurs when surface contaminants allow the voltage to "creep" over the surface of the glove and possibly jump over the rubber insulating glove to the arm. The minimum distance between the rubber insulating glove bead and leather protector cuff is determined by the rubber insulating glove class. **See Figure 9-16.**

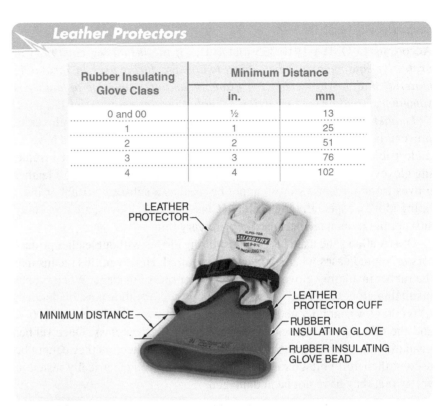

Rubber Insulating Glove Class	Minimum Distance	
	in.	mm
0 and 00	½	13
1	1	25
2	2	51
3	3	76
4	4	102

Figure 9-16. A minimum distance must be maintained between the rubber insulating glove bead and leather protector cuff to prevent shock from voltage creep or flash over.

Rubber Insulating Blankets and Sleeves

According to OSHA 1910.335(a)(2)(ii), *"Protective shields, protective barriers, or insulating materials shall be used to protect each employee from shock, burns, or other electrically related injuries while that employee is working near exposed energized parts which might be accidentally contacted or where dangerous electric heating or arcing might occur. When normally enclosed live parts are exposed for maintenance or repair, they shall be guarded to protect unqualified persons from contact with the live parts."*

Protective barriers or insulating materials include rubber insulating blankets and rubber insulating sleeves. A *rubber insulating blanket* is a blanket that is used to insulate and/or isolate energized conductors and circuit parts. *Rubber insulating sleeves* are sleeves that are worn to protect the upper arms and shoulders from coming into contact with exposed energized conductors and circuit parts.

According to OSHA 1910.269(l)(4)(i), *"When an employee uses rubber insulating gloves as insulation from energized parts (under paragraph (l)(3)(iii)(A) of this section), the employer shall ensure that the employee also uses rubber insulating sleeves. However, an employee need not use rubber insulating sleeves if: (A) Exposed energized parts on which the employee is not working are insulated from the employee; and (B) When installing insulation for purposes of paragraph (l)(4)(i)(A) of this section, the employee installs the insulation from a position that does not expose his or her upper arm to contact with other energized parts."*

Rubber insulating sleeves are required for most overhead line work where the shoulder and upper back could be exposed to contact with an energized line. However, no sleeves are needed if the conductors or circuit parts are insulated to the voltage or if there is no possible contact with energized conductors or circuit parts.

Rubber Insulating Blanket and Sleeve Inspection. The inspection points for rubber insulating blankets and sleeves are the same as for rubber insulating gloves. The main difference is that blankets and sleeves cannot be inflated to stretch the rubber to make defects more visible. Instead, rubber insulating blankets and sleeves are inspected by rolling the rubber along a flat surface, such as a tabletop, and inspecting the edge under a strong light. **See Figure 9-17.** The blanket or sleeve is then turned to a 90° angle and rolled and inspected again.

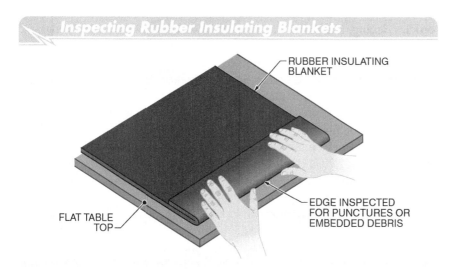

Figure 9-17. To inspect a rubber insulating blanket, roll the rubber along a flat surface and inspect the edge under a strong light. Then turn the blanket 90° and roll and inspect again.

Rubber Insulating Blanket and Sleeve Testing. Blankets and sleeves are to be tested every twelve months. ASTM F496, *Standard Specification for In-Service Care of Insulating Gloves and Sleeves,* contains testing procedures for blankets and sleeves that are more involved than the electrical testing for gloves.

Salvaging and Repairing Rubber Insulating Blankets. According to OSHA 1910.137(c)(2)(x)(B), *"Rubber insulating blankets may be salvaged by severing the defective area from the undamaged portion of the blanket. The resulting undamaged area may not be smaller than 560 mm by 560 mm (22 inches by 22 inches) for Class 1, 2, 3, and 4 blankets."*

Rubber insulating blankets that have been damaged can still be used if the damaged area is removed. The remaining undamaged area must have an area of at least 22″ × 22″.

According to OSHA 1910.137(c)(2)(x)(D), *"Rubber insulating gloves and sleeves with minor physical defects, such as small cuts, tears, or punctures, may be repaired by the application of a compatible patch. Also, rubber insulating gloves and sleeves with minor surface blemishes may be repaired with a compatible liquid compound. The patched area shall have electrical and physical properties equal to those of the surrounding material. Repairs to gloves are permitted only in the area between the wrist and the reinforced edge of the opening."*

Using a vulcanizing patch on rubber insulating equipment may not provide a safe repair if the patch or adhesive is not compatible with the electrical properties of the rubber insulating PPE. If rubber insulating PPE must be repaired, it is recommended that a professional perform the repair. Usually, the best course of action is to replace the equipment rather than try to repair it.

Eye Protection

Eye protection must be worn to prevent eye or face injuries caused by flying particles, arcing, and radiant energy. Eye protection includes safety glasses, goggles, arc-rated face shields, and arc-rated hoods.

Safety glasses are an eye protection device with special impact-resistant glass or plastic lenses, reinforced frames, and side shields. Plastic frames are designed to keep the lenses secured in the frame if an impact occurs and to minimize the shock hazard when working with electrical equipment. Side shields provide additional protection from flying objects.

Goggles are an eye protection device with a flexible frame that is secured on the face with an elastic headband. Goggles fit snugly against the face to seal the areas around the eyes, and may be used over prescription eyeglasses. Goggles protect against small flying particles or splashing liquids.

An *arc-rated face shield* is an eye and face protection device that covers the entire face with a plastic shield and has a chin cup and side shields that extend its protective coverage to the chest and beyond the ears. An arc-rated face shield provides protection from flying objects and projectiles as well as from the incident energy of an arc. An *arc-rated hood* is an eye and face protection device that covers the entire head and is used for protection from an arc flash. It also can provide one or two breaths of uncontaminated air. Arc-rated hoods have a built-in arc-rated window. **See Figure 9-18.**

According to OSHA 1910.335(a)(1)(v), *"Employees shall wear protective equipment for the eyes or face wherever there is danger of injury to the eyes or face from electric arcs or flashes or from flying objects resulting from electrical explosion."*

When work is performed on or near exposed energized electrical conductors or circuit parts there is a danger of electric arcs, flashes, or flying objects. OSHA requires equipment to be worn that protects the eyes and face from these hazards.

According to OSHA 1910.133(a)(1), *"The employer shall ensure that each affected employee uses appropriate eye or face protection when exposed to eye or face hazards from flying particles, molten metal, liquid chemicals, acids or caustic liquids, chemical gases or vapors, or potentially injurious light radiation."*

UV radiation is always a hazard when working around energized electrical equipment, so electrical workers are required to have and use safety glasses or goggles that are designed to protect against UVA and UVB. It is important

Shermco Industries

Figure 9-18. An arc-rated hood is an eye and face protection device that covers the entire head.

for employees to check the original packaging for safety glasses or goggles because there are rarely any markings on the safety glasses or goggles to indicate whether they provide protection against UV radiation. This information is typically located on the original packaging for the equipment.

Eye Protection Inspection. Safety glasses, goggles, arc-rated face shields, and arc-rated hoods are required to be maintained in a safe and reliable condition. Each piece of eye protection equipment should be inspected for excessive scratching of the lenses or window. If the scratching is severe enough that it interferes with the employee's vision, the lenses or window must be replaced. Additional inspection points for arc-rated face shields and arc-rated hoods include the following:

- Check the arc rating on the window or face shield. Most arc-rated hood windows will not have the arc rating on them. In these cases, the manufacturer should be contacted with the part number for the window to verify that it is the appropriate window for the arc rating of the hood.
- Ensure that all fasteners are securely holding an arc-rated face shield in place. **See Figure 9-19.** Also, if the chin cup and extenders are separate pieces, ensure that they are secure. Never use an arc-rated face shield without the chin cup.
- Ensure that the retainers on arc-rated face shields are properly positioned.
- Inspect for cuts, rips, or tears in the fabric of arc-rated hoods.
- Inspect for grease or oil stains that are larger than 2″ in diameter. Grease and oil increase heat transfer through fabric and can cause a burn to occur where it normally would not.
- Inspect for loose or open seams.
- Inspect the Velcro or seal around the window. There should be no gaps or openings between the arc-rated hood and its window.
- Inspect the seal between the layers of a two-layer arc-rated window.
- Check the security of the hood to the hard hat.
- Inspect for evidence of prior exposure to an arc flash.

SAFETY FACT

Safety glasses or goggles are considered primary eye protection, whereas face shields and hoods are considered secondary protection. Safety glasses or goggles must be worn under face shields and hoods, even when the face shields and hoods are arc-rated.

Face Shield Inspection

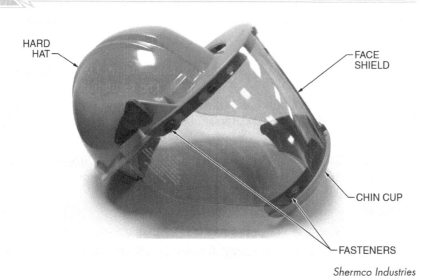

Shermco Industries

Figure 9-19. Arc-rated face shields and other accessories must be attached securely in place.

Arc Flash Protective Clothing

Arc flash protective clothing is clothing that provides protection from exposure to the extreme temperatures that occur during an arc flash. **See Figure 9-20.** Arc flash protective clothing must be used when working with energized low-voltage and high-voltage electrical circuits. Any time an employee is inside the Arc Flash Boundary, arc-rated clothing and PPE are required, regardless of voltage. Any part of the body that may be inside the Arc Flash Boundary must also be protected. This would include the hands if they are exposed to the arc flash hazard.

Arc-rated clothing characteristics are covered in NFPA 70E 130.7(C)(11). Arc-rated clothing should not have any portion of it, including fabrics, zipper tape, and findings (all the parts that hold it together), made from fabrics that can ignite or melt, unless they are treated and certified to meet the requirements of ASTM F1506, *Standard Performance Specification for Flame Resistant and Arc-Rated Textile Materials for Wearing Apparel for Use by Electrical Workers Exposed to Momentary Electric Arc and Related Thermal Hazards.* All parts of the arc-rated clothing must be constructed from arc-rated materials. This does not apply to patches or logos, which should be kept to a minimal size to prevent injury from an electric arc.

According to OSHA 1910.269(l)(8)(iii), *"The employer shall ensure that each employee who is exposed to hazards from flames or electric arcs does not wear clothing that could melt onto his or her skin or that could ignite and continue to burn when exposed to flames or the heat energy estimated under paragraph (l)(8)(ii) of this section. Note: This paragraph prohibits clothing made from acetate, nylon, polyester, rayon and polypropylene, either alone or in blends, unless the employer demonstrates that the fabric has been treated to withstand the conditions that may be encountered by the employee or that the employee wears the clothing in such a manner as to eliminate the hazard involved."*

Even though OSHA 1910.269 applies specifically to utilities and utility-like facilities, OSHA expects the same level of protection for all electrical workers. Clothing that melts and increases the severity of an injury is fundamentally inappropriate when working on or near exposed energized conductors or circuit parts. OSHA's list of fabrics (acetate, nylon, polyester, and rayon) is not intended to be all inclusive. If the clothing can melt, it is inappropriate for electrical workers to wear it. It should be noted that the OSHA regulations do not specify what should be worn instead. The regulations only provide examples of what should not be worn. NFPA 70E, however, is more helpful in this area because it is specific about what can and cannot be worn by electrical workers.

Arc Flash Protective Clothing Inspection. To thoroughly inspect arc flash protective clothing, it should be laid flat on a tabletop. Items to inspect for include the following:
- Inspect for cuts, rips, or tears to the fabric.
- Inspect for grease or oil spots larger than 2″ in diameter.
- Ensure that the zipper or Velcro® fastener is sewn into the clothing along its entire length. Occasionally, the sewing will run just along the edge of the Velcro fastener, and it will pull away and curl. **See Figure 9-21.** Curled Velcro will not seal properly. Check both sides to ensure proper sewing.

Figure 9-20. Nylon, rayon, acetate, polyester, spandex, and any other meltable fibers are prohibited from being used by electrical workers unless they are used as part of a blend that meets ASTM F1506 or F1891.

- Inspect the Velcro for lint, dirt, and other debris. If it is packed with debris, it will not seal properly. Most lint or other debris can be cleaned from Velcro using a clean, soft toothbrush.
- Check that seams in the clothing are sewn together tightly, particularly in high-stress areas such as under the arms and crotch. All seams should be sewn with a double-row stitch or better. Any seams that start to come loose can be sewn back together with an arc-rated thread of the same type the clothing is made from.
- Inspect the inside of the clothing with the same thoroughness that the outside was inspected with. The inside layer can have the same issues as the outside layer. Normally PPE Category 3 or 4 arc-rated clothing has multiple layers. In order to reach its arc rating, the clothing must have all of its layers fully intact and defect-free.
- Inspect for evidence of prior exposure to an arc flash. The clothing will discolor from the heat of an electrical arc. It is important to replace the entire arc flash suit, not just the affected piece, if there was prior exposure to an arc flash. There is no way to determine just how much heat the clothing received, and it is possible that it could break down at a lower incident energy than its rating if it has been involved in a prior arc flash.

Arc Flash Protective Clothing Inspection

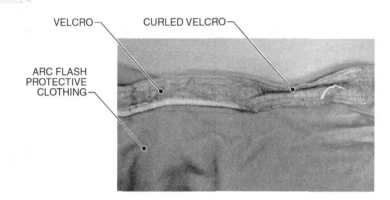

Shermco Industries

Figure 9-21. Arc flash protective clothing must be inspected to ensure that the Velcro is not curled. Curled Velcro will not provide a proper seal.

In addition, the specifications label on the inside of the clothing should be inspected. **See Figure 9-22.** The specifications label on the inside of the clothing should state the following:
- the clothing meets the requirements of NFPA 70E and ASTM F1506
- the arc rating in cal/cm^2

    Note: The clothing or equipment must have an arc rating equal to or greater than the expected hazard.
- the name of the manufacturer that made the clothing
- the type of fabric that was used to make the clothing
- the tracking ID for tracking the fabric lot for recalls

Arc Flash Protective Clothing Specifications Label

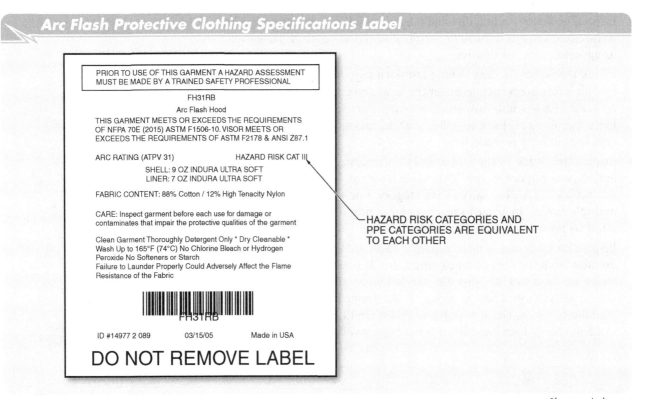

Shermco Industries

Figure 9-22. The specifications label inside arc flash protective clothing must be checked to ensure that the appropriate attire is being used for the job.

SAFETY FACT

Some manufacturers of arc flash protective clothing may have an arc rating on an outside shoulder or breast patch that is different from the arc rating on the inside specifications label. As long as the arc rating on the label is higher than that on the outside patch, employees may wear the clothing. If the arc rating on the label is lower than that on the outside patch or there is no arc rating, employees may not wear the clothing.

Wearing Arc Flash Protective Clothing. Arc flash protective clothing must be worn and stored properly in order to function properly. It is important to ensure that all ignitable clothing is covered by the arc-rated clothing. Meltable materials must not be worn underneath arc-rated garments or arc flash suits. A suit can allow energy through that is sufficient to melt polyester and other meltable materials onto a person's skin. The flaps on the arc-rated hood must be flat against the chest and back. If not, the arc blast is more likely to lift up the hood flap, exposing the worker to the heat of the arc. All of the Velcro flaps should be closed flat, without any wrinkles that could let heat through.

Arc-rated clothing should fit loosely, but it should not be so large that it causes a trip hazard or creates additional problems while an employee is working. Loose clothing allows for thermal shrinkage of the garment and may allow greater protection. NFPA 70E 130.7(C)(9)(d) requires that arc-rated clothing used for daily wear must have the shirt tucked into the pants; the shirt and coverall sleeves fastened;

Chapter 9 – Choosing and Inspecting Personal Protective Equipment

and shirts, coveralls, and jackets fastened at the neck. All these are necessary to prevent the heat from the arc from contacting bare skin. If an employee is wearing cold or wet weather gear, the outer-most layer of clothing must be arc-rated per NFPA 70E 130.7(C)(9)(b). Also, underlayers made of materials that can melt are prohibited from being worn according to NFPA 70E 130.7(C)(9)(c).

Storing Arc Flash Protective Clothing. After each use, employees should take a few minutes to inspect their arc flash protective clothing and PPE for defects. If the clothes are dirty or greasy, they should be laundered. When the clothing is stored, all of the zippers should be zipped up and all Velcro fasteners should be sealed, including those on the arms, neck, and legs. This prevents dirt and lint from collecting on the Velcro. The clothing should be neatly folded if it is to be stored in a bag or other container. If the clothing is hung in a locker, it should be neatly hung from hangers that are designed to support its weight.

The arc-rated hood is stored by folding the front flap over the window and then folding the rear flap forward so it also covers the window. **See Figure 9-23.** This will help prevent scratching of the window. Arc-rated face shields are typically purchased with a storage bag. Face shields should be stored in the storage bag when they are not in use.

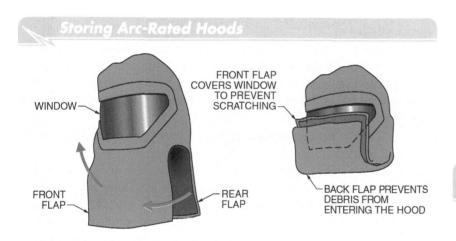

Figure 9-23. An arc-rated hood should be stored by folding both the front flap and back flap over the front of the hood to prevent the window from being scratched and debris from entering the hood.

Improperly stored PPE creates a hazard for workers using it. Some workers do not consider what improper storage will do to their equipment, but they should. PPE and arc-rated clothing that is improperly stored may be exposed to damage and can gather dirt that may be conductive. **See Figure 9-24.**

Repairing Arc Flash Protective Clothing. Repairs made to arc-rated clothing must be performed using the same type of thread that the clothing is made from. Cotton or nylon thread must never be used for repairs.

If any piece of arc-rated PPE or clothing is damaged, it should be removed from service until it is repaired properly. Arc-rated PPE and clothing that have been exposed to a prior arc flash cannot be reused. In the event that one piece

Shermco Industries

Figure 9-24. Arc rated clothing that is improperly stored may not provide the protection required.

of clothing, such as the jacket of a flash suit, is damaged, all associated pieces of arc-rated PPE that were worn at the same time would need to be retired and disposed of. There is no method to determine how much incident energy the other pieces of equipment or clothing were subjected to, and therefore, no method to determine if they still retain their full arc rating. The best practice is to retire all clothing involved in an arc event, whether it shows heat damage or not.

PPE PER NFPA 70E

Standards for determining proper PPE are mentioned throughout chapter 1 of NFPA 70E. The following are standards in NFPA 70E where PPE must be determined or identified:

- NFPA 70E 130.1 requires that all of the requirements in Article 130 apply to work involving electrical hazards, regardless of whether the table method or an incident energy analysis is used to determine PPE. Some workers rely on the tables or the labels alone to choose their arc-rated PPE and clothing. Other factors must be considered.
- NFPA 70E 130.2 requires that before an employee performs work within the Limited Approach Boundary or interacts with equipment in a manner that could cause failure, such as inserting or removing circuit breakers, the equipment must be placed in an electrically safe work condition.
- The Exception to NFPA 70E 130.2 allows a switch or disconnecting means supplying the equipment to be placed into an electrically safe work condition to deenergize a piece of equipment if it has been properly installed and maintained. A risk assessment has to be performed first and there can be no unacceptable risks.
- NFPA 70E 130.2(B)(1) requires that an energized electrical work permit be used when a person is exposed to the shock hazard (within the Restricted Approach Boundary) or when the worker interacts with equipment in a manner that could increase the likelihood of failure. This primarily means racking circuit breakers or installing or removing MCC buckets, but it could apply to other tasks as well. Previously, the implication was that the permit was only required for shock hazards, but it must be used for any electrical hazard the worker is exposed to.
- NFPA 70E 130.2(B)(2)(4)(d) and (5)(b), "Elements of Work Permit," requires that the PPE necessary to protect a worker from the hazards be identified.
- NFPA 70E 130.5, "Arc Flash Risk Assessment," requires that the arc flash risk assessment determine the Arc Flash Boundary and the necessary PPE. The incident energy analysis must be updated when changes are made to the electrical system that affect the incident energy, and it must be reviewed at least every five years. The design, operating time, and condition of maintenance of OCPDs are also to be considered. The "condition of maintenance" requirement is important. If the electrical system is not properly engineered, installed, and maintained, there is no way to determine how it will respond to a short circuit. It may or may not trip.
- NFPA 70E 130.5(C)(2) "Arc Flash PPE Categories Method," allows the use of the tables in 130.7 instead of a detailed arc flash risk assessment.

- NFPA 70E 130.5 Informational Note No.1 states that improper or inadequate maintenance causes the operating time of OCPDs to increase, which also increases incident energy. NFPA 70E Annex D shows that incident energy is proportional to the time of exposure. For example, if the operating time of an OCPD is doubled, the incident energy to the employee is doubled. The amount of time is a few cycles, so any increase in operating time will have a major effect on incident energy.
- NFPA 70E 130.5, "Arc Flash Risk Assessment," requires an arc flash hazard analysis that determines the Arc Flash Boundary and the PPE required.
- The Informational Note to NFPA 70E 130.5(C)(1) refers the reader to Informative Annex D for incident energy estimation and to Table H.3(b) in Informative Annex H for arc-rated clothing and PPE.

NFPA 70E 130.2(A)(3)

NFPA 70E 130.2(A)(3) explains that electrical conductors and circuit parts may need to be deenergized, even if they operate at less than 50 V. This could be due to factors such as the overcurrent protection used, the capacity of the circuit, and whether there is an increased risk of exposure to arc burns or explosions. An example is a 48 V battery bank. Despite relatively low voltage, the battery bank has a lot of short-circuit energy. If it is shorted, it could explode, creating a blast hazard and caustic fluids hazard.

Protective Clothing and Other PPE for Application with an Arc Flash Risk Assessment—NFPA 70E 130.5

The appropriate arc flash protective clothing and PPE are chosen by conducting an incident energy analysis or by using the table method. When performing an incident energy analysis, incident energy is calculated and documented. Incident energy is recorded in cal/cm^2 and is based on the working distance from a potential arc source to the face and chest area of the employee.

When using the table method, arc-rated clothing and PPE are chosen based on the incident energy associated with specific types of equipment and their available short circuit current and OCPD operating time. Hands and other body parts that are closer to the arc source than the chest require additional arc-rated PPE. Incident energy decreases by the inverse square of the distance as the employee moves away from a potential arc source. Incident energy increases by the square of the distance as an employee moves closer to the potential arc source. This is why body position is important when working on or near exposed energized electrical conductors or circuit parts. If employees are closer than the 18″ listed as the working distance on arc flash hazard warning labels, they would receive more incident energy than expected.

NFPA 70E 130.5 Informational Note No. 2. Short-circuit currents that are larger and smaller than the assumed maximum short-circuit currents for Tables 130.7(C)(15)(A)(a), 130.7(C)(15)(A)(b), and 130.7(C)(15)(B) can cause higher incident energy than larger short circuit currents under some circumstances. Informational Note No. 2 in 130.5 explains that if the short-circuit current increases without a reduction in the operating time of an OCPD, the incident energy will increase. If the operating time increases due to a lower short-circuit current, it can also result in higher incident energy exposure to the worker, even if the short-circuit current decreases. An example of this is a circuit protected by an OCPD using an instantaneous trip function that protects equipment at the end of a long run of cable.

Impedance is increased when a long run of cable exists between the OCPD and the load it is feeding/protecting. This impedance reduces the short-circuit current that the OCPD will detect. If the reduced short-circuit current is below the instantaneous setpoint, the OCPD will detect this as a large overcurrent condition instead of a fault. The OCPD will operate quickly, but not as fast as it would if the instantaneous function were to operate. Sometimes, less short-circuit current is worse than more.

NFPA 70E 130.5 Informational Note No. 3. Information Note No. 3 in 130.5 explains that an arcing fault inside an enclosure creates a variety of hazards not seen with a bolted fault. An arcing fault will cause arc flash and arc blast, although there are equipment designs available to reduce the risk to a worker working on or operating the equipment. A short list of available options that reduce the risk (arc-resistant switchgear, remote racking, etc.) is presented and refers the user to Informative Annex O, "Safety Related Design Requirements," for more information.

Personal and Other Protective Equipment—NFPA 70E 130.7

Any employee working where an electrical hazard potentially exists must be provided with, and must use, the appropriate PPE or other protective equipment appropriate for the hazard. The protective equipment must be designed and constructed for the specific parts of the body to be protected and for the work to be performed. While the tables in NFPA 70E 130.7 are intended to protect an employee from arc flash and shock hazards, an injury, such as a burn to the skin, can still occur even with the proper protective equipment. The rated incident energy for 0.1 sec can result in the onset of a second-degree burn on bare skin underneath arc-rated clothing and PPE. Also, the tables do not address the arc blast hazard (pressure wave) or any other hazards other than the thermal effects of an arc flash.

When the incident energy is above 40 cal/cm^2, NFPA 70E directs that extra effort must be taken to deenergize conductors or circuit parts. The pressure wave created by an arc blast is not proportional to time. It is caused by the near-instantaneous rise in air temperature in the enclosure. The tremendous expansion rate of ionizing air, which turns into an arc plasma jet, and the temperature rise to 5,000°F to 10,000°F, almost instantaneously, causes the arc blast. Since there are no industry-recognized equations to calculate the arc blast hazard and it is not known where the arc blast hazard becomes greater than the arc flash hazard, an incident energy of 40 cal/cm^2 was chosen by the NFPA 70E Committee task group as a relatively conservative value to protect workers from the arc blast hazard. Circuits or equipment rated above 40 cal/cm^2 can still be worked on, but it is important to consider using alternate methods for working on them instead.

Care of Equipment. PPE and other protective equipment must be maintained in a safe, reliable condition. PPE and other protective equipment must also be inspected before each use and must be stored in a manner that will not cause damage. Specific testing requirements for electrical PPE are addressed in NFPA 70E 130.7(C)(7)(a) and Tables 130.7(C)(7)(c) and 130.7(F).

Personal Protective Equipment (PPE)

NFPA 70E 130.7(C) requires that for work performed within the Arc Flash Boundary, employees use PPE to protect against electrical arc flash hazards. The PPE must meet the requirements of NFPA 70E 130.5. All body parts that are inside the Arc Flash Boundary must be protected. This includes hands, feet, or the back of the head if it is exposed to electrical hazards.

The 2012 edition of NFPA 70E clarified the need for arc-rated PPE and clothing in 130.7(A) Informational Note No. 2. This Informational Note stated that normal operation of low-voltage equipment (< 600 V) that is properly installed, maintained, and operated by a qualified person is not likely to pose a hazard to the operator. Normal operation means opening and closing or starting. Normal operation does not include any operation where the electrical connection between the main bus and the device is made or broken, such as inserting or removing a circuit breaker or motor control center (MCC). In the 2015 edition of NFPA 70E, this was further reinforced by adding 130.2(A)(4), which states that normal operation is permitted and gives a listing of requirements for normal operation. The requirements that must all be met include the following:

- properly installed
- properly maintained
- all covers and doors in place and properly secured
- no evidence of impending failure

The Informational Note attached to NFPA 70E 130.2(A)(4) defines what is meant by properly installed, properly maintained, doors and covers in place and properly secured, and evidence of impending failure.

Movement and Visibility. When arc-rated clothing is required, it must cover all ignitable clothing. Tall employees must verify that the cuffs of their pants are covered by the arc-rated clothing. If they are not covered, the exposed pant cuffs could ignite and burn under the arc-rated clothing.

Arc-rated clothing should also be loose fitting. **See Figure 9-25.** When arc-rated clothing fits tightly to an employee's body, heat transfer through the fabric increases and can burn the skin where it normally would not if the clothing were loose. Arc-rated PPE such as arc-rated face shields and windows must allow visibility.

Head, Neck, and Chin (Head Area) Protection. Nonconductive hard hats must be used wherever there is danger of contact with exposed energized conductors or circuit parts. Arc-rated PPE for the face, neck, and head area must be nonconductive as well. If hairnets or beard nets are required while performing electrical work, they must be nonmelting and made from arc-rated material. Even when they are worn under an arc-rated face shield or hood, the heat from an electrical arc could melt these items if they are not made from arc-rated materials.

According to NFPA 70E 130.7(C)(10)(b)(1) and (2), if the back of a worker's head is within the Arc Flash Boundary, an arc-rated balaclava or an arc-rated hood is needed. If the incident energy is expected to exceed 12 cal/cm^2, an arc-rated hood is required. According to NFPA 70E 130.7(C)(10)(c), arc-rated face shields must have a wrap-around guard that protects the face, chin, ears, forehead, and neck.

Eye Protection. UV-rated safety glasses or goggles must be worn anytime there is danger of an injury from arcs, flashes, or flying objects due to an electrical arc. When working on or near electrical circuits and devices there is always a danger of flying objects resulting from an explosion. For this reason, safety glasses or goggles should be worn at all times.

Hearing Protection. Hearing protection is any device worn to limit the noise entering the ear. According to NFPA 70E 130.7(C)(5), hearing protection is required anytime a worker is inside the Arc Flash Boundary. Rubber inserts are one form of hearing protection.

Salisbury

Figure 9-25. Arc-rated clothing should be loose fitting. When arc-rated clothing fits too tightly, heat transfer through the fabric increases during an arc flash.

SAFETY FACT

PPE manufacturers are developing ways to increase the light transmission through arc-rated windows, but workers should plan on using auxiliary lighting when wearing them.

Body Protection. If the incident energy exposure to the worker exceeds 1.2 cal/cm^2, arc-rated clothing is required. Arc-rated clothing can be any combination of arc-rated shirts, pants, coveralls, flash suits, or other appropriate equipment for the hazard. All parts of the body must be protected. Heavier fabrics generally provide more protection than lighter fabrics, but layering arc-rated clothing over flammable, nonmelting clothing is also acceptable. Employees should be aware that flammable, nonmelting clothing will not increase the arc rating of a clothing system. Flammable, nonmelting clothing only reduces the heat received by the body when it is worn under arc-rated clothing. It does not reduce the risk of ignition, which can negate the protection of the arc-rated material when the protection level of the clothing is exceeded.

According to NFPA 70E 130.7(C)(9), non-arc-rated clothing cannot be used to increase the arc rating of clothing or a clothing system. Under no circumstances, except those allowed by ASTM F1506 for manufacturing clothing, can flammable clothing be used to establish the arc rating of a clothing system.

Hand and Arm Protection. Hand and arm protection must be provided in accordance with NFPA 70E 130.7(C)(7)(a), (b), and (c). Rubber insulating gloves and leather protectors are required whenever there is a danger of injury from electric shock. Rubber insulating gloves must be rated for the voltage they will be exposed to. Arc-rated gloves that meet ASTM F2675, *Standard Test Method for Determining Arc Ratings of Hand Protective Products Developed and Used for Electrical Arc Flash Protection,* are now available, but these gloves are typically not voltage-rated gloves and are only useful for equipment operation when no shock hazard is present.

If rubber insulating gloves must be worn without leather protectors, as allowed by OSHA 1910.137, the requirements of ASTM F496 must be met. When rubber insulating gloves are used without leather protectors, the gloves must be derated until they are proven to be safe through electrical testing. For example, Class 00 gloves can only be used to 250 V, Class 0 are derated to 500 V, and the other glove classes are derated by one glove class. **See Figure 9-26.**

SAFETY FACT

According to OSHA 1910.137 and ASTM F496, Class 0 rubber insulating gloves are the only glove class that can be used without leather protectors and not be derated.

Derating Rubber Insulating Gloves

Class	Maximum Use Voltage	
	With Leather Protectors*	Without Leather Protectors*
00	500	250
0	1000	1000
1	7500	1000
2	17,000	7500
3	26,500	17,000
4	36,000	26,500

*in V

Figure 9-26. When rubber insulating gloves are used without leather protectors the gloves must be derated until they are proven to be safe through electrical testing. Using rubber insulating gloves without leather protectors is not recommended.

Electrical protective equipment must be maintained in a safe, reliable condition and inspected for damage or defects before use each day and after any incident that may have caused the equipment to be damaged. An air test is required as part of the inspection process for rubber insulating gloves. Insulating protective equipment must be tested in accordance with OSHA 1910.137 and ASTM F496.

Foot Protection. According to NFPA 70E 130.7(C)(8), dielectric overshoes must be worn when there is a possibility of step or touch potentials. Dielectric overshoes are rubber shoes worn directly over the top of normal footwear to provide protection from electric shock. Insulated soles, also known as electrical hazard (EH) shoes, cannot be used as the primary protection from step and touch potentials.

Standards for Personal Protective Equipment. A list of the current ANSI and ASTM standards that govern electrical PPE is provided in NFPA 70E Table 130.7(C)(14). For clothing material that may not be used, employees should refer to NFPA 70E 130.7(C)(11) and (12).

NFPA Table 130.7(C)(15)(A)(a), Arc Flash Hazard Identification for Alternating Current (ac) and Direct Current (dc) Systems

Table 130.7(C)(15)(A)(a) is used to determine if arc-rated clothing and PPE is mandated. There are certain tasks for which arc-rated clothing and PPE are always required, such as racking in circuit breakers or removing/installing bolted panels or covers.

Many other tasks may not require arc-rated clothing and PPE when certain conditions are met. A worker who opens doors or removes panels as part of doing infrared testing would be required to wear arc-rated clothing. Even though it is not mandated by NFPA 70E, arc-rated clothing for thermography is still recommended since equipment could be on the verge of failure. All equipment must be properly installed, be properly maintained, have all covers and doors in place and secured, and not show evidence of impending failure. For infrared thermography, the Restricted Approach Boundary cannot be crossed. The user of the table method should always be aware that NFPA 70E provides only the minimum acceptable safe work practices, not the best safe work practices. Just because arc-rated clothing and PPE are not mandated, does not mean there is no arc flash risk.

There are many circumstances that would create a need for arc-rated clothing and PPE, such as equipment that has a history of improper operation, equipment with illegible labeling, or equipment with higher voltages or larger ampacities. For example, wearing arc-rated clothing and PPE would not be reasonable for a 240 V, two-pole, molded-case circuit breaker, but it should be worn for any circuit breaker larger than a 600 A trip size because the potential for injury is too great.

Each technician must assess the risks involved for every specific piece of equipment and situation he or she may encounter, and the technician should not trust his or her safety to a label or table. If, for any reason, that person does not feel comfortable performing a task without arc-rated clothing and PPE, regardless of what the tables indicate, it should be worn. Many companies require arc-rated clothing and PPE to be worn while operating electrical equipment, such as circuit breakers, switches, or MCCs. This is perfectly acceptable as they are required to meet or exceed the requirements of NFPA 70E.

The next step in the process is to use Table 130.7(C)(15)(A)(b) to determine if the equipment involved is within the limits (estimated available short-circuit current and maximum fault clearing times) of the table method and which arc flash PPE category of clothing and PPE is required. The limits for the various PPE categories are the same as the old hazard/risk categories (HRCs), except that there is no PPE Category 0 in the new table. HRC-0 was cotton clothing and eye and ear protection and was intended to be worn when outside of the Arc Flash Boundary. Since Table 130.7(C)(15)(A)(b) is only intended for arc-rated clothing and PPE, there

---

SAFETY FACT

Using the table method to determine PPE is a clean sheet concept originally recommended by David Wallis, who at the time was OSHA's Director of the Office of Engineering Safety and OSHA's principle representative to the NFPA 70E Committee. He suggested that the table method should follow the same reasoning that OSHA field safety compliance officers do when performing audits.

> **SAFETY FACT**
>
> *One concern regarding the elimination of the HRC/PPE Category 0 designation was that some workers may believe that wearing permanent press or other meltable fabrics would be acceptable. This is not the case. Section 130.1 states that all requirements of Article 130 must be met, whether the table method is used or an incident energy analysis is performed.*

is no need for a PPE Category 0. Table 130.7(C)(15)(A)(b) also provides the Arc Flash Boundary for the appropriate PPE categories. Just as when using the old tables, the limits of the new table cannot be exceeded. If either limit is exceeded, an incident energy analysis must be performed.

The final step is to refer to Table 130.7(C)(16) and determine the arc-rated clothing and PPE recommended. The NFPA 70E Committee believes the new table method will enhance safety by simplifying the process of selecting arc-rated clothing and PPE and by clarifying the differences between tasks that have arc-rated clothing and PPE mandated and those that do not.

Regardless of whether the table method or an arc flash risk assessment is used to select PPE, the unique hazards and risks associated with each specific task and piece of equipment must be determined prior to beginning any work. Although the NFPA 70E Committee task group believes the table method to be reasonable, there are many factors that the NFPA 70E Committee could not evaluate. These factors include the age, condition, environment, maintenance, and history for each specific piece of equipment.

Using NFPA Table 130.7(C)(15)(A)(a). Table 130.7(C)(15)(A)(a) is used to determine if arc flash clothing and PPE are required. The following are five example tasks from Tables 130.7(C)(15)(A)(a) and (b). These tasks are typical of those performed in the industry each day.

Example task 1 is normal operation of a circuit breaker (CB), switch, contactor, or starter. No arc-rated clothing or PPE is mandated when the following conditions are met:

- The equipment is properly installed.
- The equipment is properly maintained.
- All equipment doors are closed and secured.
- All equipment covers are in place and secured.
- There is no evidence of impending failure.

When any of the conditions are not met, or when the worker is uncertain whether any one of them is true, arc-rated clothing and PPE must be worn. For example, when there is no data available to show that maintenance has been performed on the equipment in the past three years, such as a calibration and test label, test sheets, or personal knowledge of the electrical system and practices of the facility, then arc-rated clothing and PPE is necessary because it cannot be determined that the equipment has been properly maintained. When there is evidence that the equipment was not properly installed, such as lack of a properly sized grounds, misaligned panels, or enclosure doors that are unable to be opened to remove circuit breakers or access equipment, then arc-rated clothing and PPE is necessary.

Even when all the conditions are met, it may be advisable for the worker to wear arc-rated clothing and PPE. Smaller circuit breakers, switches, etc. present a small risk of injury if they fail. Once the continuous current rating of the device exceeds about 800 A to 1,000 A, that risk increases substantially. Equipment failure is fairly rare, but the consequences of such failure could be drastic. The worker should always assess the conditions at the time of operation and determine if arc-rated clothing and PPE is needed.

Example task 2 is removal or installation of CBs or switches. This particular task requires arc-rated clothing and PPE, regardless of conditions. It is considered to be a high-risk, potentially high-hazard task. The worker should consult Table 130.7(C)(15)(A)(b) to determine the appropriate PPE category.

Example task 3 is performing infrared thermography and other noncontact inspections outside the restricted approach boundary. This activity does not include opening doors or covers. Table 130.7(C)(15)(A)(a) shows that arc-rated clothing and PPE are not mandated without conditions for this task. Thermographers should be aware that when covers are off, and the equipment is having a thermographic survey performed on it due to problems, there could be an arc flash risk. This task is a good example of when arc-rated clothing and PPE might be reasonable in some situations.

Example task 4 is insertion or removal of individual cells or multicell units of a battery system in an enclosure for DC systems. The key to this task is whether or not the battery system is in an enclosure. When it is in an enclosure, arc-rated clothing and PPE are always necessary. When the same task is performed on an open rack, no arc-rated clothing and PPE are mandated.

Example task 5 is work on control circuits with exposed energized electrical conductors and circuit parts, 120 volts or below, without any other exposed energized equipment over 120 volts. This task, because of the very small arc flash risk, does not mandate the use of arc-rated clothing and PPE. This type of task is performed inside an annunciator or control panel, where all the conductors and circuit parts are operating at 120 V or less. In the past, unless an assessment was performed, a minimum of HRC/PPE Category 1 clothing and PPE would be required. Since many companies use HRC/PPE Category 2 daily work wear, they would be using that level of protection.

NFPA Table 130.7(C)(15)(A)(b), Arc-Flash Hazard PPE Categories for Alternating Current (ac) Systems

When Table 130.7(C)(15)(A)(a) indicates arc-rated clothing and PPE are necessary, the next step is to determine whether the equipment exceeds the limits of the table method, the appropriate PPE category, and the Arc Flash Boundary. Table 130.7(C)(15)(A)(b) is used to determine if the available short circuit current and OCPD operating time are within the limits of the table method, the appropriate PPE category, and the Arc Flash Boundary.

Regardless of the task, the PPE category that is needed is based on the equipment on which work is performed. This is a major change to the table method. The table method in previous editions of NFPA 70E assessed the risk associated with the task. The HRC number could have been reduced by 1, 2, or 3 numbers depending on the perceived risk of the task. This potentially could leave some workers underprotected, although the NFPA 70E Committee never received a substantiated report of this happening. With the new table method, when arc-rated clothing and PPE are required, it is always specified at the full rating for the equipment being worked on. The following five examples are the same tasks used previously.

Example task 1 is normal operation of a circuit breaker (CB), switch, contactor, or starter. When the conditions listed in Table 130.7(C)(15)(A)(a) are met, no arc-rated clothing or PPE is mandated. When one or more of the conditions are not met, when the worker is unsure, or when the worker would be more comfortable using arc-rated clothing or PPE, the type of arc-rated clothing and PPE needed is determined based on the equipment. When the equipment is a panelboard or other equipment rated greater than 240 V and up to 600 V, has a maximum of 25 kA short-circuit current available, and has a maximum of 0.03 sec (2 cycles) fault clearing time at a working distance of 18″, PPE Category 2 arc-rated clothing and PPE are indicated.

When the equipment is a 600 V class motor control center (MCC), the limits are a maximum of 65 kA available short-circuit current and a maximum of 0.03 sec (2 cycles) fault clearing time at a working distance of 18″. This is for tasks performed inside the MCC bucket, and PPE Category 2 is indicated. However, when the task is removing or installing an MCC bucket onto the main bus, the limits are a maximum of 42 kA available short-circuit current with a maximum of 0.33 sec (20 cycles) fault clearing time with a working distance of 18″. PPE Category 4 is indicated.

When performing tasks on an MCC, it is necessary to determine if the task is inside the MCC bucket or is exposing or interacting with the bus that feeds the MCC bucket. Tasks performed inside the bucket are typically PPE Category 2, and tasks that expose the bus feeding the MCC bucket, or removing or installing a bucket, would usually be PPE Category 4. Tasks on MCCs will always require arc-rated clothing and PPE, since MCCs are almost always fed by high-energy 208 V or 480 V bus.

As a last consideration, medium-voltage equipment is not specified any differently than low-voltage equipment using the table method. When using the table method in previous editions of NFPA 70E, operating a circuit breaker or fused switch on equipment rated 2.3 kV to 15 kV would require HRC/PPE Category 2. This would still be the best work practice. Even though the table method does not mandate arc-rated clothing and PPE, it would be prudent to wear PPE Category 2 as a precaution.

Example task 2 is removal or installation of CBs or switches. Since arc-rated clothing and PPE are mandated by the table method, the available short circuit current and the operating time of the OCPD must be estimated. The table method is used when one limit is not known with certainty. When the equipment is a panelboard or other equipment rated 240 V and below, it has limits of a maximum of 25 kA available short-circuit current and a maximum of 0.03 sec (2 cycles) fault clearing time at a working distance 18″. PPE Category 1 is mandated by the table method.

Example task 3 is performing infrared thermography and other noncontact inspections outside the restricted approach boundary. This activity does not include opening doors or covers. Arc-rated clothing and PPE are not mandated using the table method. However, the technician performing the thermographic survey must assess the conditions at the time of the task and determine whether or not it is safe to proceed without arc-rated clothing and PPE. When there is only a small distance between the equipment being scanned and a wall, especially with equipment rated 2.3 kV to 15 kV, arc-rated clothing and PPE may be warranted even though the table method indicates otherwise.

Example task 4 is insertion or removal of individual cells or multicell units of a battery system in an enclosure for DC systems. When working on DC systems, Table 130.7(C)(15)(B) must be consulted. The use of this table is similar to Table 130.7(C)(15)(A)(b), but the limits are related to batteries and other DC power systems.

If a battery bank is rated greater than 100 V but less than 250 V, the limits would be a maximum of 250 V, a maximum arc duration of 2 sec, and a working distance of 18″. The Arc Flash Boundary for DC systems is determined based on the available short circuit current. Less than 4 kA short circuit current requires PPE Category 1, and the Arc Flash Boundary is 3′. From 4 kA to 7 kA, PPE Category 2 is required, and the Arc Flash Boundary is 4′. From 7 kA to 15 kA, PPE Category 3 is required, and the Arc Flash Boundary 6′. If no engineering support is available, it may be easier to get the available short circuit current from the manufacturer or the installer of the battery system rather than calculating it in-house.

An issue that all employees working with these systems should be aware of is that the operating time of the OCPD could vary due to the state of charge of the battery system. A battery system that is fully charged will operate more quickly than one that is partially charged. This is why there is such a long maximum operating time for the limitations in the table. Note 1 for Table 130.7(C)(15)(B) requires that the need for acid-resistant PPE be evaluated in accordance with ASTM F1296, *Standard Guide for Evaluating Chemical Protective Clothing.* Note 2 requires that clothing and PPE be arc-rated in accordance with ASTM F1891, *Standard Specification for Arc Rated and Flame Resistant Rainwear,* or equivalent.

Example task 5 is work on control circuits with exposed energized electrical conductors and circuit parts, 120 volts or below, without any other exposed energized equipment over 120 volts. This task does not require arc-rated clothing and PPE, but some PPE is still necessary. A long-sleeved cotton shirt and long work pants are needed. Section 130.7(C)(12) prohibits wearing any fabric that could melt. This is another example where all requirements of Article 130 must be met, regardless of the method used to select arc-rated clothing and PPE.

NFPA Table 130.7(C)(16), Personal Protective Equipment (PPE)

The last step in the table method is to refer to Table 130.7(C)(16). Table 130.7(C)(16) lists the clothing and PPE required for each PPE category. The new table method does have an additional step, but it also clarifies the extent of the arc flash risk and provides relief in instances where arc flash clothing and PPE are not required. However, every technician must assess the conditions of the equipment and task each and every time the task is performed to ensure the risk is low. Each section in Table 130.7(C)(16) contains all necessary PPE and clothing for that particular PPE category and the minimum arc rating of the specified clothing and PPE. The following acronyms may be listed after the PPE in the table:

- AN = As Needed—This means that use of the PPE may be required if there is an exposure to a hazard. Leather gloves may be needed to protect the hands from an arc flash, even though rubber insulating gloves may not be needed.
- AR = As Required—This means that these items may be required in certain circumstances, such as when arc-rated coveralls should be worn instead of arc-rated shirts and pants.
- SR = Selection Required—This means that a worker must choose either safety glasses or safety goggles. Either type of eye protection must be rated to protect against UV-A and UV-B.

The notes at the end of this table can also directly affect the selection of PPE and clothing. Understanding their meaning is important.

- Note 1—The arc rating of PPE and clothing is defined in Article 100. The arc rating is either the arc thermal performance value (ATPV) or the energy of breakopen threshold (E_{BT}) value.
- Note 2—The arc-rated face shield that is used must protect the face, forehead, ears, neck, and chin areas per NFPA 70E 130.7(C)(10)(c). An alternative to an arc-rated face shield would be to wear an arc-rated hood.
- Note 3—If rubber insulating gloves with leather protectors are used, additional leather or arc-rated gloves are not required. The combination of

rubber insulating gloves with leather protectors satisfies the arc flash protection requirement. Salisbury by Honeywell conducted testing of rubber insulating gloves and leather protectors and found that they provide a high level of protection to the hands from arc flash.

130.7(C)(16) Informational Note No. 1. Informative Annex H in NFPA 70E contains a simplified two-category system for using arc flash PPE and clothing. Many facilities use this two-category system or some derivative of it. PPE Category 2 daily workwear is worn for most exposures along with an appropriate arc-rated face shield and other PPE. A PPE Category 4 flash suit and PPE are used for higher exposures. This system provides maximum protection and flexibility.

130.7(C)(16) Informational Note No. 2. This note cautions that the PPE and clothing in Table 130.7(C)(16) are intended to make an arc flash event survivable and is not intended to prevent burns caused by an arc flash. Some minor burns could still occur even when a worker is wearing properly rated and fitted PPE. The PPE and clothing specified in Table 130.7(C)(16) are also not intended to protect workers from the arc blast hazard, which could cause substantial injury.

Factors in Selection of Protective Clothing

NFPA 70E 130.7(C)(9) discusses factors for selecting protective clothing, which include how the arc-rated garments are to be worn as well as their characteristics. PPE designed for shock and arc flash protection is required when electrical hazards are present. PPE is a combination of separate items designed to protect specific parts of the body that provide overall protection when worn together. The appropriate PPE and clothing must cover all ignitable clothing and cover all affected parts of the body, while allowing for movement and visibility. PPE and clothing are required to be maintained in a safe, reliable, and functionally effective condition.

Arc-rated clothing and PPE typically include articles of clothing such as shirts, pants, coveralls, jackets, and parkas that are normally worn by employees who may be exposed to electrical arc flash hazards. Rainwear also must be arc-rated if it is needed. NFPA 70E 130.7(C)(9)(d) requires that arc-rated shirts be tucked into the pants and that shirts, coverall sleeves, and collars be fastened.

Shermco Industries
PPE and arc-rated clothing must cover all underlayers of clothing. Meltable clothing is not allowed under arc-rated clothing.

Layering. Flammable, nonmelting underlayers can be worn with arc-rated clothing to reduce the heat received by the body. Each layer of clothing that is under arc-rated clothing typically reduces the heat to the body, but only arc-rated systems count toward protection since in some layering systems this is not true. Many garment manufacturers have done research on their layered systems and provide arc ratings for the systems. If flammable, nonmelting layers are worn under arc-rated clothing, the arc rating of the clothing system must be high enough to prevent the innermost layer of arc-rated clothing from failing (breakopen), so the flammable underlayers do not ignite.

Cotton t-shirts, shorts, shirts, pants, and/or arc-rated coveralls are often used as underlayers. Additional arc-rated clothing and/or PPE may be required to provide the level of protection necessary for a specific task or to eliminate the possibility of ignition of under layers. Typically, arc-rated undergarments are recommended for the upper torso in medium-voltage work.

Workers must be aware that the face and neck area must also be protected to the same level as the rest of the body. Also, arc-rated clean room clothing cannot be worn with other types of arc-rated clothing, and its arc-rating cannot be added to other arc-rated clean room clothing.

Outer Layers. The outermost layer of clothing (parkas, rainwear, etc.) must be arc-rated clothing when worn over arc-rated clothing. If meltable fabric clothing is worn over arc-rated clothing and subjected to an electrical arc, it will ignite or melt. Arc-rated clothing is not intended to provide long-term protection from heat. A burning and melting outerlayer can cause the arc-rated clothing underneath to fail, causing substantial burns.

Underlayers. Acetate, nylon, polyester, polypropylene, spandex, or any other meltable fabric is not allowed to be worn as an underlayer next to the skin. A small amount of elastic used to hold up underlayers such as shorts or socks is allowed. Arc-rated underlayers provide additional protection from an electrical arc. There are several thin, lightweight arc-rated undergarments available. When arc-rated underlayers are worn, they may provide a higher arc rating for the clothing system than flammable, nonmelting underlayers. Arc-rated underlayers eliminate the risk of ignition and remove the employee's responsibility to wear nonmelting undergarments, so some companies mandate them.

Clothing that is not arc-rated cannot increase the arc-rating of a clothing system. Flammable, nonmelting underlayers reduce the heat received by the body, but can ignite and burn if the arc-rated clothing outerlayers break open.

The main concern with meltable fiber fabrics is the melting temperature, rather than the ignition temperature. Ignition temperatures may range from 600°F to 650°F, but melting temperature can be as low as 180°F for materials such as spandex. This temperature is low enough that the clothing may melt under the arc-rated clothing and cause severe burns where there normally would not be an injury. This aspect of meltable fiber clothing can be a hazard to female technicians because most of their underlayers are made from meltable synthetics. Additionally, most wicking T-shirt materials on the market today are made from meltable materials like polyester and polypropylene and should not be worn as underlayers.

SAFETY FACT

Layered clothing will always provide a reduction of heat to the body. Wearing arc-rated underlayers provides additional protection, but the arc ratings may not necessarily be added together, since the underlayer can still transfer the residual heat. Best practice is to use PPE with the appropriate arc rating for the task being performed.

SAFETY FACT

The elastic in waistbands and socks may melt during an electrical arc and cause injury. Although the injury received will be relatively minor, it is recommended that employees tuck a cotton t-shirt under the elastic waistband of shorts.

Coverage. Arc-rated clothing must cover all ignitable clothing and as much of the exposed body as possible. Shirts must be fastened at the wrists, and both shirts and jackets must be fastened at the neck.

Fit. Arc-rated clothing should be loose-fitting so as to provide air spaces between layers, but not so loose that they interfere with the task or are a hazard. When arc-rated clothing fits tightly, the heat transfer through the fabric is increased, increasing the chances of a burn.

This is especially true with balaclavas. Often, technicians wear the balaclava tight-fitting. Wearing balaclavas in this manner will cause the heat transfer through the balaclava to increase, reducing its rating. The air space between the face and balaclava is necessary to achieve the arc rating of the balaclava.

Interference. PPE and arc-rated clothing must provide adequate protection for the task, but not interfere with the performance of the task. Several factors can influence PPE and clothing choices, such as the work method, location, and the specific task being performed.

Arc Flash Protective Equipment

Arc flash protective equipment is designed to be worn whenever a worker may be exposed to the hazard of arc flash. Arc flash protective equipment per NFPA 70E 130.7(C)(10) includes arc flash suits, face protection, hand protection, and foot protection.

Arc Flash Suits. Arc flash suits must be designed to allow easy and rapid removal. The arc flash window and all other parts of the flash suit and hood must have an arc rating equal to or greater than the expected incident energy exposure. When air pumps are used to supply exterior air into the hood, the air hoses and pump housing must either be made from nonmelting, nonflammable material or covered in arc-rated material.

Face Protection. Arc-rated face shields and windows must have an arc rating equal to or greater than the expected incident energy exposure. If a face shield does not have an arc rating, it cannot be used unless the manufacturer is contacted with the part number of the window to verify that it is the appropriate window for the arc rating of the hood. UV-rated safety glasses or goggles are required under arc-rated face shields or hoods.

Arc-rated hoods are required to be rated as a product. Rating of fabric without a rating of the full system as a product is not allowed per ASTM F2178, *Standard Test Method for Determining the Arc Rating and Standard Specification for Eye or Face Protective Products*. Data for the hood style used is required. Testing of a window separate from the hood has been shown in some cases to be an insufficient system due to afterflaming characteristics of the fabric.

Arc-rated face shields and windows provide increased protection, but they can also interfere with color perception and vision. Additional lighting, such as light stands, is usually needed when wearing arc-rated PPE for the face and head. If the expected incident energy exceeds 12 cal/cm^2, an arc-rated hood must be worn instead of an arc-rated face shield.

Hand Protection. When arc flash protection is required for the hands, leather or arc-rated gloves must be worn. Leather protectors must be worn over rubber insulating gloves when they are worn for shock protection. Rubber insulating gloves, when worn with leather protectors, provide protection from an arc flash. **See Figure 9-27.** Arc-rated gloves also provide protection to the hands from an arc flash in accordance with their ratings, but they do not provide protection for the shock hazard. Heavy-duty leather gloves (with a minimum thickness of 0.076 cm) are suitable for PPE Category 2 tasks. Leather gloves will shrink during an arc flash and will then provide a reduced level of protection. According to the Informational Note for NFPA 70E 130.7(C)(10)(d)(1), heavy-duty leather gloves meeting the specification have a minimum ATPV of 10 cal/cm^2.

Foot Protection. Heavy-duty leather work shoes also provide some protection from the heat of an arc flash and must be used for PPE Category 2 and/or 4 cal/cm^2 and higher tasks. Steel-toed shoes are not considered a shock hazard as long as the steel toe remains covered. If the steel toe is exposed, the shoes need to be replaced.

Figure 9-27. When arc flash protection is required for the hands, leather protectors must be worn over rubber insulating gloves to provide protection from an arc flash.

Clothing Material Characteristics

Clothing and PPE are required to meet the requirements of NFPA 70E 130.7(C)(10) and (11). Arc-rated materials made from flame-resistant-treated cotton provide thermal protection from an arc flash, as long as they meet ASTM F1506.

Arc-rated fabrics can reduce a possible burn injury by providing a thermal barrier between the worker and the arc flash. Materials that are not arc-rated materials, such as cotton, polyester blends (permanent press, antiwrinkle, antistain, wash-and-wear, or other types of fabrics), wool, silk, and rayon, are flammable and could ignite during an arc flash and may continue to burn, causing a more serious injury than the one that would be caused by the arc flash alone.

Synthetic materials, such as acetate, nylon, polyester, polyethylene, polypropylene, and spandex, in blends or as single materials, melt below 600°F (315°C) and cannot be worn by electrical workers. Clothing made from these materials, as well as those that may not be listed, will melt during an arc flash and may increase the extent of the injury from the arc flash.

According to NFPA 70E 130.7(C)(11), "Exception," the use of flammable and meltable fabrics is allowed if they meet the requirements of ASTM F1506.

Clothing and Other Apparel Not Permitted

Clothing items such as hard hat liners, beard nets, and hairnets are not permitted to be worn by employees working on electrical equipment unless the items meet the requirements of ASTM F1506 and NFPA 70E 130.7(C)(11). Nondurable, flame-resistant-treated clothing and flame-resistant-treated acrylic materials are not recommended to be worn as protective clothing by employees working on electrical equipment. Flammable, nonmelting clothing is allowed as underlayers in accordance with Article 130.7(C)(12) Exception No. 1. When non-arc-rated PPE, such as a respirator, is required to protect the worker, it can be worn if the risk assessment identifies the protection offered by it as adequate for the level of protection from the arc flash hazard. This applies to other types of PPE that are not available as arc-rated. Some types of non-arc-rated PPE can be covered with an arc-rated balaclava or hood for protection, if necessary.

SAFETY FACT

The materials that are chosen to be worn by an employee working on electrical equipment are crucial to the employee's safety. For this reason, arc-rated daily workwear is recommended. Materials such as untreated cotton or wool, regardless of fabric weight, will not protect a worker from the effects of an arc flash.

The heat during an arc flash is extremely high, but its duration is short and should not cause heavier PPE to fail or melt. An extended arc flash, however, could create enough heat for a period of time that is long enough to melt non-arc-rated PPE and other equipment and worsen the injury to an employee. This is why a risk assessment is required for non-arc-rated PPE.

Care and Maintenance of Arc-Rated Clothing and Arc-Rated Flash Suits

Arc-rated clothing and PPE must be inspected before each use. Contaminated or damaged flash suits or arc-rated clothing and PPE cannot be used if their protective qualities are reduced. Arc-rated clothing and PPE that are contaminated with grease, oil, or flammable or combustible liquids cannot be used. NFPA 70E requires that the user of arc-rated clothing and PPE follow the care and maintenance instructions provided by the garment manufacturer.

It is also recommended that the user visit the fabric manufacturer's website and download their laundering and care requirements. These requirements are often more detailed and can offer additional insight. For example, many garment manufacturers offer arc-rated flash suits and provide guidance on the care, maintenance, and storage of the clothing. However, the fabric manufacturer may have more complete laundering and care procedures on its website than the clothing manufacturer. Arc-rated clothing and PPE must be stored in a way that protects them from any physical damage, including moisture, dust, and contamination by flammable or combustible materials or any other agent that could cause deterioration. If clothing is stored in a closet or locker, hangers that are designed to support the weight of the clothes should be used. **See Figure 9-28.**

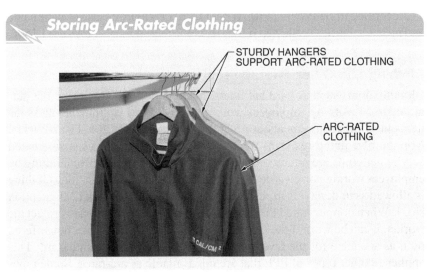

Figure 9-28. Protective clothing that is hung in lockers or closets must be hung on hangers designed to support the weight.

The instructions from the manufacturer must be followed when cleaning arc-rated clothing and PPE or it may lose some of its protecting qualities. Arc-rated clothing must be repaired with the same thread and materials that were used in the original manufacturing process. Name tags, company logos, trim, or any other additions to the clothing or PPE must meet the requirements of ASTM F1506.

SUMMARY

- OSHA specifies that workers must be trained in the use of PPE, temporary insulating and shielding materials, and insulated hand tools.
- Employers must determine the correct PPE for their workers, supply the PPE, and ensure it is being used when needed.
- Two of the most important parts of PPE maintenance is inspecting the equipment before each use and properly storing the PPE. Each part of the meaning of the phrase "use of" becomes critical when using PPE and arc-rated clothing.
- Leather protectors worn over rubber insulating gloves prevent damage to the rubber insulating glove.
- The tables and notes located in NFPA 70E 130.7(C) provide an alternative method to a full arc flash hazard analysis for determining PPE.
- In cases where the tables from NFPA 70E 130.7(C) cannot be used, NFPA 70E requires an arc flash hazard analysis to properly assess hazards and determine PPE. This occurs whenever the maximum short-circuit current or OCPD operating time specified in the notes for the tables is exceeded.
- Synthetic materials, such as acetate, nylon, polyester, polyethylene, polypropylene, and spandex, will melt during an arc flash and increase the extent of the injury from an arc flash. Clothing made from meltable fibers cannot be worn when working on or near exposed energized electrical conductors or circuit parts unless it meets the requirements of ASTM F1506 and NFPA 70E.
- Arc-rated clothing and PPE must be stored in a way to protect it from physical damage, including moisture, dust, and contamination by flammable or combustible materials or any other agent that could cause deterioration.

Digital Learner Resources
ATPeResources.com/QuickLinks
Access Code: 705798

9 CHOOSING AND INSPECTING PERSONAL PROTECTIVE EQUIPMENT

Review Questions

Name _____   Date _____

_____ 1. The acronym ATPV stands for ___.
   A. available to protect vehicles
   B. arc temporary performance value
   C. arc thermal performance value
   D. associated temporary protection value

_____ 2. The ___ is the point at which a fabric allows a 1″ crack or a ½″ hole, but no burn is registered.
   A. energy breakopen threshold
   B. arc thermal performance value
   C. incident energy exposure
   D. Category 1 PPE category

_____ 3. Arc flash protective clothing and PPE are chosen by conducting an incident energy analysis or ___.
   A. by completing an energized electrical work permit
   B. by using the table method
   C. an arc flash hazard analysis
   D. job briefing

_____ 4. NFPA 70E Table ___ is used to determine the PPE category for a specific task for AC electrical systems.
   A. 130.7(C)(14)
   B. 130.7(C)(15)(A)(b)
   C. 130.7(C)(15)(b)
   D. 130.7(C)(16)

T   F   5. Per NFPA 70E, normal operation of electrical equipment includes inserting or removing a circuit breaker.

_____ 6. The NFPA 70E standard that covers the selection of PPE and equipment is Article ___.
   A. 100
   B. 110
   C. 120
   D. 130

263

_____ 7. Improper storage of rubber insulating gloves includes ___.
   A. storing them in areas with moderate temperatures
   B. storing them out of direct sunlight
   C. folding the glove cuffs to prevent dirt and debris from entering
   D. placing them in a storage bag cuff down

_____ 8. OSHA 1910.335 refers to ___ PPE.
   A. arc flash
   B. shock
   C. arc blast
   D. all electrical

T F 9. It is acceptable to store rubber insulating gloves inside out.

T F 10. Safety glasses for electrical workers must be UV rated.

T F 11. Using NFPA 70E Table 130.7(C)(15)(A)(a), the PPE category for operating a 480 V MCC with the doors closed is PPE Category 4.

_____ 12. Using NFPA 70E Table 130.7(C)(15)(A)(a), the PPE category for inserting or removing a 480 V MCC bucket or cubicle is PPE Category ___.
   A. 1
   B. 2
   C. 3
   D. 4

_____ 13. NFPA 70E Table ___ lists the clothing and PPE required for each PPE category.
   A. 130.7(C)(14)
   B. 130.7(C)(15)(A)(b)
   C. 130.7(C)(16)
   D. 130.7(C)(17)

_____ 14. Items of inspection for rubber insulating gloves include all of the following except ___.
   A. UV damage
   B. punctures
   C. embedded or foreign material
   D. open or loose seams

_____ 15. Required PPE for PPE Category 2 tasks includes all of the following except ___.
   A. safety glasses
   B. a long-sleeve cotton or arc-rated shirt and pants
   C. an arc-rated face shield
   D. hearing protection

10
GUIDELINES FOR COMMON ELECTRICAL TASKS

Choosing proper personal protective equipment (PPE) is critical to worker safety. It is easy for employees to get overwhelmed by the amount of information needed to plan a task. Determining the PPE for electrical tasks can be time consuming and cumbersome, but it must be accomplished. In addition, equipment condition and reliability must be evaluated before a task is performed. Safe approach distances must be observed for shock and arc flash hazards and a safe work zone must be established whenever exposed energized conductors or circuit parts are present. Every electrical task may have unique challenges, but common guidelines can be followed to perform each task safely.

OBJECTIVES

- Explain risk assessment for common electrical tasks.
- Describe the task of removing and inserting low- or medium-voltage drawout-type circuit breakers.
- Identify the hazards involved with operating medium-voltage air-break switches.
- Explain the task of inserting and removing motor control center buckets.
- List the recommended PPE for troubleshooting circuits rated 120 V and less.
- Describe the unique challenges involved when troubleshooting AC drives.
- Identify the risks involved with operating equipment rated 240 V and less and equipment rated 240 V to 600 V.
- Explain the hazards involved with removing covers and panels from electrical enclosures.
- Describe the risks involved with replacing light ballasts.
- Explain the task of replacing low-voltage motors.
- Explain the task of replacing medium-voltage motors.

RISK ASSESSMENT FOR COMMON ELECTRICAL TASKS

Each task and type of equipment presents unique challenges, but they all share one common requirement: a risk assessment must be conducted before any work is performed. This applies no matter if the tables in NFPA 70E are used or if an incident energy analysis has been performed and the equipment has been labeled with an arc flash hazard warning label. The new table method in Article 130 requires that a determination be made as to whether or not arc flash clothing and PPE are required. Generally, no arc-rated clothing and PPE are mandated if the equipment meets the following conditions:

- Equipment is properly installed.
- Equipment is properly maintained.
- All covers and doors are in place and properly secured.
- There is no evidence of impending failure.

If any of the above conditions are not met, or if the technician performing the task cannot say positively that they have all been met, then arc-rated clothing and PPE must be worn. The condition that is most difficult to verify is the condition of maintenance. This is especially true for technicians who do not work at a specific location or facility on a continual basis and cannot know for certain that proper maintenance has been performed. NFPA 70B, *Recommended Practice for Electrical Equipment Maintenance,* Section 11.27, provides a method for applying decals or labels that provide the information needed to verify the condition of maintenance.

SAFETY FACT

It is critical for employee safety that a risk assessment be performed prior to starting work on electrical equipment, even if the NFPA 70E Table method is used. Regardless of whether the NFPA 70E Tables are used to choose arc flash protective clothing and PPE or an incident energy analysis has been performed, NFPA 70E 130.1 requires that all requirements of Article 130 be met.

Another important aspect of the new table method is that just because arc-rated clothing and PPE are not mandated by the table, it does not mean that they are not required in all circumstances. The qualified person must assess the conditions and circumstances at the time of performing the task and determine at that time whether arc-rated clothing and PPE are needed.

Some tasks, such as applying personal protective grounds, are performed after absence-of-voltage testing has been completed. Even when the circuit or equipment has been tested and found deenergized, there is still a small risk of an arc flash involved in applying grounds. This residual risk must be evaluated before the task is attempted. If it is determined that the risk is too high, additional safety measures to reduce the risk must be implemented or other means of completing the task must be discussed.

Arc Flash Hazard Warning Labels

If an incident energy analysis has been performed, there will be arc flash hazard warning labels affixed to the electrical equipment. If a risk assessment is performed and the employee believes the equipment will function in accordance with the manufacturer's specifications, the employee would dress to the level of protection indicated on the label. This includes the proper class of rubber insulating gloves as indicated by the nominal voltage and arc-rated clothing and PPE that has an arc rating equal to or greater than the incident energy indicated on the arc flash hazard warning label.

Arc flash hazard warning labels provide information on two separate hazards: electric shock and arc flash. These hazards are addressed individually. This can lead to some odd-looking labels where the arc flash hazard is so high that certain tasks would be unsafe but the shock hazard information is provided as if the arc flash were not an issue. **See Figure 10-1.**

Common Electrical Tasks

Table 130.7(C)(15)(A)(a) is used to determine whether arc-rated clothing and PPE are mandated. If so, Table 130.7(C)(15)(A)(b) is used to determine the appropriate PPE category and the Arc Flash Boundary for AC systems. Table 130.7(C)(15)(B) is used for DC systems. In order to use the table method to determine the PPE needed, the available short circuit current and the operating time of the OCPD must be within the limits given in Table 130.7(C)(15)(A)(b).

Tasks that used to be included in Table 130.7(C)(15)(A) of the previous edition or other tasks that are not listed in the current tables can be determined by the more general task categories listed. For example, there is no specific task listed for replacing light ballasts. This task now falls under the task category "Work on exposed energized electrical conductors and circuit parts directly supplied by a panelboard or motor control center." Table 130.7(C)(16) is used to identify the specific protective clothing and PPE to be used with each PPE category. Some of the tasks listed in Table 130.7(C)(15)(A)(a) include the following:

- removing and inserting low- or medium-voltage drawout-type circuit breakers
- operating medium-voltage air-break switches
- removing and inserting motor control center (MCC) buckets
- troubleshooting circuits rated 120 V or less
- troubleshooting circuits rated 240 V to 600 V

- testing for the absence of voltage
- operating equipment rated 240 V or less
- operating equipment rated 240 V to 600 V
- operating NEMA E2 (fused contactor) motor starters rated 2.3 kV to 7.2 kV
- removing covers and panels from electrical enclosures

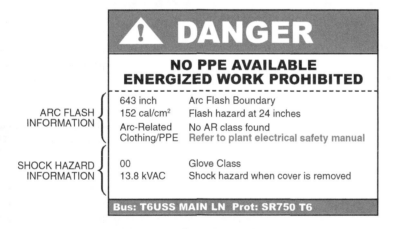

Figure 10-1. Arc flash hazard warning labels provide information on both shock and arc flash hazards.

REMOVING AND INSERTING LOW- OR MEDIUM-VOLTAGE DRAWOUT-TYPE CIRCUIT BREAKERS

The task of removing and inserting drawout-type circuit breakers is commonly known as racking. All drawout-type circuit breakers basically function in the same way when it comes to racking them in or out. There are differences from manufacturer to manufacturer, but the process is usually the same.

SAFETY FACT

The physical size of a drawout-type circuit breaker makes it difficult for a single person to safely remove or insert a circuit breaker without mechanical aid. A mobile cable and pulley mechanism is typically used to raise or lower the circuit breaker and guide it into place.

Medium-voltage drawout-type breakers have some minor differences, but rack in or out in the same way low-voltage breakers do.

Low-voltage circuit breakers of the drawout-type design can be referred to by different names, such as drawout or air-frame, but they are most commonly referred to as low-voltage power circuit breakers. This name differentiates them from other types of low-voltage circuit breakers, such as molded-case circuit breakers and insulated-case circuit breakers. Low-voltage power circuit breakers are typically of the drawout-type design, but on rare occasions they may be bolted directly to the bus.

The task of racking a low-voltage power circuit breaker is inherently hazardous because the process involves making and breaking a load-carrying electrical connection without arc chutes. An *arc chute* is an element inside a circuit breaker that extinguishes an arc when the contacts are opened.

If an arc were to occur during the racking operation, it would start at the bus-to-breaker connections and would have to be extinguished by the upstream breaker (the breaker that is feeding the bus). The upstream breaker may not see the short circuit as such due to the settings of its overcurrent protective device (OCPD). Instead, depending on its settings, the upstream breaker may see it as a large overcurrent. The circuit breaker will trip, probably in one or two seconds. However, it may not use its instantaneous function to trip the breaker open, which would take no more than 6 cycles (0.10 sec). One or two seconds is a long time for an arc to exist. Since incident energy is proportional to time, the incident energy from this situation would be much greater than the incident energy that would be calculated by an arc flash hazard analysis. **See Figure 10-2.**

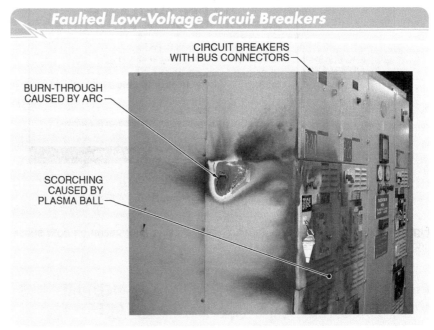

DuPont Company

Figure 10-2. Circuit breaker failure at the bus connection produces a high incident energy that can cause major damage to equipment and serious injury to nearby employees.

Removing Low-Voltage Power Circuit Breakers

Listed below are general guidelines for removing low-voltage power circuit breakers. "General" implies that the procedure may be incomplete, depending on the specific installation, or it may not apply at all, depending on the design of the breaker and enclosure. The instructions provided by the manufacturer for the specific drawout-type circuit breaker should always be followed.

- Verify that the correct circuit breaker is about to be operated.
- If possible, disconnect the load fed by the circuit breaker before operating. This reduces the chance of a failure.
- If possible, remotely open the circuit breaker. If it is not possible to open the circuit breaker remotely, open the circuit breaker locally while standing at the hinge-side of the enclosure.
- Note the position of the OPEN/CLOSED indicator. It should indicate that it is in the OPEN position. **See Figure 10-3.**

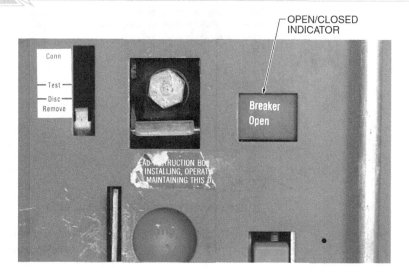

Shermco Industries

Figure 10-3. Before removing or inserting drawout-type circuit breakers, it is important to verify that the OPEN/CLOSED indicator is in the open position.

- Standing next to the hinge side of the cubicle, manually depress the TRIP pushbutton, even if the indicator shows that it is in the OPEN position. If there is an arc flash in the circuit breaker cell, the handle side will almost certainly open. The hinge side will usually remain connected to the cubicle. The door will provide some amount of protection if an arc flash were to occur. Even if the door has louvers or vents, it should help divert some of the heat of the arc flash away from the employee's body.
- If possible, rack the circuit breaker in or out of its cubicle with the door closed. Cubicle doors on old switchgear may not have access for racking built into them, but they can be modified by using a circular saw and cutting a hole in-line with the racking screw. **See Figure 10-4.**

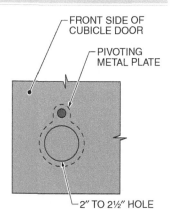

Figure 10-4. Cubicle doors on old switchgear may not have access for racking built into them, but they can be modified.

A teardrop-shaped piece of flat sheet metal can be attached to the door from the inside to cover the opening when not in use. Only qualified persons should attempt to make such modifications and only after discussing it with the engineering department, supervisor, or any other authority having jurisdiction (AHJ). Also, the location of any control wiring or components must be noted to prevent damaging them or creating an unsafe condition. All required safe work practices must be observed while planning and performing this type of modification.

- Insert the racking handle while observing that the racking interlock depresses the TRIP pushbutton. **See Figure 10-5.** This is a common feature of all but the oldest circuit breakers. This interlock feature ensures that the circuit breaker does not get racked when the circuit breaker is still closed.

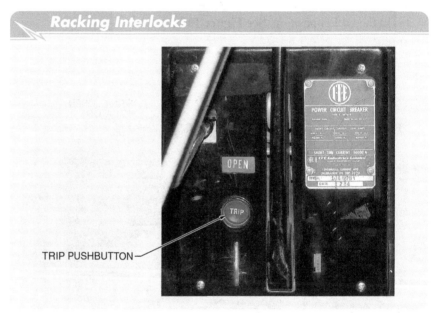

Shermco Industries

Figure 10-5. Most low-voltage circuit breakers have an interlock feature to ensure that the circuit breaker cannot be racked into or out of the cubicle when it is closed.

- Slowly turn the racking handle to remove the circuit breaker. Perform this task slowly, as it is best practice to listen for any arcing that may occur when the primary disconnects on the circuit breaker just begin to part from the bus stabs. **See Figure 10-6.** If the breaker is racked in the CLOSED position, an arcing sound will be heard. If this happens, discontinue racking out immediately and rack the breaker back in. It is likely that a mechanical failure has occurred, possibly with the safety interlock mechanisms. On medium-voltage circuit breakers it is not uncommon for a slight buzzing sound to be heard when the primary stabs part. This is much different from an arcing sound, which indicates that the breaker is still closed. If in doubt, stop racking and reinsert the circuit breaker to the bus.

- Once the stabs are clear of the bus, the arc flash hazard is almost nonexistent. Rack the circuit breaker through the TEST position, unless there is an operational problem with the circuit breaker. The TEST position is used to electrically cycle the circuit breaker to check for proper functioning.

Bus Stabs

Shermco Industries

Figure 10-6. As the racking handle is slowly turned, the primary disconnects on the circuit breaker begin to part from the bus stabs.

- Continue racking until the circuit breaker is in the DISCONNECTED position. The door may be opened at this point and the extension arms (found on low-voltage circuit breakers and some medium-voltage vacuum circuit breakers) can be installed or extended from the cubicle. Verify that the extension arms are locked into place and able to support the weight of the circuit breaker before proceeding.
- Ensure that the circuit breaker's rollers are squarely on the extension arms and all components appear to be in place. Use a flashlight if needed to see the front rollers.
- Slowly pull the circuit breaker out of the cubicle and onto the extension arms. Pull until the rear rollers of the circuit breaker can be seen. Again, ensure that the rollers are squarely on the extension arms.
- Slowly pull the circuit breaker completely out of the cubicle until it reaches the stops on the extension arms. **See Figure 10-7.** Do not allow the circuit breaker to move too quickly. It can become dislocated from the extension arms when it reaches the stops.
- If removing the circuit breaker from the extension arms, use the spreader bar from the manufacturer. Make certain the spreader bar is properly attached to the circuit breaker frame. Often, one side will come loose when the cable is tightened. This can cause the circuit breaker to be dumped on the floor. Do not use rope or wire slings strung through the frame of the circuit breaker. These will squeeze the sides of the frame, possibly causing damage to the arc chutes and/or operating mechanism.
- If using a cable and pulley mechanism, be certain to inspect the cable reel for any crossed turns before lifting the circuit breaker. If there are crossed turns, they must be corrected by uncrossing the cables. Crossed cables on the reel will cause the circuit breaker to bounce when lowering it, possibly causing the cable to snap under the instantaneous weight/force increase.

Low-Voltage Circuit Breaker on Extension Arms

Shermco Industries

Figure 10-7. The rollers on the circuit breaker must be squarely on the extension arms when the circuit breaker is pulled out of the cubicle.

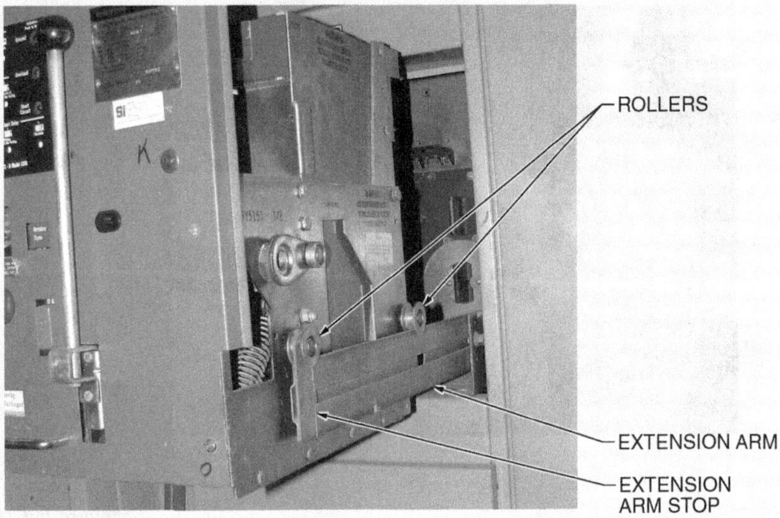

- If using a lift platform to remove the circuit breaker, be certain it is completely under the circuit breaker frame before lifting.
- Slowly lower the circuit breaker onto the floor and remove the spreader bar. **See Figure 10-8.**

Cable and Pulley Mechanism

Shermco Industries

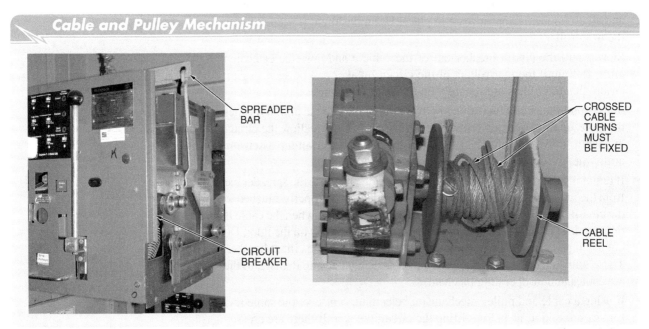

Figure 10-8. When using a cable and pulley mechanism, the cable reel must be checked for any crossed cable turns before the circuit breaker is lifted.

Inserting Low-Voltage Power Circuit Breakers

The instructions provided by the manufacturer for the specific drawout-type circuit breaker being inserted into a cubicle should always be followed. The general guidelines that should be used to insert a circuit breaker into its cubicle include the following:

- When placing the circuit breaker onto the extension arms, be sure that the spreader bar is secure before lifting.
- Raise the circuit breaker and position it so that the rollers will contact the extension arms squarely. Because the circuit breaker can rotate due to the cable tension, be sure to continuously guide it until it rests on the extension arms.
- Push the circuit breaker into the cell until it is in the DISCONNECTED position.
- Manually trip the circuit breaker using the TRIP pushbutton, even if the circuit breaker indicator shows it as open.
- Close and secure the cubicle door if possible.
- Rack the circuit breaker through the TEST position, unless it is necessary to test operate the circuit breaker. Be certain to push the TRIP pushbutton before continuing to rack the circuit breaker in from this point.
- Rack the circuit breaker into the CONNECTED position. Do not apply too much torque to the racking mechanism. Once the breaker is firmly seated onto the bus connections, discontinue racking.

The shock approach boundaries for low-voltage power circuit breakers include the Limited Approach Boundary at 42" and the Restricted Approach Boundary at 12". The Arc Flash Boundary can be 36" to 84" or more, depending on the available short-circuit current and the trip function (short-time delay or instantaneous) being used for short-circuit protection. A circuit breaker is installed in a cubicle, which is essentially a metal box. An arc flash will reflect off the back and sides of the cubicle and focus the arc plasma and heat toward the front opening, which is where an employee would be standing.

The following is an example of the effect that time has on incident energy. A low-voltage power circuit breaker has 42 kA of available short-circuit current with an instantaneous trip (4 cycles or 0.07 sec) and an Arc Flash Boundary of 30.5". However, if that same circuit breaker is using a short-time delay function of 0.33 seconds (nominal), the Arc Flash Boundary increases to 66.3". The short-time delay function can have a nominal delay of up to 0.5 sec, which would cause the Arc Flash Boundary to increase to 81.6".

Recommended PPE for Removing and Inserting Low-Voltage Power Circuit Breakers

To remove and insert low-voltage power circuit breakers, there are two separate tasks that must be performed. The first task is operating the circuit breaker to ensure it is open. The second task is racking the circuit breaker in or out of its cubicle.

First, the heading "Normal operation of a circuit breaker (CB), switch, contactor, or starter" is located in Table 130.7(C)(15)(A)(a). Arc-rated clothing and PPE are not mandated if the equipment condition meets all of the following conditions:

- Equipment is properly installed.
- Equipment is properly maintained.

- All covers and doors are in place and secured.
- There is no evidence of impending failure.

However, if any one of the conditions is not met, then arc-rated clothing and PPE are mandated. If the technician is not certain whether these conditions are or are not met, arc-rated clothing and PPE must be worn. It is also recommended that arc-rated clothing and PPE be worn when operating a low-voltage drawout circuit breaker. The risk of operating smaller-frame circuit breakers is low, and the consequences of a failure are less severe. However, when the circuit breaker is 600 A or larger, the consequences of a failure become much more substantial.

Next, the heading "Insertion or removal (racking) of CBs or starters from cubicles, doors open or closed" is located. This task always requires the use of arc-rated PPE, regardless of equipment condition. The recommended arc-rated clothing and PPE for inserting and removing low-voltage power circuit breakers is PPE Category 4 and includes the following:

- an arc-rated clothing system, which includes an arc-rated flash suit, pants, and hood with a minimum arc rating of 40 cal/cm^2
- a Class E or G hard hat
- an arc-rated hard hat liner, as required
- UV-rated safety glasses or goggles
- ear canal inserts for hearing protection
- arc-rated gloves or Class 2 rubber insulating gloves and leather protectors
- heavy-duty leather work shoes

The total arc-rating of the clothing system may be achieved or increased by layering arc-rated clothing, such as arc-rated daily workwear coveralls or shirt and pants, and wearing arc-rated underclothing. Wearing flammable, nonmelting underlayers, such as cotton or wool t-shirts and shorts, can decrease the heat that may reach the body and decrease the likelihood of a burn. However, flammable, nonmelting clothing worn under arc-rated clothing and PPE does not increase the arc rating of the clothing system.

Eye protection must be UV-rated to comply with OSHA 1910.133(a)(1), which states, *"The employer shall ensure that each affected employee uses appropriate eye or face protection when exposed to eye or face hazards from flying particles, molten metal, liquid chemicals, acids or caustic liquids, chemical gases or vapors, or potentially injurious light radiation."*

Removing and Inserting Medium-Voltage Drawout-Type Circuit Breakers

The general guidelines for racking medium-voltage drawout-type circuit breakers are similar to the guidelines for racking low-voltage power circuit breakers. However, there are a few differences. Some of the primary differences include the following:

- Floor trippers are used to open the circuit breaker and discharge the closing springs when the circuit breaker is racked into or out of its cubicle. **See Figure 10-9.** A floor tripper should trip the circuit breaker before the bus connections (stabs) touch when the circuit breaker is racked in. It should also trip the circuit breaker before the bus connections separate when the circuit breaker is racked out.

Floor Trippers

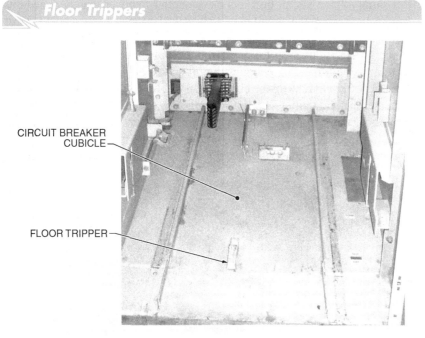

Shermco Industries

Figure 10-9. Floor trippers open the circuit breaker and discharge the closing springs when the circuit breaker is being racked into or out of its cubicle.

- On some medium-voltage vacuum circuit breakers there are two racking screws that rack the circuit breaker in or out. **See Figure 10-10.** If these racking screws are not cleaned and lubricated as part of a regular maintenance program, the circuit breaker will often bind during racking. This binding is caused by the circuit breaker not racking into the cubicle squarely. The racking screws will not push the circuit breaker forward evenly, moving one side ahead of the other. When the racking screws are properly maintained, there are fewer problems when racking the circuit breaker.

Racking Screws

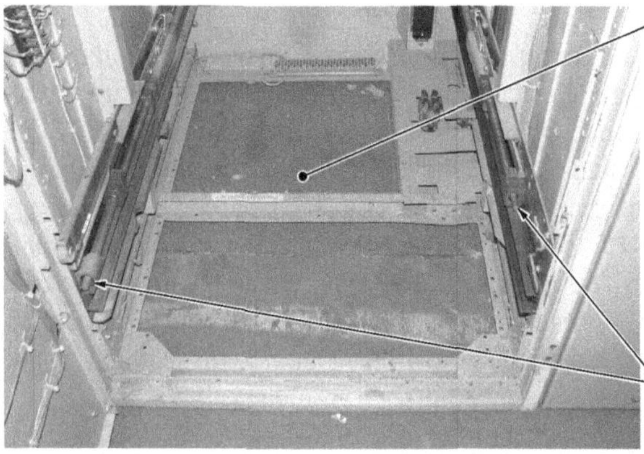

Shermco Industries

Figure 10-10. The racking screws on medium-voltage vacuum circuit breakers must be properly maintained to prevent the circuit breaker from binding when it is racked in or out.

- Medium-voltage circuit breakers use a shutter to cover the energized bus connections when the circuit breaker is racked out. **See Figure 10-11.** Most companies do not allow any work to be performed within the cubicle, even if the shutter is covering the energized bus connections. This is due to the extremely high incident energies that may be generated if there is an arc. Do not attempt to remove the circuit breaker from its cubicle or perform any other tasks if the shutter does not cover the bus connections when the circuit breaker is racked out. Rack the circuit breaker back into place. Under no circumstances should any attempt be made to fix the problem until the entire bus can be deenergized and placed in an electrically safe work condition.

Bus Connection Shutters

UNCOVERED BUS CONNECTIONS — EXPOSED ENERGIZED BUS CONNECTIONS

COVERED BUS CONNECTIONS — SHUTTER

Shermco Industries

Figure 10-11. A shutter is used to cover the energized bus connections on medium-voltage circuit breakers and to protect employees from an arc flash.

Some modern low-voltage drawout-type circuit breakers are also equipped with shutter mechanisms and the same precautions must be taken when racking them out as well. Operating these circuit breakers when the shutter mechanism is defective or inoperable violates the instructions provided by the manufacturer and constitutes an unsafe act.

Other components to consider when racking in medium-voltage circuit breakers are the stab finger cluster (connector) assemblies, sometimes referred to as tulips, which connect the circuit breaker to the main bus. Although these clusters are lubricated at the factory, they require periodic relubrication. If installed without lubricant, they sometimes tilt and then bind when sliding onto the bus. **See Figure 10-12.** If they are not lubricated, they will not slide onto the bus and can break apart. This can cause an arc flash at the bus-to-circuit breaker connection. The shock approach boundaries for medium-voltage drawout-type circuit breakers (ranging from 2.3 kV to 15 kV) include the Limited Approach Boundary at 60″ and the Restricted Approach Boundary at 26″.

The Arc Flash Boundary can be several feet. Most medium-voltage circuit breakers in this voltage range tend to have incident energies of about 30 cal/cm^2 to 40 cal/cm^2. Some medium-voltage circuit breakers can have extremely high incident energies, such as circuit breakers feeding arc-furnace transformers or other high-demand loads.

Stab Finger Clusters

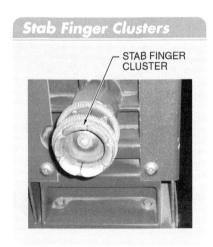

Shermco Industries

Figure 10-12. Stab finger clusters must be lubricated when a circuit breaker is connected onto a bus.

Chapter 10 – Guidelines for Common Electrical Tasks

Shermco Industries
This medium-voltage circuit breaker was severely damaged when the finger cluster broke apart due to lack of lubrication after it was inserted into the cubicle. The circuit breaker was damaged so severely it could not be repaired.

Recommended PPE for Inserting and Removing Medium-Voltage Drawout-Type Circuit Breakers

The recommended PPE for inserting and removing medium-voltage drawout-type circuit breakers is the same as for racking low-voltage circuit breakers. However, some medium-voltage systems can have incident energies well above those found at the 480 V level and there may not be limits for short-circuit current or operating time of the OCPD per any notes. In this situation, more protection is better. Per Table 130.7(C)(15)(A)(b), the recommended PPE category for this task is PPE Category 4. The recommended PPE for inserting and removing medium-voltage drawout-type circuit breakers includes the following:

- an arc-rated clothing system, which includes an arc-rated flash suit, pants, and hood with a minimum arc rating of 40 cal/cm^2
- a Class E or G hard hat
- an arc-rated hard hat liner, if needed
- UV-rated safety glasses or goggles
- ear canal inserts for hearing protection
- arc-rated gloves or Class 2 rubber insulating gloves and leather protectors
- heavy-duty leather work shoes

OPERATING MEDIUM-VOLTAGE AIR-BREAK SWITCHES

A *medium-voltage air-break switch* is a switch that uses air as the interrupting medium. Medium-voltage air-break switches are often used to feed a power transformer and usually contain a main contact assembly, an arcing contact assembly, arc chutes, and current-limiting fuses.

Due to their design, medium-voltage air-break switches can present unique hazards when they are operated. One of the most common problems with these switches is a lack of lubrication. As current passes through the contact assembly, heat is generated due to the copper losses (I^2R), which then dries out the lubricant in the contact pivot point. As the lubricant dries out, it becomes thick and gummy, eventually flaking off and exposing the pivot point to metal-to-metal wear. The amount of heat increases as larger loads are put on a switch, which causes the lubricant to deteriorate faster.

This creates a hazardous situation where the main contacts are open but the phase unit is still energized because the arcing contact remains closed. An employee who does not carefully apply safe work procedures, which would include live-dead-live testing for the absence of voltage, could easily miss this and be electrocuted. A viewing/inspection window allows the contacts to be inspected and visually verified as being open. **See Figure 10-13.** However, these windows can become discolored with weathering, age, and usage. The windows can also be painted over by employees who are unaware of their safety purposes.

Medium-voltage air-break switches have two contacts per phase. The larger contact is the main contact. The main contact carries the load current and is constructed of an alloy that consists primarily of copper and silver. The smaller contact is known as the arcing contact. An *arcing contact* is a thin, blade-like contact that is designed to interrupt an arc when the switch is opened or closed. The arcing contact moves in and out of the arc chute, which contains, stretches, cools, and extinguishes the arc. **See Figure 10-14.** Arcing contacts are often made of an alloy such as copper with zinc or tungsten added to withstand the heat of an electric arc.

Shermco Industries

Figure 10-13. A viewing/inspection window is used to visually verify that medium-voltage air-break switches are open.

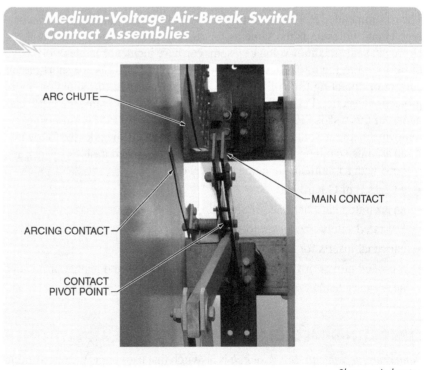

Shermco Industries

Figure 10-14. The arcing contact is designed to interrupt an arc when opening or closing the switch.

The arcing contact has a spring connected to it that gives it extra force when opening. The main contact carries the load while the switch is closed. As the main contact opens, the load is transferred to the arcing contact. After the arc has been interrupted and cleared, both the arcing contact and main contact are open. **See Figure 10-15.**

Arc Interruption Sequence for Medium-Voltage Air-Break Switches

CLOSED   INTERRUPTION (NO ARCING ON MAIN BLADE)   OPEN

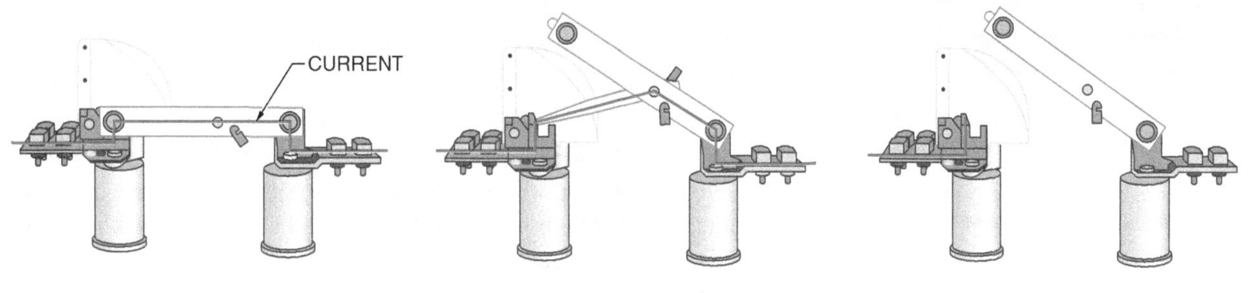

Figure 10-15. As the main contact of a circuit breaker opens, the load is transferred to the arcing contact.

Due to the location of the operating handle on most of these switches, it is difficult for an employee to stand to the hinge side when operating them. **See Figure 10-16.** This exposes the employee to greatly increased arc flash and blast hazards.

The shock approach boundaries for medium-voltage air-break switches include the Limited Approach Boundary at 60″ and the Restricted Approach Boundary at 26″. The Arc Flash Boundary can be several feet, although often times medium-voltage air-break switches will have lower incident energies available when they are equipped with current-limiting fuses.

Recommended PPE for Operating Medium-Voltage Air-Break Switches—Door Open or Closed

There is not a specific equipment category for medium-voltage air-break switches, but the hazards and risks involved with operating them would be similar to the task category "Normal operation of a circuit breaker (CB), switch, contactor, or starter." Per Table 130.7(C)(15)(A)(a), arc-rated clothing and PPE are not mandated if the equipment condition meets all of the following requirements:
- Equipment is properly installed.
- Equipment is properly maintained.
- All covers and doors are in place and secured.
- There is no evidence of impending failure.

When operating medium-voltage equipment, including circuit breakers, switches, and starters, PPE Category 2 arc-rated clothing and PPE should be worn, even though it is not expressly required by Table 130.7(C)(15)(A)(a). This is due to the increased risk involved with operating medium-voltage equipment.

Shermco Industries

Figure 10-16. Sometimes the location of the operating handle for a medium-voltage air-break switch prevents the employee from standing to the hinge side of the door where it is safer.

Shermco Industries
Due to their design, medium-voltage air-break switches can present unique hazards. As illustrated, the B-phase arcing contact is still connected even though the main contacts are not.

The recommended arc-rated clothing and PPE for operating a medium-voltage air-break switch when the door is closed includes the following:
- an arc-rated clothing system, which includes an arc-rated flash suit, pants, and hood with a minimum rating of 8 cal/cm^2
- a Class E or G hard hat
- an arc-rated balaclava with a minimum arc raiting of 8 cal/cm^2
- an arc-rated face shield with a minimum rating of 8 cal/cm^2 or an arc-rated hood
- UV-rated safety glasses or goggles
- ear canal inserts for hearing protection
- arc-rated gloves or Class 0 or 00 rubber insulating gloves and leather protectors
- heavy-duty leather work shoes

If all of the conditions above are not met, or if the qualified person is not certain they have all been met, arc-rated clothing and PPE are mandated by Table 130.7(C)(15)(A)(a). Table 130.7(C)(15)(A)(b) is referred to for the equipment category, which would be "Metal-clad switchgear, 1 kV through 15 kV." This equipment category requires PPE Category 4, which includes the following:
- an arc-rated flash suit and hood with a minimum arc rating of 40 cal/cm^2
- UV-rated safety glasses or goggles
- a Class E or G hard hat
- Class 2 rubber insulating gloves and leather protectors or arc-rated gloves with a minimum arc rating of 40 cal/cm^2
- heavy-duty leather work shoes
- ear canal inserts

FIELD NOTES:
ARC-RESISTANT SWITCHGEAR

Arc-resistant switchgear is used as an alternative to regular NEMA-rated switchgear. Arc-resistant switchgear is much more substantial in its construction and is designed to divert the arc flash and blast upwards through the top of the switchgear, where it can be safely exhausted. When the doors are properly closed and secured, no arc flash PPE or clothing is required to rack circuit breakers in or out of their cubicles. NFPA 70E Table 130.7(C)(15)(A)(a) also states that if the doors are opened, PPE Category 4 arc flash clothing and PPE are required. Arc-resistant switchgear can cost 15% to 20% more than standard switchgear, but it may be worthwhile to a company since it eliminates the PPE needed for operating this equipment and greatly reduces the risk for some of the more hazardous equipment and tasks. Table 130.7(C)(15)(A)(a) specifies that arc-resistant switchgear has to be in compliance with IEEE C37.20.7, *IEEE Guide for Testing Metal-Enclosed Switchgear Rated up to 38 kV for Internal Arcing Faults,* Type 1 or Type 2.

SAFETY FACT

Type 1 switchgear has an arc-resistant front only. Type 2 switchgear is arc-resistant on the sides, rear, and front.

INSERTING AND REMOVING MOTOR CONTROL CENTER BUCKETS

A *motor control center (MCC) bucket* is an assembly that connects to a main bus assembly and contains the necessary components to operate a load, usually a motor. An MCC may contain several buckets or dozens of buckets, depending on the size and processes of the facility. **See Figure 10-17.**

Shermco Industries

Figure 10-17. MCC buckets do not use an arc chute to extinguish an arc when connecting to a bus.

Similar to the task of racking drawout-type circuit breakers, MCC buckets make and break a load-carrying contact. MCC buckets also do not have an arc chute at the bus connections. Since the bucket is connected to the main bus, there is a large amount of short-circuit current available and the arc will be substantial.

An MCC presents two levels of hazards. One hazard level is for work inside the bucket. The other hazard level occurs when a worker interacts with the main bus for the MCC.

Per NFPA 70E, tasks associated with MCCs in Tables 130.7(C)(15)(A)(a) and (b) are divided into two groups based on the level of hazard. Tasks performed inside the bucket use maximum limits of 65 kA short-circuit current available and 0.03 sec operating time. When tasks involve the hazards and risks presented by the main bus, such as removing and inserting the bucket from the main bus or removing and installing bolted panels, then the maximums limits for using the table are 42 kA short-circuit current available and 0.33 sec operating time. No manufacturer currently recommends removing or inserting MCC buckets while the MCC is energized. Even though it is included in the tasks listed in Table 130.7(C)(15)(A)(a), the author does not recommend performing this task while the MCC is energized.

Workers should be aware that if the line-side of the circuit breaker in the MCC bucket is not guarded, the bucket cannot be placed in an electrically safe work condition because there are still exposed energized conductors or circuit parts inside the bucket. Most older MCCs have exposed line-side connectors. If there is a guard that completely covers the line-side terminals, the hazard in the MCC bucket will appear on the load-side terminals, which will have a much lower available short-circuit current. Plastic guards are also sometimes used to cover unused bus sections to reduce the possibility of an arc flash when the bucket is not connected. **See Figure 10-18.**

Shermco Industries

Figure 10-18. Plastic guards can be used to cover unused bus sections to reduce the possibility of an arc flash when an MCC bucket is not connected.

The safest method to remove or install MCC buckets is to deenergize the entire unit. Sometimes it is not possible to deenergize the entire unit due to operational limitations. If this is the case, there are some general guidelines to follow. Since the design of MCCs vary from one manufacturer to another, these guidelines may be incomplete or may not apply at all. The instructions provided by the manufacturer for the specific MCC should always be followed. It is important to assess the condition of the equipment and circuit as well as the applicability of the following steps:

- Verify that the correct MCC bucket is being serviced. It is easy to mistake one MCC bucket for another, especially in low-voltage MCCs. NFPA 70E 130.7(E)(4) includes a specific warning concerning look-alike equipment that requires that one of the alerting techniques in 130.7(E)(1), (2), or (3) be used if there is a possibility of confusion between equipment.
- Deenergize the load before operating the MCC circuit breaker.
- Move the circuit breaker handle to the OFF position while standing to the hinge side of the enclosure. If wearing an arc-rated face shield with a balaclava, face the hazard and do not look away from the equipment.

- Visually confirm that the MCC bucket is off-line and that the equipment is not operating.
- Follow instructions from the manufacturer for the removal of the bucket from the MCC. There is typically some type of interlock that must be released before the bucket can be removed.
- Some manufacturers state that the MCC bucket is not to be removed while the MCC is energized. If this is the case, do not violate the manufacturer's procedures.
- When removing or installing an MCC bucket, the available incident energy is from the main bus of the MCC, not the bucket. Properly rated arc flash protective equipment and clothing must be worn in accordance with company procedures, a valid arc flash hazard analysis, or the table method.

The shock approach boundaries for removing or installing a 600 V MCC bucket include the Limited Approach Boundary at 42″ and the Restricted Approach Boundary at 12″.

Recommended PPE for Inserting and Removing MCC Buckets

Per Tables 130.7(C)(15)(A)(a) and (b), the recommended arc-rated clothing and PPE category for removing or installing an MCC bucket is PPE Category 4. Table 130.7(C)(15)(A)(a) requires arc-rated clothing and PPE regardless of the equipment condition. The PPE required for removing or installing an MCC bucket includes the following:

- an arc-rated clothing system, which includes an arc-rated flash suit, pants, and hood with a minimum rating of 40 cal/cm^2
- a Class E or G hard hat
- an arc-rated hard hat liner, as required
- UV-rated safety glasses or goggles
- ear canal inserts for hearing protection
- arc-rated gloves or Class 2 rubber insulating gloves and leather protectors
- heavy-duty leather work shoes

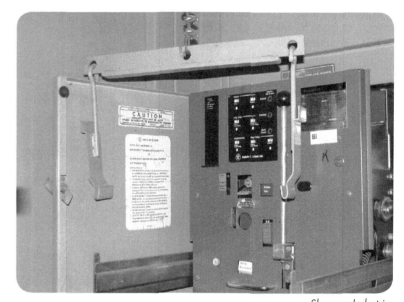

Shermco Industries
A spreader bar is used together with a cable and pulley mechanism to remove or install heavy electrical equipment without damaging it.

TROUBLESHOOTING CIRCUITS RATED 120 V AND LESS

When troubleshooting electrical circuits that are 120 V and less, employees must understand the function of every part of the circuit. More attention is required when working with low-voltage circuits, as the dangers are often downplayed or ignored. The arc flash risk is very low at this voltage range, but a significant shock risk still exists. If contact with an energized conductor or circuit part is possible, rubber insulating gloves and leather protectors are required.

Troubleshooting 120 V Circuits

Shermco Industries

Figure 10-19. CPTs are used in some MCC buckets to reduce voltage from 480 V to 120 V.

Some MCC buckets use a control power transformer (CPT) to reduce the voltage from 480 V to 120 V. Since there is 480 V inside the bucket, the shock hazard would be considered at the 480 V level, not the 120 V level. When there are multiple voltages inside a single enclosure, the highest exposed voltage is used to determine the shock hazard. **See Figure 10-19.**

For the task of troubleshooting a 120 V control circuit, the risk assessment would have to consider that the line-side of the disconnecting switch is guarded, but the line-side (top) of the fuses is exposed while the switch is in the CLOSED position. **See Figure 10-20.** Temporary rubber insulating shielding could be installed at multiple locations to guard the exposed energized elements. Although this would be necessary since the voltage is not reduced to 120 V until it reaches the load-side of the CPT, it is impractical. The risk assessment would be at the 480 V level.

Terminal Guards

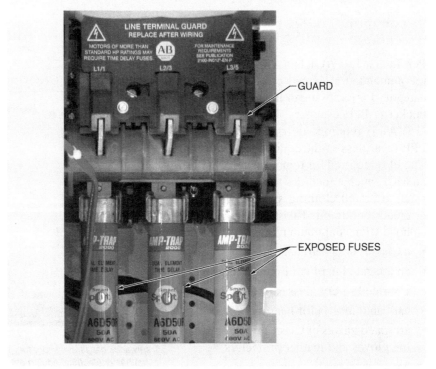

Shermco Industries

Figure 10-20. Even though the line-side of the MCC bucket terminal can be covered with a guard, the line-side of the fuse is still exposed when the switch is closed.

The incident energy would be calculated using the short-circuit current available at the main bus and the operating time of the OCPD protecting the main bus. This OCPD usually has a short-time delay function, which can delay tripping up to 0.5 sec (30 cycles). The shock approach boundaries for troubleshooting while exposed to 480 V include the Limited Approach Boundary at 42″ and the Restricted Approach Boundary at 12″.

Recommended PPE for Troubleshooting Circuits Rated 120 V and Less with Exposure to Greater than 120 V

Per Tables 130.7(C)(15)(A)(a) and (b), the recommended arc-rated clothing and PPE for tasks involving work on 120 V control circuits when there is exposure to 480 V is PPE Category 2. The PPE recommended at this level includes the following:

- an arc-rated clothing system, which includes an arc-rated shirt and pants or a coverall with a minimum rating of 8 cal/cm^2
- an arc-rated face shield and balaclava with a minimum rating of 8 cal/cm^2 or an arc-rated hood
- a Class E or G hard hat
- an arc-rated hard hat liner, as required
- UV-rated safety glasses or goggles
- ear canal inserts for hearing protection
- arc-rated gloves or Class 0 or Class 00 rubber insulating gloves and leather protectors (Since there is an exposure to 480 V for this task inside an MCC bucket, rubber insulating gloves are required and leather protectors must be worn over them. Rubber insulating gloves may not be needed when troubleshooting a lighting panel or other electrical equipment that only has an exposure up to 120 V, but leather gloves may still be required to protect the hands from the arc flash hazard.)
- heavy-duty leather work shoes

Recommended PPE for Troubleshooting Circuits Rated 120 V and Less with No Exposure to Greater than 120 V

Per Tables 130.7(C)(15)(A)(a) and (b), arc-rated clothing and PPE are not mandated for tasks involving work on 120 V control circuits when there is no exposure to voltages greater than 120 V inside an MCC bucket. However, since NFPA 70E 130.7(C)(12) prohibits the wearing of clothing that could melt, the PPE recommended at this level of hazard includes the following:

- a flammable, nonmelting long-sleeved shirt and long pants or a coverall (cotton, wool, or silk)
- UV-rated safety glasses or goggles
- ear canal inserts for hearing protection
- leather gloves, if there is a possibility of an arc flash

Recommended PPE for Troubleshooting Circuits Rated 240 V and Less

For tasks such as voltage testing or troubleshooting inside panelboards or other equipment rated 240 V or less, the recommended arc flash clothing and PPE category per Tables 130.7(C)(15)(A)(a) and (b) is PPE Category 1. This includes equipment connected to the panelboard or utilization equipment such as lighting, motors, or other devices. The PPE required for voltage testing or troubleshooting conductors or circuit parts rated 240 V and less includes the following:

- an arc-rated clothing system, which can include an arc-rated shirt and pants or a coverall with a minimum rating of 4 cal/cm^2
- an arc-rated face shield with a minimum rating of 4 cal/cm^2

- an arc-rated hood (minimum rating of 4 cal/cm^2) could be worn instead of the arc-rated face shield
- a Class E or G hard hat
- UV-rated safety glasses or goggles
- ear canal inserts for hearing protection
- arc-rated gloves or Class 0 or Class 00 rubber insulating gloves and leather protectors
- leather work shoes, as needed

TROUBLESHOOTING CIRCUITS RATED GREATER THAN 240 V AND UP TO 600 V

Troubleshooting is performed with the circuits energized. The shock approach boundaries for troubleshooting circuits greater than 240 V and up to 600 V include the Limited Approach Boundary at 42″ and the Restricted Approach Boundary at 12″.

The Arc Flash Boundary is calculated using the highest short-circuit current available exposure within the enclosure. The operating time is determined from the OCPD supplying power to the equipment being troubleshot.

Recommended PPE for Troubleshooting Circuits Rated Greater than 240 V and Up to 600 V

Per the task located in the equipment category "Panelboards or other equipment rated > 240 V and up to 600 V," found in Table 130.7(C)(15)(A)(b), the recommended arc-rated clothing and PPE category for voltage testing and troubleshooting is PPE Category 2. The PPE required for voltage testing and troubleshooting at this level of hazard includes the following:
- an arc-rated clothing system, which includes an arc-rated shirt and pants or a coverall with a minimum rating of 8 cal/cm^2
- an arc-rated face shield and balaclava with a minimum rating of 8 cal/cm^2 or an arc-rated hood
- a Class E or G hard hat
- an arc-rated hard hat liner, as required
- UV-rated safety glasses or goggles
- ear canal inserts for hearing protection
- arc-rated gloves or Class 0 or Class 00 rubber insulating gloves and leather protectors
- heavy-duty leather work shoes

TROUBLESHOOTING AC DRIVES

Large AC drives present unique concerns due to the combination of low-energy control circuits, high-energy electronic components such as capacitors and power supplies, and high-energy power components such as primary motor power circuits. For example, a circuit board containing AC drive controls has a low hazard from shock or arc flash, but there may be high-energy primary circuits operating at 480 V in close proximity to it. **See Figure 10-21.**

Since the AC drive has low-energy control circuits located in close proximity to high-energy circuits, employees are required to follow one of the two following rules:
- Wear PPE rated for the highest exposure in the enclosure.
- Guard the main power circuit of the motor, which is usually the line side and load side of the circuit breaker, and any other exposed energized conductors or circuit parts operating at the higher voltage/energy level.

AC Drive Controls

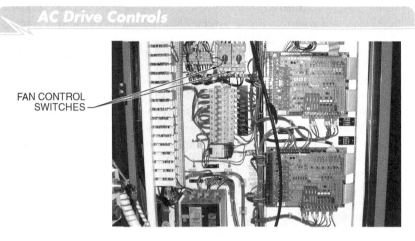

Shermco Industries

Figure 10-21. Large AC drives present unique concerns due to the combination of low-energy control circuits, high-energy electronic components, and primary motor circuits.

The high-energy electronic components can present a serious electric shock hazard. However, it is not practical to use Class 2 or 4 rubber insulating gloves and leather protectors when troubleshooting these circuits. Test equipment designed for these tasks have high-voltage test probes adequate for the troubleshooting. Leather gloves may be needed if there is an arc flash hazard present. The arc energy from such components is usually fairly low, while the shock hazard can be high. The reasons for choosing the specific PPE should be documented when assessing the hazards.

Troubleshooting Low-Voltage AC Drives

Low-voltage AC drives should be evaluated as any other equipment operating at its voltage and available short-circuit current levels. The shock approach boundaries for low-voltage AC drives include the Limited Approach Boundary at 42″ and the Restricted Approach Boundary at 12″.

Recommended PPE for Troubleshooting Low-Voltage AC Drives

There is no specific task category for AC drives in the new tables. The tasks would be under various categories, such as "Normal operation of a circuit breaker (CB), switch, contactor, or starter" and "Work on control circuits with exposed energized electrical conductors and circuit parts, greater than 120 V" in Table 130.7(C)(15)(A)(a). It is necessary to determine the appropriate equipment type in Table 130.7(C)(15)(A)(b). The closest equipment type is "Other 600 V-class equipment."

When working with low-voltage AC drives, the following should be considered:
- Some printed circuit boards (PCBs) are not only in close proximity to power circuits, but they are directly connected to power components and relatively high voltage. For instance, gate-driver boards are typically directly wired to the insulated gate bipolar transistors (IGBTs) in the inverter (output) sections of AC drives. In addition, connections to DC buses are usually made to PCBs for voltage sensing. Fortunately, the gate-driver boards are typically not as accessible as control boards.
- With regard to the relatively high voltage of the DC bus, the bus bars are sometimes insulated with removable covers. These covers should remain in place during troubleshooting. There are typically other test points that can be used for directly measuring DC bus voltage with a suitable digital multimeter (DMM).
- Although low-voltage AC drives are not treated differently than other low-voltage AC apparatuses with regard to short-circuit current, very large AC drives store a tremendous amount of energy in DC bus capacitors. A short circuit or touch incident directly across large DC bus bars can be fatal. This makes it especially important for a technician to assess the risk properly and use a higher PPE category than mandated if he or she feels more comfortable doing so.
- When repairing AC drives with failed insulated gate bipolar transistors (IGBTs) in the output (inverter) sections, technicians can often use manufacturer's procedures for either bench testing the electronic gate signals or testing them without DC power applied. Improper switching of the IGBTs can result in catastrophic failure with arc flash, flying shrapnel, and bent DC bus bars. This is a good example of an instance where a qualified person would require specific safety and procedure training for the task. This type of procedure is typically only taught in factory training classes.

As long as the available short circuit current and OCPD operating time do not exceed the maximum values given, the table method can be used. Table 130.7(C)(15)(A)(b) recommends PPE Category 2 arc-rated clothing and PPE for tasks involving this equipment. The recommended PPE for low-voltage AC drives includes the following:
- an arc-rated clothing system, which includes an arc-rated shirt and pants or a coverall with a minimum rating of 8 cal/cm^2
- an arc-rated face shield and balaclava with a minimum rating of 8 cal/cm^2 or an arc-rated hood
- a Class E or G hard hat
- arc-rated hard hat liner, as required
- UV-rated safety glasses or goggles
- ear canal inserts for hearing protection
- arc-rated gloves or Class 0 or Class 00 rubber insulating gloves and leather protectors
- heavy-duty leather work shoes

Troubleshooting Medium-Voltage AC Drives

Medium-voltage AC drives are usually divided into a low-voltage control section and a medium-voltage power section. The medium-voltage power section is secured by a trapped-key interlock system. **See Figure 10-22.** To access the

Shermco Industries

Figure 10-22. A trapped-key interlock system is used to secure the medium-voltage section of an AC drive.

medium-voltage section, the main circuit breaker must be turned off, which releases a key. That key is then used to release a second key, which unlocks the medium-voltage compartment as well as any other compartments. Without both keys, the compartment door will not release.

The shock approach boundaries for medium-voltage AC drives include the Limited Approach Boundary at 60″ and the Restricted Approach Boundary at 26″. The Arc Flash Boundary must be calculated for medium-voltage AC drives that have unguarded medium-voltage circuits. Energized work is not recommended on medium-voltage circuits due to the close proximity of the enclosure and other parts.

Recommended PPE for Troubleshooting Medium-Voltage AC Drives

If medium-voltage circuits are not guarded by an enclosure and are exposed, the PPE category recommended is PPE Category 4 for tasks such as absence-of-voltage testing and grounding. However, since the medium-voltage circuits are usually guarded by an enclosure and not accessible, troubleshooting the low-voltage control circuits is a PPE Category 2 task. This is similar to the task "Work on control circuits with exposed energized electrical conductors and circuit parts, greater than 120 V." The recommended PPE for troubleshooting the low-voltage control circuits in medium-voltage AC drives includes the following:

- an arc-rated clothing system, which includes an arc-rated shirt and pants or a coverall with a minimum rating of 8 cal/cm^2
- an arc-rated face shield and balaclava with a minimum rating of 8 cal/cm^2 or an arc-rated hood
- a Class E or G hard hat
- an arc-rated hard hat liner, as required
- UV-rated safety glasses or goggles
- ear canal inserts for hearing protection
- arc-rated gloves or Class 0 or Class 00 rubber insulating gloves and leather protectors
- heavy-duty leather work shoes

TESTING FOR THE ABSENCE OF VOLTAGE

Circuits rated greater than 240 V up to 600 V require arc-rated clothing and PPE that is PPE Category 2. Circuits rated 120 V or less that do not have an exposure to voltages greater than 120 V do not have arc-rated clothing and PPE mandated. Medium-voltage circuits rated from 2.3 kV to 15 kV require PPE Category 4.

Employees should always be aware of their working distance when performing tasks while exposed to energized conductors or circuit parts. Low-voltage electrical systems typically have a working distance of 18″. Medium-voltage electrical systems typically have a working distance of 36″. The working distance can vary, depending on the actual available clearances. Employees must be certain to verify the working distance before beginning any task.

Recommended PPE for Absence-of-Voltage Testing

The PPE category for testing low-voltage control circuits in medium-voltage equipment depends on the specific equipment configuration. If there is no

exposure to medium-voltage conductors or circuit parts, the task would most likely require PPE Category 2. The recommended PPE for PPE Category 2 tasks includes the following:
- an arc-rated clothing system, which includes an arc-rated shirt and pants or a coverall with a minimum rating of 8 cal/cm^2
- an arc-rated face shield and balaclava with a minimum rating of 8 cal/cm^2 or an arc-rated hood
- a Class E or G hard hat
- arc-rated hard hat liner, as required
- UV-rated safety glasses or goggles
- ear canal inserts for hearing protection
- arc-rated gloves or Class 0 or Class 00 rubber insulating gloves and leather protectors
- heavy-duty leather work shoes

If there is exposure to medium-voltage conductors or circuit parts, the recommended arc-rated clothing and PPE category is PPE Category 4. The recommended PPE for PPE Category 4 tasks includes the following:
- an arc-rated clothing system, which includes an arc-rated flash suit, pants, and hood with a minimum rating of 40 cal/cm^2
- a Class E or G hard hat
- arc-rated hard hat liner, as required
- UV-rated safety glasses or goggles
- ear canal inserts for hearing protection
- arc-rated gloves or Class 2 rubber insulating gloves and leather protectors
- heavy-duty leather work shoes

For the equipment category "NEMA E2 (fused contactor) motor starters, 2.3 kV through 7.2 kV," the recommended arc-rated clothing and PPE category is PPE Category 2 if the circuits or parts are rated greater than 120 V. The recommended PPE for PPE Category 2 includes the following:
- an arc-rated clothing system, which includes an arc-rated shirt and pants or a coverall with a minimum rating of 8 cal/cm^2
- an arc-rated face shield and balaclava with a minimum rating of 8 cal/cm^2 or an arc-rated hood
- a Class E or G hard hat
- arc-rated hard hat liner, as required
- UV-rated safety glasses or goggles
- ear canal inserts for hearing protection
- arc-rated gloves or Class 0 or Class 00 rubber insulating gloves and leather protectors
- heavy-duty leather work shoes

All tasks that expose an employee to medium-voltage circuits are rated PPE Category 4. The recommended PPE for PPE Category 4 tasks includes the following:
- an arc-rated clothing system, which includes an arc-rated flash suit, pants, and hood with a minimum rating of 40 cal/cm^2
- a Class E or G hard hat

- arc-rated hard hat liner, as required
- UV-rated safety glasses or goggles
- ear canal inserts for hearing protection
- arc-rated gloves or Class 2 rubber insulating gloves and leather protectors
- heavy-duty leather work shoes

Absence-of-voltage testing inside panelboards rated 240 V or below or on utilization equipment fed by those panelboards is a PPE Category 1 task. The recommended PPE for PPE Category 1 tasks includes the following:

- an arc-rated clothing system, which can include an arc-rated shirt and pants or a coverall with a minimum rating of 4 cal/cm^2
- an arc-rated face shield with a minimum rating of 4 cal/cm^2
- an arc-rated hood (minimum 4 cal/cm^2) could be worn instead of the arc-rated face shield
- a Class E or G hard hat
- UV-rated safety glasses or goggles
- ear canal inserts for hearing protection
- arc-rated gloves or Class 0 or Class 00 rubber insulating gloves and leather protectors
- leather work shoes, as required

OPERATING EQUIPMENT RATED 240 V AND LESS

Generally, equipment rated 240 V and less is low energy and therefore low hazard. This is true as long as the maximum available short-circuit current and operating time in NFPA 70E Table 130.7(C)(15)(A)(b) are not exceeded. Failure to read and understand these limits puts the employee and others at risk for injury. Some 120 V to 208 V power systems with large ampacities can sustain an arc that causes massive heat and damage to equipment and personnel. The shock approach boundaries for operating equipment rated 240 V and less include the Limited Approach Boundary at 42″ and the Restricted Approach Boundary at 12″.

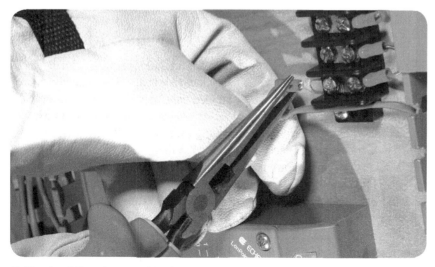

Rubber insulating gloves and leather protectors are required when working on electrical systems 240 V and less if it is likely that contact may occur.

The words "avoid contact" mean that if work is being performed on circuits or components using insulated tools and no physical contact will occur, rubber insulating gloves are not required. If contact could be made or is likely to be made, rubber insulating gloves are required. Both OSHA and NFPA 70E agree that it is unacceptable at any time to make contact with an energized circuit or component with bare, unprotected skin.

When it is calculated, the Arc Flash Boundary for low-voltage, low-energy systems will be small. The calculated Arc Flash Boundary for a 240 V, three-phase system with a 10 kA available short-circuit current protected by a molded-case circuit breaker with an operating time of 0.03 sec (2 cycles) is 6.9". At 208 V, using the same current and operating time, the Arc Flash Boundary is 6.4", which is about the distance from the knuckles of the hand to the cuff of a leather work glove. This is the reason leather work gloves are worn when working on this type of equipment. Even on low-voltage, low-energy circuits, the hands can be considerably injured from the arc flash.

Recommended PPE for Operating Equipment Rated 240 V and Less

Table 130.7(C)(15)(A)(a) shows arc-rated clothing and PPE is not mandated for operating equipment rated 240 V and less. However, there may be circumstances where it may be advisable to wear PPE, especially when operating larger 208 V circuit breakers or switches. Even though the voltage is low, when the continuous current rating of the circuit breaker approaches 800 A, the worker could be at an increased risk. It is always better to wear arc-rated clothing and PPE if there are any concerns. The recommended PPE for operating equipment rated 240 V and less includes the following:

- a flammable, nonmelting long-sleeved shirt and long pants or a coverall (cotton, wool, or silk)
- UV-rated safety glasses or goggles
- ear canal inserts for hearing protection
- leather gloves if there is a possibility of an arc flash
- heavy-duty leather work shoes

OPERATING EQUIPMENT RATED 240 V TO 600 V

Operating panelboards and switchboards rated for 240 V to 600 V with the doors closed or the covers installed is a fairly low-risk task. These panelboards or switchboards are usually protected by molded-case circuit breakers with an available short-circuit current of 25 kA or less. As long as the limits for the maximum available short-circuit current and operating time of the OCPD listed in Table 130.7(C)(15)(A)(b) are not exceeded, arc-rated clothing and PPE are not mandated.

When the door is open or when covers are not installed the PPE category increases to PPE Category 2. The higher voltages involved make an arc more likely, and if there is an arc, it is more likely to continue until an upstream OCPD clears it.

Operating an MCC rated for 600 V or less with the covers on is also listed as not requiring arc-rated clothing or PPE. Many of the larger MCCs are outside the limits given in Table 130.7(C)(15)(A)(b). NFPA 70E states that for tasks not found in Table 130.7(C)(15)(A)(a), or when the maximum available short-circuit current or operating time exceeds the limits given in Table 130.7(C)(15)(A)(b), an incident energy analysis must be performed. The shock approach boundaries for operating equipment 240 V to 600 V include the Limited Approach Boundary at 42″ and the Restricted Approach Boundary at 12″.

Table 130.4(D)(a), column 1, gives the phase-to-phase voltage for these approach distances as 151 V to 750 V. A 277 V circuit or equipment would be 480 V phase-to-phase, which would require the use of rubber insulating gloves with leather protectors.

Recommended PPE for Operating Equipment Rated 240 V to 600 V

Arc-rated clothing and PPE are not mandated for operating equipment rated up to 600 V. There is considerably more arc energy available for equipment rated 240 V to 600 V. The arc flash hazard can be considerable even when the limits for available short-circuit current and OCPD operating time in Table 130.7(C)(15)(A)(b) are not exceeded.

For example, the limits provided for the equipment category "600-V class motor control centers (MCCs)" lists the maximum short-circuit current available as 65 kA, the OCPD operating time as a maximum of 0.033 sec (2 cycles), and the potential arc flash boundary as 5′. The Arc Flash Boundary can be an indication of the available arc energy that may be encountered when working on a particular piece of equipment.

The recommended PPE for operating equipment rated 240 V to 600 V when the doors are open or covers are not installed is PPE Category 2. The recommended PPE for operating equipment 240 V to 600 V includes the following:

- an arc-rated clothing system, which includes an arc-rated shirt and pants or a coverall, with a minimum rating of 8 cal/cm^2
- an arc-rated face shield and balaclava with a minimum rating of 8 cal/cm^2 or an arc-rated hood with a minimum arc rating of 8 cal/cm^2
- a Class E or G hard hat
- UV-rated safety glasses or goggles
- ear canal inserts for hearing protection
- arc-rated gloves or Class 0 or 00 rubber insulating gloves and leather protectors
- heavy-duty leather work shoes, as required

OPERATING NEMA E2 (FUSED CONTACTOR) MOTOR STARTERS — 2.3 kV TO 7.2 kV

NEMA E2 motor starters present an increased hazard for both electric shock and arc flash due to the operating voltages that range from 2300 V to 7200 V. The shock hazard is obvious, but the arc flash hazard at these voltages is very substantial as well. The available short-circuit current can still be high, and the higher voltage gives it a greater ability to arc over to grounded metal.

Per NFPA 70E Table 130.7(C)(15)(A)(a), arc-rated clothing or PPE is not mandated for operating NEMA E2 motor starters when the doors are closed, the equipment is properly installed, the equipment is properly maintained, and there is no evidence of inpending failure. PPE Category 4 is used when doors are open. Past editions of the table method in NFPA 70E allowed the level of PPE to be reduced by 1, 2, or 3 categories based on assumed risk.

The new table method does not allow this. If a worker is exposed to an arc flash hazard, that worker must wear full-rated PPE and clothing based on the incident energy. This makes the new table method easier to apply in the field and provides adequate protection for the arc flash hazard.

If there is any doubt as to whether the equipment is functioning properly, the best course of action is to deenergize the equipment and correct the problem. If the equipment cannot be deenergized, then the employer or management must make a decision based on the residual risk and determine how that risk is to be handled.

At the very least, PPE Category 4 PPE and arc-rated clothing are necessary. Other steps may be required, such as using remote operation, setting up a safe work zone, or using a safety backup, to help reduce the residual risk.

There is another concern with this type of equipment when troubleshooting the 480 V control circuit. When troubleshooting with the motor starter racked in and energized, an employee will be working in a cramped area while wearing arc-rated clothing and PPE, including a face shield, a balaclava and hard hat, and rubber insulating gloves with leather protectors. Often, the CPT is not clearly visible because it is located between the starter and the side of the enclosure. This situation would be considered blind reaching. If the employee trying to troubleshoot the 480 V system inadvertently connects a voltmeter to the 4160 V side of the CPT, the results could be fatal.

Per NFPA 70E 130.6(C)(2), if there is an obstruction or if for any other reason hazards cannot be clearly seen and identified, work cannot be performed. Many motor starters have the CPT mounted to the side of the starter. Because the CPT is only a few inches away from the side of the enclosure, it can be difficult for an employee to see where the low-voltage terminals are located while wearing arc flash protective PPE. Other motor starters may have the CPT mounted to the inside of the enclosure, providing easier access and safer troubleshooting. **See Figure 10-23.** The shock approach boundaries for operating NEMA E2 motor starters include the Limited Approach Boundary at 60″ and the Restricted Approach Boundary at 26″.

CPT Mounting Locations

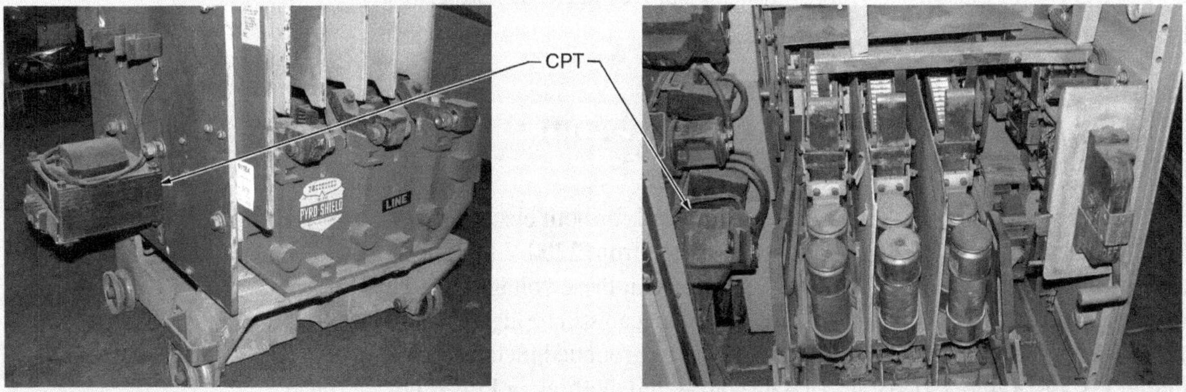

MOTOR STARTER WITH CPT MOUNTED ON THE SIDE OF A STARTER

MOTOR STARTER WITH CPT MOUNTED INSIDE THE ENCLOSURE

Shermco Industries

Figure 10-23. The CPT for motor starters can be mounted to the outside of the starter frame or inside of the enclosure.

Recommended PPE for Operating NEMA E2 (Fused Contactor) Motor Starters — 2.3 kV to 7.2 kV

The task of operating NEMA E2 motor starters that are rated at 2.3 kV to 7.2 kV and have doors closed, equipment properly installed, equipment properly maintained, and no evidence of impending failure is covered under "Normal operation of a circuit breaker (CB), switch, contactor, or starter" in Table 130.7(C)(15)(A)(a) as not requiring arc-rated clothing and PPE. However, due to the higher dielectric stresses imposed on the insulation system, the author considers it prudent to wear PPE Category 2 while operating this equipment. The recommended PPE for operating NEMA E2 motor starters rated at 2.3 kV to 7.2 kV with doors closed includes the following:

- an arc-rated clothing system, which includes an arc-rated shirt and pants or a coverall, with a minimum rating of 8 cal/cm^2
- an arc-rated face shield and balaclava with a minimum rating of 8 cal/cm^2 or an arc-rated hood with a minimum arc rating of 8 cal/cm^2
- a Class E hard hat
- UV-rated safety glasses or goggles
- ear canal inserts for hearing protection
- arc-rated gloves or Class 0 or 00 rubber insulating gloves and leather protectors
- heavy-duty leather work shoes, as required

Operating NEMA E2 motor starters rated at 2.3 kV to 7.2 kV with doors open is listed as a PPE Category 4 task. The recommended PPE for PPE Category 4 tasks includes the following:

- an arc-rated clothing system with a minimum rating of 40 cal/cm^2
- an arc-rated hood with a minimum arc rating of 40 cal/cm^2
- a Class E hard hat
- an arc-rated hard hat liner, as required
- UV-rated safety glasses or goggles
- ear canal inserts for hearing protection
- arc-rated gloves or Class 2 rubber insulating gloves and leather protectors
- heavy-duty leather work shoes

SAFETY FACT

The National Electrical Manufacturers Association (NEMA) was created in 1926 and has had a hand in creating thousands of electrical standards that contribute to making safe electrical products. NEMA motor starters are conservatively rated compared to the rating system of the International Electrotechnical Commission (IEC). Because of this, NEMA motor starters are favored by most employers despite being significantly more expensive than IEC motor starters.

REMOVING COVERS AND PANELS FROM ELECTRICAL ENCLOSURES

This task of removing covers from electrical enclosures is complicated by the fact that there often are no handles on the covers. The covers are sometimes heavy and awkward to handle when only one employee is removing them. There is also a risk that the employee may drop the cover into an energized bus.

The best solutions for safely removing covers and panels from a switchgear enclosure include the following:
- Use two people to remove and install enclosure covers.
- Install permanent handles. If permanent handles cannot be installed, there are temporary handles that use suction cups to attach to the panel.
- If possible, deenergize the equipment inside the enclosure that will have its cover removed.

Recommended PPE for Removing Covers and Panels from Electrical Enclosures

Table 130.7(C)(15)(A)(a) shows the task of removing covers and panels from electrical enclosures as always presenting an arc flash hazard. This is due to the high risk of the cover or tools making contact with the exposed energized bus. Covers and panels are awkward to handle and can slip from a worker's grip. Permanent handles that aid in removing covers and panels reduce the risk. The appropriate level of arc-rated clothing and PPE is selected from Table 130.7(C)(15)(A)(b) based on the equipment involved.

REPLACING LIGHT BALLASTS

A *light ballast* is a small transformer that takes 120 V or 277 V and steps it up to several thousand volts. This high voltage is needed to ionize the gas inside standard fluorescent light tubes, causing them to glow. Over time, light ballasts may burn out and require replacement.

One of the changes made to the 2015 edition of NFPA 70E was in Table 130.4(C)(a), which is now Table 130.4(D)(a). Due to the low risk presented by 120 V single-phase circuits, the committee added a voltage range of 50 V to 150 V to accommodate work on these single-phase circuits. The Limited Approach Boundary remains at 42″, but the Restricted Approach Boundary now indicates to "Avoid Contact." Replacing a light ballast is the type of task the committee had in mind for this change. The risk of contact is low and it is often inconvenient to deenergize an entire room to change out one ballast, especially in an interior room that has no ambient lighting or when work is being performed in the room.

Working off a ladder using light stands presents its own set of risks, so the qualified person must assess which method presents the least risk under the circumstances he or she is working. However, a 277 V single-phase circuit would still be under the 151 V to 750 V phase-to-phase voltage range. The Limited Approach Boundary would be 42″ and the Restricted Approach Boundary would be 12″, which requires the use of Class 0 or 00 rubber insulating gloves and leather protectors.

The arc flash hazard from this type of equipment is very low. With a little advance planning, there should be no reason to change light ballasts while they are energized. Circuit breakers and wall switches can be locked out, the ballast wiring can be connected to quick-disconnect plugs, or the task can be performed after hours.

Recommended PPE for Replacing Light Ballasts

Per Table 130.7(C)(15)(A)(b), under the equipment category "Panelboards or other equipment rated >240 V and up to 600 V," PPE Category of 2 is recommended. The PPE recommended for PPE Category 2 includes the following:

- an arc-rated clothing system, which includes an arc-rated shirt and pants or a coverall with a minimum rating of 8 cal/cm^2
- an arc-rated face shield and balaclava, both with a minimum arc rating of 8 cal/cm^2, or an arc-rated hood with a minimum arc rating of 8 cal/cm^2
- a Class E or G hard hat
- arc-rated hard hat liner, as required
- UV-rated safety glasses or goggles
- ear canal inserts for hearing protection
- arc-rated gloves or Class 0 or 00 rubber insulating gloves and leather protectors
- heavy-duty leather work shoes

For 120 V lighting, the equipment category in Table 130.7(C)(15)(A)(b) is "Panelboards or other equipment rated 240 V and below." In this case, PPE Category 1 is recommended. The recommended PPE for PPE Category 1 includes the following:

- an arc-rated shirt and pants or coveralls with a minimum arc rating of 4 cal/cm^2
- a Class E or G hard hat
- an arc-rated face shield or hood with a minimum arc rating of 4 cal/cm^2
- UV-rated safety glasses or goggles
- leather work shoes
- ear canal inserts

SAFETY FACT

When testing motors for the absence of voltage, always test at the motor. The basic rule is to test at the point of contact.

REPLACING LOW-VOLTAGE MOTORS

Replacing low-voltage motors is usually a low-risk task, but there are safe work practice guidelines that must be followed. General guidelines for replacing low-voltage motors include the following:

- When performing lockout/tagout, shut down the controller in accordance with the company's written procedures.
- Verify that the lockout/tagout was successful by attempting to operate the equipment using its normal operating controls. Be certain to turn the controls back to the OFF position when finished.
- Apply locks and tags as required.
- Remove the terminal cover on the motor. If desired, an absence-of-voltage test can be performed at the MCC or controller. Whether an absence-of-voltage test is performed at that location, a test for the absence of voltage must be performed at the point of contact, which for this task would be the motor.
- If the connections are taped, use a proximity-type voltage tester to test for the presence of voltages.
- If no voltage is detected, wear the proper PPE and cut the taped connections open to provide access.
- Using a direct-contact voltmeter, test phase-to-phase and phase-to-ground voltages on each phase.

The arc flash hazard from low-voltage motors may be low, but large, three-phase motors can have considerable available short-circuit current. The shock approach boundaries for replacing 480 V low-voltage motors include the Limited Approach Boundary at 42″ and the Restricted Approach Boundary at 12″.

Recommended PPE for Replacing Low-Voltage Motors

For panelboards or other equipment rated at 240 V and less, the PPE category is 1. For equipment rated between 240 V and 600 V, the PPE category is 2. The recommended PPE for PPE Category 1 tasks includes the following:
- an arc-rated clothing system, which includes an arc-rated shirt and pants or a coverall with a minimum rating of 4 cal/cm^2
- an arc-rated face shield with a minimum rating of 4 cal/cm^2 or an arc-rated hood
- a Class E or G hard hat
- UV-rated safety glasses or goggles
- ear canal inserts for hearing protection
- arc-rated gloves or Class 0 or 00 rubber insulating gloves and leather protectors
- heavy-duty leather work shoes

The recommended PPE for replacing low-voltage motors rated between 240 V and 600 V is PPE Category 2. The recommended PPE for PPE Category 2 tasks includes the following:
- an arc-rated clothing system, which in-cludes an arc-rated shirt and pants or a coverall with a minimum rating of 8 cal/cm^2
- an arc-rated face shield and balaclava, both with a minimum arc rating of 8 cal/cm^2, or an arc-rated hood
- a Class E or G hard hat
- an arc-rated hard hat liner, as required
- UV-rated safety glasses or goggles
- ear canal inserts for hearing protection
- arc-rated gloves or Class 0 or 00 rubber insulating gloves and leather protectors
- heavy-duty leather work shoes

Shermco Industries
After being replaced, used motors can be refurbished at motor rewind facilities.

REPLACING MEDIUM-VOLTAGE MOTORS

The large amount of voltage and substantial amount of available short-circuit current makes replacing medium-voltage motors hazardous. General guidelines used to replace medium-voltage motors include the following:
- Follow the written lockout/tagout procedure provided by the company.
- Test-operate the motor using the normal controls. Be certain to place the controls back in the OFF position when finished.

- Remove the terminal cover on the motor. If desired, an absence-of-voltage test can be performed at the MCC or controller. Whether an absence-of-voltage test is performed at that location, a test for the absence of voltage must be performed at the point of contact, which for this task would be the motor.
- If the connections are taped, wear the proper PPE for the hazards and use a proximity-type voltage tester to test for the absence of voltage. The use of a direct-contact voltmeter is not required because the electromagnetic field is much stronger and proximity testers are much more reliable at medium-voltages.
- Always use the live-dead-live method of voltage testing. Test the voltage detector on a known live source and ensure it is operating properly, then test for the absence of voltage, and reverify that the voltage detector is functioning properly by retesting it on a known live source.

The shock approach boundaries for replacing medium-voltage motors include the Limited Approach Boundary at 60″ and the Restricted Approach Boundary at 26″.

Recommended PPE for Replacing Medium-Voltage Motors

The only task that can be performed while equipment is energized is testing for the absence of voltage. The equipment category from Table 130.7(C)(15)(A)(b) would be "Other equipment 1 kV to 15 kV." PPE Category 4 arc-rated clothing and PPE are required for this task. The recommended PPE for PPE Category 4 includes the following:
- an arc-rated clothing system, which includes an arc-rated flash suit, pants, and hood with a minimum rating of 40 cal/cm^2
- a Class E or G hard hat
- an arc-rated hard hat liner, as required
- UV-rated safety glasses or goggles
- ear canal inserts for hearing protection
- arc-rated gloves or Class 2 rubber insulating gloves and leather protectors
- heavy-duty leather work shoes

Digital Learner Resources
ATPeResources.com/QuickLinks
Access Code: 705798

SUMMARY

- Performing a hazard/risk analysis and choosing proper PPE is critical to worker safety.
- PPE is determined by using an arc flash hazard analysis or by using the tables in NFPA 70E.
- Equipment condition and reliability must be evaluated at the time work is to be performed.
- If the ability of the equipment to function in accordance with manufacturer specifications is uncertain, the upstream OCPD must be used as the basis for protecting workers.
- Safe approach distances must be observed for shock and arc flash hazards, and a safe work zone must be established whenever exposed energized conductors or circuit parts are present.

10 GUIDELINES FOR COMMON ELECTRICAL TASKS

Review Questions

Name _____ Date _____

_____ 1. The task of removing and inserting drawout-type circuit breakers is commonly known as ___.
   A. operating
   B. mounting
   C. racking
   D. setting

_____ 2. ___ are used to open a medium-voltage air circuit breaker and discharge the closing springs when the circuit breaker is racked into or out of its cubicle.
   A. Floor trippers
   B. Racking screws
   C. Shutters
   D. Finger clusters

_____ 3. A Limited Approach Boundary of ___" is used when operating medium-voltage air-break switches.
   A. 7
   B. 26
   C. 42
   D. 60

_____ 4. When removing or installing MCC buckets, the arc-rated clothing and PPE category recommended by NFPA 70E Tables 130.7(C)(15)(A)(a) and (b) is PPE Category ___.
   A. 1
   B. 2
   C. 3
   D. 4

T   F   5. Flammable, nonmelting clothing worn under arc-rated clothing and PPE increases the arc rating of the clothing system.

T   F   6. The Limited Approach Boundary for removing or installing an MCC bucket is 48".

_____ 7. When troubleshooting conductors or circuit parts rated 240 V and less, an arc-rated clothing system with a minimum rating of ___ cal/cm^2 is recommended by NFPA 70E.
   A. 4
   B. 8
   C. 10
   D. 40

_____ 8. When testing for the absence of voltage on circuits rated 120 V and less with no exposure to greater than 120 V, the recommended arc-rated clothing and PPE is ___.
   A. PPE Category 1
   B. PPE Category 2
   C. PPE Category 3
   D. not mandated

T   F   9. When there are multiple voltages inside a single enclosure, the highest exposed voltage is used to determine the shock hazard.

T   F   10. When operating equipment rated 240 V to 600 V, an arc-rated clothing system with a minimum rating of 40 cal/cm² is required.

_____ 11. A Limited Approach Boundary of ___″ is used when operating equipment rated 240 V and less.
   A. 12
   B. 24
   C. 36
   D. 42

_____ 12. For tasks not found in NFPA 70E Tables 130.7(C)(15)(A)(a) and (b), or when the maximum available short-circuit current or operating time exceeds the limits given in the tables, a(n) ___ must be performed.
   A. energized work permit
   B. incident energy analysis
   C. arc flash hazard analysis
   D. short circuit study

_____ 13. According to NFPA 70E, the PPE category for operating equipment rated 240 V to 600 V is PPE Category ___ when enclosure doors are open and covers are not in place.
   A. 0
   B. 1
   C. 2
   D. 3

_____ 14. Safely removing panels and covers from a switchgear enclosure involves ___.
   A. using at least two people
   B. installing handles
   C. deenergizing equipment inside the enclosure
   D. all of the above

T   F   15. Removing covers and panels from electrical enclosures has an HRC rating of 0.

T   F   16. The Limited Approach Boundary for replacing low-voltage motors is 42″.

APPENDIX

Energized Electrical Work Permit < 600 Volts 304

Lockout/Tagout Switching and Grounding Procedure. 305

Risk Assessment Matrix 306

Risk Assessment Form 307

Grounding of Equipment for Personal Safety Policy and Procedure 309

Industry and Standards Organizations. 310

Basis for Arc Flash Protection Distance Boundaries. 311

Electrical/Electronic Abbreviations/Acronyms 312

Ohm's Law 313

Voltage Drop Formulas—1ϕ, 3ϕ 313

AC/DC Formulas 314

NEMA Enclosure Selection. 314

Fuses and ITCBs 314

ENERGIZED ELECTRICAL WORK PERMIT < 600 VOLTS

Customer: _____ **Location:** _____ **Job #:** _____ **Date:** _____

Equipment/circuit: _____

The goal is to perform work deenergized. When deenergizing use remote switching and racking, and when grounding use longest hot stick as practical.

Justification of why the equipment/circuit cannot be deenergized:
- ☐ Troubleshooting
- ☐ Operational limits
- ☐ Other _____
- ☐ Energized work is being performed to deenergize?
- ☐ Process or life support equipment

1. **MINIMUM REQUIRED PPE:** Hard hat, safety glasses, ear plugs, 8 cal FR clothing and ER leather work boots.

2. **SHOCK HAZARD ANALYSIS:** (After voltage test, guard all exposed energized conductors if possible.)

    Voltage: _____ ☐ DC ☐ AC ☐ Use appropriate voltage detector: _____

    Working distance: _____ ☐ feet ☐ inches ☐ from body ☐ from hand

    Specific hazards and surrounding work area: _____

3. **SHOCK PROTECTION BOUNDARY AND PPE:**

    a. > 50 V and < 600 V. Voltage rated gloves when the working body is less than 4 feet of exposed to bare energized parts. In addition, wear arc rated face shield and balaclava for associated flash hazard.
    - ☐ If you performed a hazard analysis you may not have to wear a face shield and balaclava.
    - ☐ For control circuits, heater circuits, lighting circuits, no face shield or balaclava necessary.
    - ☐ Gloves class 00 (500 V)   ☐ Class 0 (1000 V)   ☐ Other _____

4. **ARC FLASH RISK ASSESSMENT:**

    Note: When racking, removing or replacing covers, or applying grounds wear a minimum 40 cal suit. Where at all possible, a remote racking device will be used for inserting or removing drawout-type circuit breaker.
    - ☐ 40 cal suit w/hood

    This is in effect if you cannot perform the below calculations. The preferred option is to perform the calculations.

    Bus/circuit rating amps: _____ Short circuit amps: _____ Fuses/C.B./relay clearing time cycles: _____

    Default clearing time: ☐ 0.1 sec   ☐ 2 sec _____

    Working distance: ☐ 18"   ☐ _____ cal/cm² _____

    Condition of equipment: ☐ Clean/excellent   ☐ Medium/good   ☐ Dirty/poor   Last Calibration Date: _____

5. **FLASH PROTECTION BOUNDARY AND PPE:** When inside FPB _____ ☐ Default option 10'.

    PPE Category ☐ 1   ☐ 2   ☐ 3   ☐ 4

6. **ADDITIONAL PPE BASED ON ABOVE ANALYSIS**
    - ☐ Insulated tools
    - ☐ Floor mat
    - ☐ 40 cal flash suit w/hood
    - ☐ 10 cal arc flash face shield
    - ☐ 8 cal balaclava
    - ☐ Rubber blanket
    - ☐ Other _____
    - ☐ No additional PPE needed because of the low energy available

7. **GROUND CLUSTERS:**
    - ☐ Inspected _____
    - ☐ Size _____
    - ☐ Inspection date _____

    * Use hot stick to install or remove grounds.
    * Refer to ASTM F855 tables for proper ground size.

I UNDERSTAND THE HAZARDS ABOVE AND THE WORK CAN PROCEED SAFELY.

_____ / _____   _____ / _____
TECHNICIAN (Print and Sign)   CUSTOMER AUTHORIZED REPRESENTATIVE (Print and Sign)

_____
SUPERVISOR

© 2012 by Gary Donner and Tony Demaria

LOCKOUT/TAGOUT SWITCHING AND GROUNDING PROCEDURE

Customer: _____   Date: _____

Job Location: _____   Job #: _____

WARNING: THE STEPS GIVEN BELOW ARE FOR A TYPICAL RADIAL POWER SYSTEM AND MAY NOT BE THE SAME AS YOUR SYSTEM. HAVE A CURRENT ONE-LINE DIAGRAM ON HAND AND DEVELOP A SWITCHING PROCEDURE THAT IS SPECIFIC TO YOUR SYSTEM.

NOTES:
- A GOOD WAY TO ACHIEVE AN ELECTRICALLY SAFE WORK CONDITION IS THE EXAMPLE BELOW.
- IT IS ALWAYS BEST TO USE AN UPDATED ONE-LINE DIAGRAM.
- WHENEVER POSSIBLE, VISUALLY VERIFY THAT ALL BLADES OF THE DISCONNECTING DEVICES ARE FULLY OPEN OR DRAWOUT-TYPE CIRCUIT BREAKERS ARE WITHDRAWN TO THE FULLY DISCONNECTED POSITION.

	SWITCHING STEPS	POSITION	SAFETY
☐	1. Open all load side breakers.	1	
☐	2. Open 480 volt main breaker.	2	
☐	3. Open high voltage main breaker.	3	
☐	4. Open high voltage switch.	4	
☐	5. Rack out the main circuit breaker.	5	
☐	6. Open the metering and control power low voltage fuse or circuit breaker.	6	
☐	7. Open the power and control transformer trunnion.	7	
☐	8. Test of high voltage.	4, 9	
☐	9. Apply high voltage grounds.	4, 9	
☐	10. Test for voltage on line and load sides of the transformer.	1, 10	
☐	11. Apply low voltage grounds depending on work location.	1, 10	
☐ 12.			
☐ 13.			
☐ 14.			
☐ 15.			

(PRINT NAME CLEARLY BELOW)   DATE   DATE

TECHNICIAN: _____   **AUTHORIZED CUSTOMER REP:** _____

FOREMAN: _____   **OTHER:** _____

© 2012 by Gary Donner and Tony Demaria

RISK ASSESSMENT MATRIX

Customer: _____ Date: _____

Location: _____ Job #: _____

Task: _____

ASSIGN A, B, AND C BELOW WITH A NUMBER 1 THROUGH 5; 1 BEING THE LOWEST AND 5 BEING THE MAXIMUM

A. *FREQUENCY OF PERFORMING TASK*

☐ 1. Very infrequently _____
☐ 2. _____
☐ 3. _____
☐ 4. _____
☐ 5. Constantly/several times per day _____

B. *PROBABILITY OF BEING HARMED*

☐ 1. No probability of being harmed _____
☐ 2. _____
☐ 3. _____
☐ 4. _____
☐ 5. You will be harmed _____

C. *POSSIBLE DEGREE OF HARM*

☐ 1. No degree of harm _____
☐ 2. _____
☐ 3. _____
☐ 4. _____
☐ 5. Death _____

RISK ASSESSMENT = **A** + **B** + **C** = _____

RISK ASSESSMENT = _____ + _____ + _____ = _____

COMMENTS: _____
_____
_____

RISK ASSESSMENT RATING MATRIX

☐ **1-5** Proceed with the task
☐ **6-9** Proceed with caution
☐ **10-11** Reassess the plan to see if there is a better plan
☐ **12-15** Make a new plan

TOTAL RISK = _____

(PRINT NAME CLEARLY BELOW)   DATE   DATE

TECHNICIAN: _____ **AUTHORIZED CUSTOMER REP:** _____

FOREMAN: _____ **OTHER:** _____

© 2012 by Gary Donner and Tony Demaria

RISK ASSESSMENT FORM ☐ General ☐ Job specific

Customer: _____ Location: _____ Job #: _____ Date: _____

Check in w/operations, operator(s) name(s): _____ Phone #: _____

Daily work scope: _____ Permit #: _____

_____

_____

Required Energized Work Permit (EWP)? ☐ Y ☐ N   Any * below must have EWP   Lockout/tagout required? ☐ Y ☐ N

THIS SIDE FOCUSES ON ELECTRICAL HAZARDS — BACK SIDE FOCUSES ON ALL OTHER HAZARDS
IF THERE ARE NO, OR MINOR, ELECTRICAL HAZARDS, DRAW A LINE THROUGH THIS SECTION

TASK/WORK DESCRIPTION	HAZARDS (INCLUDING ENVIRONMENTAL); WHAT WILL THIS CIRCUIT TRIP?	ELIMINATE/MINIMIZE
☐ Office work/shop work	Very low shock hazard and no flash hazard	Be aware
☐ Megger, hi-pot, PF testing, etc.	High shock hazard, but no flash hazard	Voltage rated gloves and barricade
☐ Working around large energized equipment w/covers closed	No shock hazard, possible flash hazard	Do not work in front of gear and/or use flash blanket, etc.
☐ 120/240 panels max 100 A or other low energy circuits	Low shock and no flash hazard	Voltage rated gloves
☐ Protective device testing	Inadvertent tripping of process equipment	Review prints and obtain customer/operator(s) permission
☐ * Voltage testing and phasing	Usually higher shock hazard and lower flash hazard	Voltage gloves and face shield
☐ * Switching energized equipment	Possible flash hazard	Remote switching/body position
☐ * Racking energized equipment	High flash hazard	Remote racking
☐ * Removing covers	Shock and arc flash	Wear appropriate PPE
☐ * Applying grounds	Shock and arc flash	40 cal full FR suit, longest hot stick possible and position body away from direct flash
☐ * Contact work on bare energized conductors	High shock hazard and high flash hazard	Do not perform the work or fill out Energized Electrical Work Permit and get permission
☐		
☐		
☐		

The foreman, supervisor or second replacement foreman are responsible for safety, job performance and customer relations.

Foreman's Name _____   Supervisor or Second Replacement Foreman _____

IF CONDITIONS ARE UNCLEAR, YOU MUST OBTAIN ADDITIONAL INFORMATION *(PRINT ALL NAMES CLEARLY)*
ALL PERSONS WORKING ON THE ABOVE WORK SCOPE MUST HAVE THEIR NAMES BELOW

_____ _____ _____

_____ _____ _____

CUSTOMER REPRESENTATIVE
(PRINT NAME CLEARLY) _____ DATE _____

© 2012 by Gary Donner and Tony Demaria

☐ **Standard industrial PPE**: Hard hat, hearing protection, safety glasses, appropriate gloves, FR clothing, and required boots, etc. _____

Special safety equipment: ☐ Pocket voltage detector ☐ Goggles ☐ Personal H2S monitor ☐ _____
☐ Arc flash suit ☐ Face shield ☐ Floor mat ☐ Insulated tools ☐ _____ ☐ _____

Fill out job safety inspection sheet if necessary (on large jobs) ☐ Y ☐ N By: _____
Review evacuation plan: _____
Location of first aid kit: _____
Location of the nearest hospital: _____
Cell phone use allowed at job site? ☐ Y ☐ N EXPLAIN: _____ EXEMPTIONS: _____
Near misses during the day: _____
Spot-check safety audit during the day @ _____ ☐ AM ☐ PM By: _____

MAJORITY OF ALL ACCIDENTS ARE CAUSED BY THE BELOW—CHECK MARK BELOW WHERE APPROPRIATE

TASK/WORK DESCRIPTION	HAZARDS	ELIMINATE/MINIMIZE
☐ Walking around job site	Slipping, tripping, or falling	Housekeeping, awareness
☐ Hand tools, lifting and climbing	Cuts, pinch points	Work gloves
☐ Other people working in the same area	Traffic, cranes, etc.	Communicate clearly at all times
☐ Production equipment running	Noise	Ear plugs, double protection
☐ Cords, hoses	Tripping, leaks, etc.	Out of path, reroute, etc.
☐ Chemicals, cleaners, solvents	Fumes, skin irritation	Ventilation, gloves, respirators
☐ Ladders and scaffolding	Failing	Tie down ladder, relocate as needed, body harness, short tie offs
☐ Lifting heavy objects	Back injury	Use legs, ask for help, etc.
☐ Awkward body positions	Body injury	Change body positions
☐ Other people working above	Falling objects, materials, tools, etc.	Wear hard hat, move, wait
☐ Working outside in the sun, heat	Heat stroke, heat exhaustion	Drink fluids, get shade, take breaks
☐		
☐		
☐		
☐		

© 2012 by Gary Donner and Tony Demaria

GROUNDING OF EQUIPMENT FOR PERSONAL SAFETY
POLICY AND PROCEDURE

Customer: _____ Location: _____ Date: _____

Equipment/circuit: _____

Any circuits over 600 V or where possible stored electrical energy may exist must be grounded before any direct contact. Circuits less than 600 V may be grounded if deemed necessary.

- ☐ Deenergize the proper circuit using the LOCKOUT/TAGOUT SWITCHING AND GROUNDING PROCEDURE.
- ☐ When opening a door or removing a cover, remember that the equipment is considered energized until proven otherwise. Wear proper PPE while performing this type of work and when applying grounds.
- ☐ Test the circuit to confirm that all conductors are deenergized. Use an adequately rated voltage detector to test each circuit for NO VOLTAGE. BEFORE and AFTER testing the affected conductors, determine that the voltage detector is operating satisfactorily by proving it on an energized source.
- ☐ When capacitors are involved, they should be grounded and shorted to drain off any stored charge.
- ☐ Clamps should be of proper size to fit the conductors and have adequate capacity for the fault current.
- ☐ Solid metal-to-metal connections are essential between grounding clamps and the deenergized conductors. The clamps should be slightly tightened in place, given a slight rotation on the conductors to provide cleaning action by the serrated jaws, and then be securely tightened. Ground clamps that attach to the steel tower, switchgear, or station ground bus are equipped with pointed or cupped set screws that should be tightened to ensure penetration through corrosion and paint to provide adequate connections.
- ☐ Grounding cable should be no longer than is necessary to keep resistance as low as possible and to minimize slack in cables to prevent their violent movement under fault conditions.
- ☐ Grounding cables should be connected between phases to the grounded structure and to the system neutral (when available) to minimize the voltage drop across the work area if inadvertent reenergization should occur.
- ☐ Prior to installing grounding equipment, inspect it for broken strands in the conductors, loose connections to the clamp terminals, and defective clamp mechanisms. Defective equipment should not be used.
- ☐ Verify and test the point(s) where you apply grounds (Is it a good ground?).
- ☐ Grounding equipment must be installed at each point where work is being performed on deenergized equipment. Often it is advisable to install grounding equipment on each side of a work point or at each end of a deenergized circuit.
- ☐ Always install a tag on the grounds identifying TDE and yourself.
- ☐ One end of the grounding "down level" must be connected to the metal structure of ground bus of the switchgear before connecting the other end of a phase conductor of the deenergized equipment. Then, and only then, the grounding cables should be connected between phase conductors.
- ☐ When removing grounding equipment, the above installation procedure must be reversed by first disconnecting the cables between phases, then disconnecting the "down lead" from the phase conductor and, finally, disconnecting the "down lead" from the metal structure or ground bus.
- ☐ Do not remove any grounds without proper permission. **AUTHORIZED BY:** _____
- ☐ Have customer witness, if appropriate or required. **CUSTOMER NAME:** _____
- ☐ It is very important that all personnel be notified of grounds being removed.
- ☐ When hi-potting, meggering, or other testing on bus, only remove 1∅ at a time.
- ☐ On motors or transformers all 3∅ must be removed and proper gear worn.

(PRINT NAME CLEARLY BELOW)   DATE   DATE

TECHNICIAN: _____   **AUTHORIZED CUSTOMER REP:** _____

FOREMAN: _____   **OTHER:** _____

© 2012 by Gary Donner and Tony Demaria

Industry and Standards Organizations . . .

FM Global 270 Central Avenue PO Box 7500 Johnston, RI 02919 www.fmglobal.com	Tests equipment and products to verify conformance to national codes and standards
International Association of Electrical Inspectors (IAEI) 901 Waterfall Way, Suite 602 Richardson, TX 75080 www.iaei.org	Focuses on interpretation of the National Electrical Code® and teaches safe installation and use of electricity
International Code Council (ICC) 500 New Jersey Avenue NW, 6th Floor Washington, DC 20001 www.iccsafe.org	Administers the International Building Code® including building materials, building systems, energy efficiency, fire protection systems, and structural design
National Electrical Installation Standards (NEIS) National Electrical Contractors Association 3 Bethesda Metro Center, Suite 1100 Bethesda, MD 20814 www.neca-neis.org	Standards developed by NECA in partnership with other industry organizations; all NEIS are submitted for approval to ANSI
National Electrical Manufacturers Association (NEMA) 1300 N. 17th Street, Suite 1847 Rosslyn, VA 22209 www.nema.org	Provides standards for manufacturers of electrical equipment
InterNational Electrical Testing Association (NETA) 106 Stone Street PO Box 687 Morrison, CO 80465 www.netaworld.org	Defines standards by which electrical equipment is deemed safe and reliable; also provides training for and certification of electrical testing technicians
International Electrotechnical Commission (IEC) 3, rue de Varembe' PO Box 131 CH—121 GENEVA 20 Switzerland www.iec.ch	Provides standards for most international installations and some domestic installations
National Fire Protection Association (NFPA) Batterymarch Park Quincy, MA 02269 www.nfpa.org	Provides guidance in assessing hazards of products of combustion; publishes the National Electrical Code®; develops hazardous materials information
American National Standards Institute (ANSI) 1819 L Street NW Washington, DC 20036 www.ansi.org	Coordinates and encourages activities in national standards department; identifies industrial and public needs for standards; acts as national coordinator and clearinghouse for consensus standards
The Institute of Electrical and Electronics Engineers (IEEE) 1828 L Street NW, Suite 1202 Washington, DC 20036 www.ieee.org	Provides guidance in all electrical and electronic systems including aerospace systems, computers and telecommunications, biomedical engineering, electric power, and consumer electronics

...Industry and Standards Organizations

National Institute for Occupational Safety and Health (NIOSH) 4676 Columbia Parkway Cincinnati, OH 45226 www.cdc.gove/niosh	Acts in conjunction with OSHA to develop recommended exposure limits for hazardous substances or conditions located in the workplace; recommends preventive measures to reduce or eliminate adverse health and safety effects
Underwriters Laboratories Inc. (UL) 333 Pfingsten Road Northbook, IL 60062 www.ul.com	Tests equipment and products to verify conformance to national codes and standards
Canadian Standards Association (CSA) 5060 Spectrum Way, Suite 100 Mississauga, ON L4W 5N6 www.csa.ca	Tests equipment and products to verify conformance to Canadian codes and standards

Basis for Arc Flash Protection Distance Boundaries

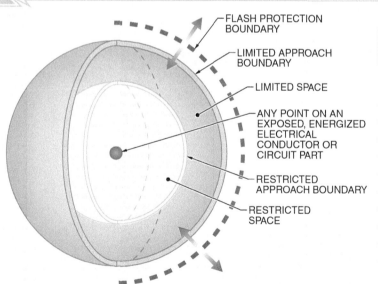

Minimum Air Insulation Distances to Avoid Flashover

Voltage	Distance
300 V and less	0′–0.03″
Over 300 V, not over 750 V	0′–0.07″
Over 750 V, not over 2 kV	0′–0.19″
Over 2 kV, not over 15 kV	0′–1.5″
Over 15 kV, not over 36 kV	0′–6.3″
Over 36 kV, not over 48.3 kV	0′–10.0″
Over 48.3 kV, not over 72.5 kV	1′–3.0″
Over 72.5 kV, not over 121 kV	2′–1.2″
Over 138 kV, not over 145 kV	2′–6.6″
Over 161 kV, not over 169 kV	3′–0.0″
Over 230 kV, not over 242 kV	4′–2.4″

Electrical/Electronic Abbreviations/Acronyms

Abbr/Acronym	Meaning	Abbr/Acronym	Meaning	Abbr/Acronym	Meaning
A	Ammeter; Ampere; Anode; Armature	FU	Fuse	PNP	Positive-Negative-Positive
AC	Alternating Current	FWD	Forward	POS	Positive
AC/DC	Alternating Current; Direct Current	G	Gate; Giga; Green; Conductance	POT.	Potentiometer
A/D	Analog to Digital	GEN	Generator	P-P	Peak-to-Peak
AF	Audio Frequency	GRD	Ground	PRI	Primary Switch
AFC	Automatic Frequency Control	GY	Gray	PS	Pressure Switch
Ag	Silver	H	Henry; High Side of Transformer; Magnetic Flux	PSI	Pounds Per Square Inch
ALM	Alarm			PUT	Pull-Up Torque
AM	Ammeter; Amplitude Modulation	HF	High Frequency	Q	Transistor
AM/FM	Amplitude Modulation; Frequency Modulation	HP	Horsepower	R	Radius; Red; Resistance; Reverse
		Hz	Hertz	RAM	Random-Access Memory
ARM.	Armature	I	Current	RC	Resistance-Capacitance
Au	Gold	IC	Integrated Circuit	RCL	Resistance-Inductance-Capacitance
AU	Automatic	INT	Intermediate; Interrupt	REC	Rectifier
AVC	Automatic Volume Control	INTLK	Interlock	RES	Resistor
AWG	American Wire Gauge	IOL	Instantaneous Overload	REV	Reverse
BAT.	Battery (electric)	IR	Infrared	RF	Radio Frequency
BCD	Binary Coded Decimal	ITB	Inverse Time Breaker	RH	Rheostat
BJT	Bipolar Junction Transistor	ITCB	Instantaneous Trip Circuit Breaker	rms	Root Mean Square
BK	Black	JB	Junction Box	ROM	Read-Only Memory
BL	Blue	JFET	Junction Field-Effect Transistor	rpm	Revolutions Per Minute
BR	Brake Relay; Brown	K	Kilo; Cathode	RPS	Revolutions Per Second
C	Celsius; Capacitance; Capacitor	L	Line; Load; Coil; Inductance	S	Series; Slow; South; Switch
CAP.	Capacitor	LB-FT	Pounds Per Foot	SCR	Silicon Controlled Rectifier
CB	Circuit Breaker; Citizen's Band	LB-IN.	Pounds Per Inch	SEC	Secondary
CC	Common-Collector Configuration	LC	Inductance-Capacitance	SF	Service Factor
CCW	Counterclockwise	LCD	Liquid Crystal Display	1 PH; 1φ	Single-Phase
CE	Common-Emitter Configuration	LCR	Inductance-Capacitance-Resistance	SOC	Socket
CEMF	Counter Electromotive Force	LED	Light Emitting Diode	SOL	Solenoid
CKT	Circuit	LRC	Locked Rotor Current	SP	Single-Pole
CONT	Continuous; Control	LS	Limit Switch	SPDT	Single-Pole, Double-Throw
CPS	Cycles Per Second	LT	Lamp	SPST	Single-Pole, Single-Throw
CPU	Central Processing Unit	M	Motor; Motor Starter; Motor Starter Contacts	SS	Selector Switch
CR	Control Relay			SSW	Safety Switch
CRM	Control Relay Master	MAX.	Maximum	SW	Switch
CT	Current Transformer	MB	Magnetic Brake	T	Tera; Terminal; Torque; Transformer
CW	Clockwise	MCS	Motor Circuit Switch	TB	Terminal Board
D	Diameter; Diode; Down	MEM	Memory	3 PH; 3φ	Three-Phase
D/A	Digital to Analog	MED	Medium	TD	Time Delay
DB	Dynamic Braking Contactor; Relay	MIN	Minimum	TDF	Time Delay Fuse
DC	Direct Current	MN	Manual	TEMP	Temperature
DIO	Diode	MOS	Metal-Oxide Semiconductor	THS	Thermostat Switch
DISC.	Disconnect Switch	MOSFET	Metal-Oxide Semiconductor Field-Effect Transistor	TR	Time Delay Relay
DMM	Digital Multimeter			TTL	Transistor-Transistor Logic
DP	Double-Pole	MTR	Motor	U	Up
DPDT	Double-Pole, Double-Throw	N; NEG	North; Negative	UCL	Unclamp
DPST	Double-Pole, Single-Throw	NC	Normally Closed	UHF	Ultrahigh Frequency
DS	Drum Switch	NEUT	Neutral	UJT	Unijunction Transistor
DT	Double-Throw	NO	Normally Open	UV	Ultraviolet; Undervoltage
DVM	Digital Voltmeter	NPN	Negative-Positive-Negative	V	Violet; Volt
EMF	Electromotive Force	NTDF	Nontime-Delay Fuse	VA	Volt Amp
F	Fahrenheit; Fast; Field; Forward; Fuse	O	Orange	VAC	Volts Alternating Current
FET	Field-Effect Transistor	OCPD	Overcurrent Protection Device	VDC	Volts Direct Current
FF	Flip-Flop	OHM	Ohmmeter	VHF	Very High Frequency
FLC	Full-Load Current	OL	Overload Relay	VLF	Very Low Frequency
FLS	Flow Switch	OZ/IN.	Ounces Per Inch	VOM	Volt-Ohm-Milliammeter
FLT	Full-Load Torque	P	Peak; Positive; Power; Power Consumed	W	Watt; White
FM	Frequency Modulation	PB	Pushbutton	w/	With
FREQ	Frequency	PCB	Printed Circuit Board	X	Low Side of Transformer
FS	Float Switch	PH; φ	Phase	Y	Yellow
FTS	Foot Switch	PLS	Plugging Switch	Z	Impedance

Ohm's Law

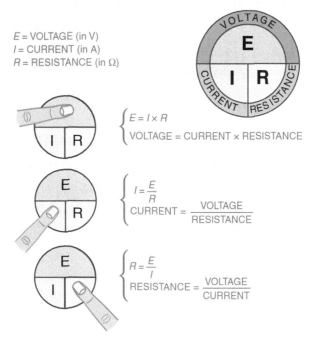

E = VOLTAGE (in V)
I = CURRENT (in A)
R = RESISTANCE (in Ω)

$E = I \times R$
VOLTAGE = CURRENT × RESISTANCE

$I = \dfrac{E}{R}$
CURRENT = $\dfrac{\text{VOLTAGE}}{\text{RESISTANCE}}$

$R = \dfrac{E}{I}$
RESISTANCE = $\dfrac{\text{VOLTAGE}}{\text{CURRENT}}$

Voltage Drop Formulas — 1ϕ, 3ϕ

Phase	To Find	Use Formula	Example Given	Find	Solution
1ϕ	VD	$VD = \dfrac{2 \times R \times L \times I}{1000}$	240 V, 40 A, 60′ L, 0.764 R	VD	$VD = \dfrac{2 \times R \times L \times I}{1000}$ $VD = \dfrac{2 \times 0.764 \times 60 \times 40}{1000}$ $VD =$ **3.67 V**
3ϕ	VD	$VD = \dfrac{2 \times R \times L \times I}{1000} \times 0.866$	208 V, 110 A, 75′ L, 0.194 R, 0.866 multiplier	VD	$VD = \dfrac{2 \times R \times L \times I}{1000} \times 0.866$ $VD = \dfrac{2 \times 0.194 \times 75 \times 110}{1000} \times 0.866$ $VD =$ **2.77 V**

Note: $\dfrac{\sqrt{3}}{2} = 0.866$

AC/DC Formulas

To Find	DC	AC		
		1ϕ, 115 or 220 V	1ϕ, 208, 230, or 240 V	3ϕ — All Voltages
I, HP known	$\dfrac{HP \times 746}{E \times Eff}$	$\dfrac{HP \times 746}{E \times Eff \times PF}$	$\dfrac{HP \times 746}{E \times Eff \times PF}$	$\dfrac{HP \times 746}{1.73 \times E \times Eff \times PF}$
I, kW known	$\dfrac{kW \times 1000}{E}$	$\dfrac{kW \times 1000}{E \times PF}$	$\dfrac{kW \times 1000}{E \times PF}$	$\dfrac{kW \times 1000}{1.73 \times E \times PF}$
I, kVA known		$\dfrac{kW \times 1000}{E}$	$\dfrac{kW \times 1000}{E}$	$\dfrac{kVA \times 1000}{1.763 \times E}$
kW	$\dfrac{I \times E}{1000}$	$\dfrac{I \times E \times PF}{1000}$	$\dfrac{I \times E \times PF}{1000}$	$\dfrac{I \times E \times 1.73 \times PF}{1000}$
kVA		$\dfrac{I \times E}{1000}$	$\dfrac{I \times E}{1000}$	$\dfrac{I \times E \times 1.73}{1000}$
HP (output)	$\dfrac{I \times E \times Eff}{746}$	$\dfrac{I \times E \times Eff \times PF}{746}$	$\dfrac{I \times E \times Eff \times PF}{746}$	$\dfrac{I \times E \times 1.73 \times Eff \times PF}{746}$

Note: *Eff* = efficiency

NEMA Enclosure Selection

Type	Use	Service Conditions	Tests	Comments	Type
1	Indoor	No unusual	Rod entry, rust resistance		1
3	Outdoor	Windblown dust, rain, sleet, and ice on enclosure	Rain, external icing, dust, and rust resistance	Do not provide protection against internal condensation or internal icing	
3R	Outdoor	Falling rain and ice on enclosure	Rod entry, rain, external icing, and rust resistance	Do not provide protection against internal condensation, or internal icing	
4	Indoor/outdoor	Windblown dust and rain, splashing water, hose-directed water, and ice on enclosure	Hosedown, external icing, and rust resistance	Do not provide protection against internal condensation or internal icing	4
4X	Indoor/outdoor	Corrosion, windblown dust and rain, splashing water, hose-directed water, and ice on enclosure	Hosedown, external icing, and corrosion resistance	Do not provide protection against internal condensation or internal icing	4X
6	Indoor/outdoor	Occasional temporary submersion at a limited depth			
6P	Indoor/outdoor	Prolonged submersion at a limited depth			
7	Indoor location classified as Class I, Groups A, B, C, or D, as defined in the NEC®	Withstand and contain an internal explosion of specified gases, sufficiently contain an explosion so an explosive gas-air mixture in the atmosphere is not ignited	Explosion, hydrostatic, and temperature	Enclosed heat-generating devices shall not cause external surfaces to reach temperatures capable of igniting explosive gas-air mixtures in the atmosphere	7
9	Indoor location classified as Class II, Groups E or G, as defined in the NEC®	Dust	Dust penetration, temperature, and gasket aging	Enclosed heat-generating devices shall not cause external surfaces to reach temperatures capable of igniting explosive gas-air mixtures in the atmosphere	9
12	Indoor	Dust, falling dirt, and dripping noncorrosive liquids	Drip, dust, and rust resistance	Do not provide protection against internal condensation	12
13	Indoor	Dust, spraying water, oil, and noncorrosive coolant	Oil explosion and rust resistance	Do not provide protection against internal condensation	

Fuses and ITCBs

Increase	Standard Ampere Ratings
5	15, 20, 25, 30, 35, 40, 45
10	50, 60, 70, 80, 90, 100, 110
25	125, 150, 175, 200, 225
50	250, 300, 350, 400, 450
100	500, 600, 700, 800
200	1000, 1200
400	1600, 2000
500	2500
1000	3000, 4000, 5000, 6000

1 A, 3 A, 6 A, 10 A, and 601 A are additional standard ratings for fuses.

GLOSSARY

A

absence-of-voltage test: A test performed to verify that a conductor or circuit part contains no voltage.

ampere-turn: The number of amperes in a circuit multiplied by the number of turns of the primary or secondary winding.

arc chute: An element inside a circuit breaker that extinguishes an arc when the contacts are opened.

arc flash: An extremely high temperature discharge produced by an electrical current that flows through air, usually to ground.

Arc Flash Boundary: The distance from exposed energized conductors or circuit parts where bare skin would receive the onset of a second-degree burn, equivalent to 1.2 cal/cm^2.

arc flash protective clothing: Clothing that provides protection from exposure to the extreme temperatures that occur during an arc flash.

arcing contact: A thin, blade-like contact that is designed to interrupt an arc when the switch is opened or closed.

arc-rated clothing: Clothing that meets ASTM F1506, *Standard Performance Specification for Flame Resistant Textile Material for Wearing Apparel for Use by Electrical Workers Exposed to Momentary Electric Arc and Related Thermal Hazards.*

arc-rated face shield: An eye and face protection device that covers the entire face with a plastic shield and has a chin cup and side shields that extend its protective coverage to the chest and beyond the ears.

arc-rated hood: An eye and face protection device that covers the entire head and is used for protection from an arc flash.

arc rating: The incident energy (in cal/cm^2) on clothing that has a 50% probability of resulting in a second-degree burn on bare skin underneath the clothing or material breakopen.

arc thermal performance value (ATPV): The incident energy that results in sufficient heat transfer through PPE to cause the crossing of the Stoll curve burn injury model, which is designed to prevent second-degree burns.

assured equipment grounding conductor program: The process of testing and verifying that a ground path is unbroken and continuous.

automated external defibrillator (AED): A portable electronic device capable of interpreting a person's heart rhythm and automatically delivering a defibrillating shock to stop an irregular heart rhythm, thereby allowing the heart to reestablish a normal rhythm.

B

backfed voltage: An unexpected voltage from another circuit that is backfed through other electrical devices. Also known as backfeed.

backfeed: *See* backfed voltage.

C

capacitor: An electric device that stores electrical energy and is used for power factor correction in electrical power systems.

conductive object: An item that is not rated for the voltage to which it is exposed.

corona: The ionization of air caused by exposure to the dielectric stress of a high-voltage electrical field.

current transformer (CT): A transformer used to step down line current to make it easier and safer to measure.

D

digital multimeter (DMM): A test instrument that can measure two or more electrical properties and displays the measured properties as numerical values.

duty cycle: The amount of time an electrical tool is carrying a specific amount of current.

E

energized electrical work permit: A document that describes the job planning needed to perform energized electrical work safely.

energized work: Work on or near exposed energized conductors or circuit parts where an employee is exposed to electrical hazards.

energy of breakopen threshold (E_{BT}): The point at which a fabric allows a 1″ crack or a ½″ hole, but no burn is registered.

equipotential zone: An area in which all conductive elements are bonded or otherwise connected together in a manner that prevents a difference of potential from developing within the area.

F

flexible cord: An assembly of two or more insulated conductors contained within a jacket and is used for the connection of equipment to a power source.

G

goggles: An eye protection device with a flexible frame that is secured on the face with an elastic headband.

ground fault circuit interrupter (GFCI): A device that protects against electrical shock by detecting an imbalance of current in the energized conductor and the neutral conductor.

H

hard hat: A protective helmet that is used in the workplace to prevent injury from the impact of falling or flying objects and electrical shock.

hard hat suspension: A shock-absorbing lining that keeps the shell away from the head to provide protection from falling objects striking the hard hat.

I

incident energy: A measurement of thermal energy (in cal/cm^2) at a working distance from an arc fault.

incident energy analysis: Part of an arc flash risk assessment; determines the incident energy a worker would receive at a specific working distance.

induced voltage: An unexpected voltage that is created by placing a deenergized conductor in the magnetic field surrounding a conductor that is energized and carrying a load.

J

job briefing: A discussion that takes place between all employees engaged in a task and occurs before work is performed on energized conductors and devices.

L

leather protectors: Gloves worn over rubber insulating gloves to prevent damage to the rubber insulating glove.

light ballast: A small transformer that takes 120 V or 277 V and steps it up to several thousand volts.

Limited Approach Boundary: The distance from an exposed energized conductor or circuit part at which a shock hazard exists; the closest distance an unqualified person can approach.

Limited Space: The area between the Limited Approach Boundary and the Restricted Approach Boundary.

lockout: The process of removing the source of electrical power and installing a lock to prevent the power from being turned ON.

M

material breakopen: The incident energy where clothing chars, crumbles, and falls apart, which allows flames and heat to penetrate it.

maximum use voltage: The maximum amount of voltage a rubber insulating glove can be exposed to without suffering dielectric deterioration over time.

medium-voltage air-break switch: A switch that uses air as the interrupting medium.

motor control center (MCC) bucket: An assembly that connects to a main bus assembly and contains the necessary components to operate a load, usually a motor.

N

National Electrical Code® (NEC®): A standard on practices for the design and installation of electrical power systems, circuits, and components.

National Fire Protection Association® (NFPA®): A national organization that provides guidance in assessing the hazards of the products of combustion.

NFPA 70E®, Standard for Electrical Safety in the Workplace®: A voluntary standard for electrical safety-related work practices.

noncontact voltage tester: A test instrument that indicates the presence of voltage when the test tip is near an energized conductor or circuit part.

O

Occupational Safety and Health Administration (OSHA): A federal government agency established under the Occupational Safety and Health (OSH) Act of 1970, which requires employers to provide a safe environment for their employees.

Ohm's law: The relationship between voltage, current, and resistance.

ozone cutting: The damage caused to natural rubber through exposure to ozone.

P

personal protective equipment (PPE): Clothing and/or equipment worn by an employee to reduce the possibility of injury in the work area.

proof test voltage: The amount of voltage applied to test rubber insulating gloves for weakening or damage.

R

remote earth: Any object that is on or in the ground outside the equipotential zone and extends into the zone.

rescue hook: A fiberglass pole with a large metal hook on the end, much like a shepherd's crook.

Restricted Approach Boundary: The distance from an exposed energized conductor or circuit part where an increased risk of electric shock exists due to the close proximity of the person to the energized conductor or circuit part.

Restricted Space: The area between the Restricted Approach Boundary and the exposed energized conductors and circuit parts.

risk assessment: A field document that accounts for all hazards.

rubber insulating blanket: A blanket that is used to insulate and/or isolate energized conductors and circuit parts.

rubber insulating gloves: Gloves made of natural latex rubber (Type I) or a synthetic ozone-resistant material (Type II) that are used to provide protection from electrical shock hazards.

rubber insulating sleeves: Sleeves that are worn to protect the upper arms and shoulders from coming into contact with exposed energized conductors and circuit parts.

S

safety glasses: An eye protection device with special impact-resistant glass or plastic lenses, reinforced frames, and side shields.

safe work zone: A barrier that is established to protect unqualified or unaware persons from entering an area that has electrical hazards.

selective coordination: The act of isolating a faulted circuit from the remainder of the electrical system, thereby eliminating unnecessary power outages.

solenoid voltage tester: A test instrument that indicates approximate voltage level and type (AC or DC) by the movement and vibration of a pointer on a scale.

step potential: A difference in voltage between each foot of a person when standing near an energized object.

T

tagout: The process of placing a danger tag on a source of electrical power, which indicates that the equipment may not be operated until the danger tag is removed.

tank-loss-index (TLI): A diagnostic test performed using an insulation power factor test set.

touch potential: A difference in voltage between an energized object and a person in contact with that object, which causes current to flow through that person's body.

turns ratio: The number of turns on the primary winding (usually one turn for a CT) of a transformer to the number of turns on the secondary winding.

U

unqualified person: A person who is not qualified to perform a specific task.

INDEX

Page numbers in italic refer to figures.

A

abbreviations, electrical, *312*
absence-of-voltage tests
  defined, 45, 122
  NFPA 70E and, 131–132, *132*
  PPE and, 289–291
  tools for, *122*, 122–124, *123*, *124*
AC/DC formulas, *314*
AC drives, 286–289, *287*, *288*
AC electric power systems, 9, *10*
acronyms, electrical, *312*
administrative controls, 32
AEDs (automated external defibrillators), 8–9, *9*, 52, 177
air-break switches, 277–280, *278*, *279*, *280*
alerting techniques, 166–168, *167*, *168*
ampere-turn, 191
ANSI/AIHA/ASSE Z10, *Occupational Health and Safety Management Systems*, 35, 107
ANSI/NETA MTS, 72
ANSI Z535, *Series of Standards for Safety Signs and Tags*, 166, *167*
apparel not permitted, 259–260
applying lockouts/tagouts, 134
apprenticeships, 40
approach boundaries, *63*, 63–65, *64*, *65*
  company, 102
  energized conductors or circuit parts and, 151–152, *153*
  unqualified persons and, 152, *153*
approach distances. *See* minimum approach distances (MADs)
arc blasts, 12–13, 43
arc chutes, 268
arc flash, 12
Arc Flash Boundary
  arc flash risk assessments and, 155
  defined, 60, 65, *311*
  equations, *71*, 71
  per IEEE 1584, 66, *67*
  per NFPA 70E, 67–73, *70*, *71*
  per OSHA, 60, 67, *68*
  uninsulated overhead lines and. *See* uninsulated overhead lines
arc flash hazards, 47–48, *84*, 84, 86–87
arc flash hazard warning labels, *84*, 84–85, 266, *267*
arc flash protective clothing, 242
  arc flash risk assessments, 247–248
  care of, *260*, 260–261

  inspection of, 242–243, *243*, *244*
  material, 259
  not permitted, 259–260
  repairing, 245–246
  selecting, *256*, 256–258
  storing, *245*, 245
  wearing, 244–245
arc flash protective equipment, 258–259, *259*
arc flash risk assessments
  EEWPs and, 73, *304*
  NFPA 70E and, 68, 153–157, *156*
  PPE and, 247–248
  risk assessments and, 83
arc flash suits, 258, 260–261
architects, 25
arcing contacts, *278*, 278–279, *279*
arcing faults, 248
arc-rated clothing. *See also* PPE (personal protective equipment)
  low-energy equipment and, *43*
  maintenance, *260*, 260–261
  ratings, 13–14, *14*, *16*
  selection, 69, *256*, 256–258
arc-rated face shields, *240*, 240, *241*, 255, 258
arc-rated hoods, *240*, 240, 245
arc ratings, 13, 255–256
arc-resistant switchgear, 280
arc thermal performance values (ATPVs), 223
arm protection, *250*, 250
assessing hazards, 87–88
assured equipment grounding conductor programs, 215
ATPVs (arc thermal performance values), 223
attachment means, lockout/tagout, 129
attachment plugs, *198*
  per NFPA, 214
  per OSHA, *207*, 207, 208–209, *209*
attendants, 167
auditing, 34–35, 98
authority and contracts, 25
automated external defibrillators (AEDs), 8–9, *9*, 52, 177

B

backfed voltages, 118–119, *119*, *120*
ball-and-socket temporary protective grounds, 182, *183*

barricades, 166
barricade tape, 77–78, *78*
battery banks, *146*
battery systems, *30*, 253, 254–255
blankets, rubber insulating, 238–240, *239*
blind reaching, 158, *159*
body protection, 250
breakopen threshold value (E_{BT}), 223
burn injuries, *10*, 10–11, *11*

C

cable sizes, 213, *214*
capacitors, 189–190, *190*
cardiopulmonary resuscitation (CPR) training, 177
CBs (circuit breakers). *See* circuit breakers (CBs)
cells of battery systems, 253, 254–255
checklists, 29
chemical damage to gloves, 232
chin protection, 249
circuit breakers (CBs)
    maintenance, 90, *91*, *92*, *93*
    normal operation, 252, 254
    operating time, 87
circuit interlocks, 129
circuits, energized work on. *See* energized work
circuits rated between 240 V and 600 V, 286
circuits rated 120 V and less, 283–286, *284*
circuits rated 240 V and less, 285–286
classes of hard hats, 229
classes of rubber insulating gloves, 235, *236*
clearance distances. *See also* minimum approach distances (MADs)
    defined, 46, *47*
    energized conductors and, 57–59, *58*, *59*, *60*
    overhead power lines and, 60–62, *61*, *62*
closed cord caps, 202, *203*
closing circuits, 162
clothing, protective. *See* PPE (personal protective equipment)
clothing not permitted, 259–260
Color Book series, 35
color-coded cords, 201, *202*
company responsibility, 76
complex lockout/tagout procedure, 125, *126*
computer-based training, 49–50
condition of maintenance, 32, *70*, 70, 86–87
conductive materials, 160, *161*
conductive work locations, 206, 214
conductors, energized work on. *See* energized work
conductors, flexible cords and, 201, *202*, 213, *214*
confined work spaces, 160
connector assemblies, *276*, 276
continuous industrial processes, 147
contract employers, 26–28, *27*
contracts, 25

control circuits, 253, 255
control devices, 129
controlling employers, *22*, 23, 25
control power transformers (CPTs)
    mounting locations, *294*, 294–295
    reducing voltage and, *284*, 284
    troubleshooting, *124*, 124
controls, 32–33, *33*
coordinating lockout/tagouts, 126, 133
copper calorimeters, 16
cord-and-plug-connected equipment, 211–213, *212*, *213*
cord caps, 202, *203*
cord connectors, 208–209
corona, 233
correcting employers, *22*, 23
Correlating Committee, 3–4
corrosion of OCPDs, 90
cotton glove liners, *232*, 232
cover removal from electrical enclosures, 295–296
CPR (cardiopulmonary resuscitation) training, 177
CPTs. *See* control power transformers (CPTs)
cranes, *171*, 171–172
creating employers, *22*, 22
current transformers (CTs), *190*, 190–191, *191*

D

data plates, *85*, 85
DC electric power systems, 9, *10*
DC formulas, *314*
deenergizing equipment
    elements of control and, 130
    per NFPA, 247
    per OSHA, 111
    procedures for, 114
    uninsulated overhead lines and, 169
    verification and, 116–117
defective equipment, 214
derricks, *171*, 171–172
digital multimeters (DMMs), 123–124, *124*
disconnecting equipment from all energy sources, 114
disconnecting means, 131
DMMs (digital multimeters), 123–124, *124*
documentation of ESPs, 28
doors, 161
dot standards, 35
drawout-type circuit breakers
    low-voltage. *See* low-voltage drawout-type circuit breakers
    medium-voltage, 274–277, *275*, *276*, *277*
    qualifications to operate, 40
duty cycle, 213

E

E_{BT} (energy of breakopen threshold), 223
EEWPs. *See* energized electrical work permits (EEWPs)

electrical abbreviations and acronyms, *312*
electrical arc flash, 12
electrical burns, *10*, 10–11, *12*, 12
electrical circuit interlocks, 129
electrical enclosures, removing covers and panels, 295
electrical hazards
  arc blast, 12–13
  arc-rated clothing, 13–14, *14*
  electrical arc flash, 12
  electric shock, *8*, 8–9, *9*, *10*, 11–12, *12*
  incident energy, 16
  NFPA 70E and, *145*, 145
  risk assessment and, *84*, 84–86, *85*
electrical injuries
  burns, *10*, 10, *11*
  shock, 6–7, *7*
  trends in, *61*, 61–62, *62*
electrical installation requirements, 41, *42*
electrically safe work conditions per NFPA 70E, 125, 145–149, *146*.
electrically safe work conditions per OSHA
  deenergizing equipment, 114
  disconnecting equipment, 114
  ghost voltages, 117–119, *118*, *119*, *120*
  lockout/tagouts, 113, 115–116, *128*
  preventing other devices from energizing equipment, 114
  reenergizing equipment, 120–121
  releasing stored electric energy, 114
  verifying equipment is deenergized, 116–117, *117*
electrical regulations per OSHA, 5, *5*
Electrical Safety: A Program Development Guideline, 36
Electrical Safety Program Book, 36
electrical safety programs (ESPs)
  auditing, 34–35
  controls, 32–33, *33*
  documenting, 28
  philosophy, 29
  policies, 35
  principles, 34
  procedures, 29–31, *30*, *32*
  responsibility, 29
  safety teams, 29
  scope, 28
  standards, 35–36
  standards of performance, 34
  training, 34
electrical tasks, general guidelines, 265–267, *267*
electrical testing, 212–213, *213*
electrical workers, 41–42, 45–47, 48
electric shock, *8*, 8–9, *9*, *10*, 11–12, *12*
electric tools, 204–205
electric tool testers, *213*, 213
electrocutions, *61*, 61
electronic tool testers, 180–181

elements of control for lockout/tagout
  accountability for personnel, 133
  coordination, 133
  deenergizing equipment, 130
  disconnecting means, 131
  grounding, 133
  lockout/tagout application, 134
  release for return to service, 134–135
  removal of lockout/tagout devices, 134
  responsibility, 131
  shift change, 133
  stored energy, *130*, 130
  temporary release for testing/positioning, 135
  testing, 131–132, *132*
  verification, 131
elimination of hazards, 32
emergency response training, 177
employee/employer responsibilities, 169
employer roles, *22*, 22–28
  contract employers, 26–28, *27*
  controlling employers, *22*, 23, 25
  correcting employers, *22*, 23
  creating employers, *22*, 22
  exposing employers, *22*, 22
  host and contract employers' responsibilities, 26–27, *27*
  host employers, 26–27, *27*
  reasonable care standard, 24
enclosed work spaces, 160
energized electrical work permits (EEWPs)
  defined, 64, 73, *304*
  exemptions to, 75
  parts of, *74*, 74–75
  required/not required, *150*, 150–151
  risk assessments and, 100–101
energized work
  less than 50 volts, *146*, 146–147
  more than 50 volts, 151
  normal operation, 148–149
  permits for, *150*, 150–152
  reasons for, 146
  recognition of, 143–144, *144*, 151
  test instruments and equipment, 149
energy of breakopen threshold (E_{BT}), 223
engineering controls, 32
engineers, 25
equations for Arc Flash Boundary, *71*, 71
equipment contact, 172
equipment failure, anticipating, 162
equipment for lockouts/tagouts, *126*, *127*, 127–129, *128*
equipment grounding, 172
equipment handling and storing, 210
equipment labeling, *156*, 156–157
equipment rated between 240 V and 600 V, 292–293
equipment rated 240 V and less, *291*, 291–292
equipotential zones, 135, *174*, 174–175, *175*

ESPs. *See* electrical safety programs (ESPs)
evidence of impending failure, 69, 70–71
existing conditions of equipment per OSHA, 177–178
exposed persons during lockout/tagout, 130
exposing employers, *22*, 22
eyeglasses, 160, *161*
eye protection, *240*, 240–241, *241*, 249

F

face protection, 258
face shields, 240, *241*, 255
failure of equipment, 162
fatalities, electrical, *61*, 61
fenced-in substations, *185*, 185
fiberglass-reinforced plastic (FRP), 165, *179*, 179
fiberglass-reinforced plastic (FRP) rods, 165
First Draft meetings, 3
First Drafts, 3
first responder training, 177
First Revisions (FRs), 3
flame-resistant (FR) clothing, *14*, 14
flammable materials, 162
flexible cables, 199, 202
flexible cords
  conductor sizing, 213, *214*
  connecting, 202, *203*
  defined, 197, *198*
  identification of, 201, *202*
  inspection, 204–205, 211–213, *213*
  repairing, 202
  types of, 200–201, *201*
  uses, *199*, 199
floor trippers, 274, *275*
foot protection, 251, 259
forms for performing risk assessments
  energized electrical work permit, 100–101, *304*
  grounding for personal safety policy and procedure, 107, *305, 309*
  lockout/tagout switching and grounding procedure, 103–105, *104*
  risk assessment, 98–99, *99, 307*
  risk assessment matrix, 105–106, *106, 306*
FR (flame-resistant) clothing, *14*, 14
frequency (Fr) risk factor, 97
FRP (fiberglass-reinforced plastic), 165, *179*, 179
FRP (fiberglass-reinforced plastic) rods, 165
FRs (First Revisions), 3
fuse and fuseholder equipment, *165*, 165
fused contactor motor starters 2.3 kV to 7.2 kV, 293–295, *294*
fuse pullers, *165*, 165
fuses, *314*

G

General Duty Clause, 88
generators, portable, 209–210, *210*

GFCI circuit breakers, 215
GFCI receptacles, *215*, 215–216, *216*
GFCIs (ground fault circuit interrupters), *215*, 215–216, *216*, *217*
GFCI testing, 215–216, *216*
ghost voltages, 117–119, *118*, *119*, *120*
gloves. *See* rubber insulating gloves
goggles, 240
ground fault circuit interrupters (GFCIs), *215*, 215–216, *216*, *217*
grounding, 172, *181*, 181–184, *183*, *184*
grounding, lockout/tagout procedures, 133, *305*
grounding adapters, *205*, 205–206
grounding of equipment for personal safety policy and procedure forms, 107, *309*
grounding procedure, 103–105, *305*, *309*
grounding sticks, *188*, 188
grounding-type portable electric equipment, *205*, 205–206, 211
ground mats, *174*, 174–175, *175*
group lockout/tagout situations, 125, *126*
guarding uninsulated overhead lines, 169

H

handlamps, 207–208, *208*
handlines, 165
handling portable electric equipment, 210
hand protection, *250*, 250, 258–259, *259*
hands-on training, 50
hand tools, insulated, *163*, *164*, 164
hard hats, 226–229, *227*, *228*, *229*, *241*
hasps, *128*
hazard assessment, 88–89, 225. *See also* risk assessments
hazard/risk analysis. *See* risk assessments
hazard/risk categories (HRCs). *See* PPE (personal protective equipment): categories of
hazards, determining, 94–95
hazards, electrical. *See* electrical hazards
hazards greater than energized electrical work, 146
head protection, 226–229, *227*, *228*, *229*, 249
hearing protection, 249
hinged panels, 161
hipot testers, *186*
hoods, arc-rated, 258
host employers, 26–27, *27*
housekeeping, 162
HRCs (hazard/risk categories). *See* PPE (personal protective equipment): categories of

I

ID numbers for gloves, *235*, 235
IEEE
  1584, *Guide for Performing Arc Flash Hazard Calculations*, 66, *67*
  3007.1, *Recommended Practice for the Operation and Management of Industrial and Commercial Power Systems*, 36

3007.2-2010, *IEEE Recommended Practice for the Maintenance of Industrial and Commercial Power Systems*, 72
3007.3, *Recommended Practice for Electrical Safety in Industrial and Commercial Power Systems*, 36
  guidelines for live-line tools, 180
illumination, 158, *159*
incident energy
  analysis, 72–73, 84, 154, 155, 247
  calculating, *68*
  defined, 16, *84*, 84
  effects of distance, 157
  effects of time, 90–92, *91*, *92*, 153
incident energy analyses, 72–73, 84, 154, 155, 247
induced voltages, 117–118, *118*, 118–119, *119*, *120*
infeasibility of deenergizing equipment, 146, 147
infrared (IR) scans, *96*, 96–97
infrared thermography, 253, 254–255
injuries, electrical. *See* electrical injuries
inspecting protective equipment
  arc flash protective clothing, 242–243, *243*, *244*
  eye protection, *241*, 241
  hard hats, *227*, 227–228, *228*
  leather protectors, 237, *238*
  rubber insulating blankets and sleeves, 238–240, *239*
  rubber insulating gloves, 230–232, *231*, *232*
inspection windows, *278*, 278
instantaneous trip functions, 92–93, *93*
instantaneous tripping, *217*, *218*
instructor-led classroom training, 48, *49*
insulated tools and equipment, *163*, 163–164, *164*
insulating equipment, 165
insulation ratings, overhead lines, 169
interference of PPE and clothing, 258
IR scans, *96*, 96–97

J

JHA (job hazard analysis). *See* risk assessments
job briefings, *178*, 178–179
JSA (job safety analysis). *See* risk assessments

L

labeling equipment, *156*, 156–157
labeling systems, *70*, 70, 72
labels, *32*, 32, 212, 266, *267*
layering protective clothing, 257
leather protectors, 237, *238*, 255, *258*, 258
light ballasts, 296–297
Limited Approach Boundaries
  defined, *63*, 63, *311*
  determining safe work zones, 77
  qualified persons, *64*, 64
  shock protection and, 151
  uninsulated overhead lines and. *See* uninsulated overhead lines
  unqualified persons and. *See* unqualified persons

Limited Space, 63, *64*, *311*
live-dead-live testing, 131
live-line tools, *179*, 179–181
local indication, *70*, 70, 72
locking-type connectors, *207*, 207
lockout/tagout
  absence-of-voltage testing, 131–132, *132*
  applying, 115–116, 127, 134
  coordination of, 126, 133
  deenergizing equipment, 130
  defined, 113
  devices and equipment, *126*, *127*, 127–129, *128*, 134
  disconnecting means, 131
  elements of control. *See* elements of control for lockout/tagout
  grounding of equipment and systems, 133
  per OSHA, 113, 115–116, 120–121, *128*
  personnel, 133
  procedures for, 125, *126*, 129–130, *305*
  removing, 134
  responsibility for, 131
  return to service, 134–135
  shift changes, 133
  stored energy, *130*, 130
  switching and grounding procedure, 103–105, *104*
  temporary release for testing/positioning, 135
  training, retraining, and documentation for, 126–127
  verifying, 131
look-alike equipment, *168*, 168
low-voltage AC drives, 287–288
low-voltage drawout-type circuit breakers, 267–274, *268*
  inserting, 273
  recommended PPE, 273–274
  removing, *269*, 269–272, *270*, *272*
low-voltage motors, 297–298, *298*
low-voltage proximity testers, *119*, *120*
lubrication for OCPDs, *89*, 89–90

M

MADs. *See* minimum approach distances (MADs)
maintenance of OCPDs
  condition of, 43, 86–87, 153
  corrosion, 90
  importance of, 86–87, 89
  incident energy and time, 90–91, *91*, *92*
  lubrication, *89*, 89–90
  STD function, 92–95, *93*, *94*
  testing and, 71–72
maintenance switches, 94
manufactured dates, 228, *229*
material breakopen, 13
material characteristics, 259
mating cords to receptacles, 214
maximum use voltages, 235, *236*
MCC buckets, *281*, 281–283, *282*

MCCs (motor control centers), 254
mechanical barricades, 166
mechanical equipment, 170–175, *171*
   equipment contact, 172
   equipment grounding, 172
   equipotential zone created by ground mats, *174*, 174–175, *175*
   step potential, 172–173, *173*
   touch potential, 172, *173*
mechanical inspections, 212
medium-voltage AC drives, *288*, 288–289
medium-voltage air-break switches, 277–280, *278*, *279*, *280*
medium-voltage drawout-type circuit breakers, 274–277, *275*, *276*, *277*
medium-voltage motors, 298–299
metal-framed eyeglasses, 160, *161*
milliamps, 8
minimum approach distances (MADs). *See also* clearance distances
   changes to, 46, *47*, 47
   per OSHA, 57–59, *58*, *59*, *60*
   unqualified persons and, 169, *170*
mobile cranes, *171*, 171–172
motor control center (MCC) buckets, *281*, 281–283, *282*, *283*
motor control centers (MCCs), 254
motors, 297–299, *298*
motor starters, 293–295, *294*
multicell battery systems, 253, 254–255
multi-employer worksites, 21, *22*
multiple roles, 26

N

nameplates, *86*, 86
National Electrical Code (NEC), 1
National Fire Protection Association (NFPA), 1
nationally recognized testing laboratories (NRTLs), *212*, 212
near-miss accidents, 4
NEC (National Electrical Code), 1
NEC-compliant installations, 41, *42*
neck protection, 249
NEMA E2 (fused contactor) motor starters 2.3 kV to 7.2 kV, 293–295, *294*
NEMA enclosures, *314*
network systems and protectors, *15*, 15
NFPA (National Fire Protection Association), 1
NFPA, *Electrical Safety Program Book*, 36
NFPA 70B, *Recommended Practice for Electrical Equipment Maintenance*, *70*, 70, 72
NFPA 70E
   Arc Flash Boundaries, 67–73, *70*, *71*
   arc flash risk assessments, 153–157, *156*
   Article 130, 43–44, 145,
   consensus process, 3–4
   defined, 1
   electrically safe work conditions, 125, 145–149, *146*.

   examples vs exceptions, 148
   history of, 2
   Informative Annex F, 96–98
   lockout/tagout. *See* lockout/tagout
   portable electric equipment. *See* portable electric equipment per NFPA
   PPE (personal protective equipment)
      arc flash protective equipment, 258–259
      arc flash risk assessments and, 247–248
      arc-rated clothing and flash suits, *260*, 260–261
      body protection, 250
      clothing and other apparel not permitted, 259–260
      eye protection, 249
      foot protection, 251
      hand and arm protection, *250*, 250
      head, neck, and chin (head area) protection, 249
      hearing protection, 249
      movement and visibility, *249*, 249
      other protective equipment, 248
      selection of, 256–258
      standards for, 246–247, 251
      table method, 69, 251–256
   risk assessments, 83–84, 96–98
   table method, 69, 251–256
   training requirements, 52–53
NITMAMs (Notices of Intent to Make a Motion), 3–4
nominal voltages, 45–46, *46*, 85, 85–86
noncontact electric arc-induced injuries, 6–7, *7*, 61–62, *62*
noncontact inspections, 253–254, *254*
noncontact voltage testers (proximity testers), 119, *120*, *122*, 122–123
normally operating equipment, 43–44, 148–149, 249
normal operation of a circuit breaker (CB), switch, contactor, or starter, 252, 253–254
Notices of Intent to Make a Motion (NITMAMs), 3–4
Notices to Appeal, 4
NRTLs (nationally recognized testing laboratories), *212*, 212

O

obstructed views of work areas, 158, *159*
occupational categories, 44
Occupational Health and Safety Management Systems, 35
Occupational Safety and Health Administration. *See* OSHA (Occupational Safety and Health Administration)
OCPDs. *See* overcurrent protective devices (OCPDs)
Ohm's law, 8, *313*
on-the-job training (OJT), 40, 48, *49*
opening and closing circuits, 162
OSHA (Occupational Safety and Health Administration)
   Arc Flash Boundary, 67, *68*
   computer-based training, 49–50
   defined, 1–2
   electrically safe work conditions. *See* electrically safe work conditions per OSHA

electrical regulations, 5
employer responsibility for PPE, 88–89
examples vs exceptions, 148
General Duty Clause, 88
lockout/tagout, 113, 115–116, 120–121, *128*
minimum approach distances (MADs), 57–59, *58*, *59*, *60*
multi-employer worksites and, 26
portable electric equipment. *See* portable electric equipment per OSHA
PPE and, 224–226
qualified electrical workers, 41–43, 45–47, 48
qualified workers, 39
Table R-6, 46–47, 58–59, *59*, *60*
Table R-7, *59*
Table S-4, 44–45
Table S-5, 59, *60*
training of qualified workers, 39–40
outer layers of arc-rated clothing, 257
overcurrent protection modification, 216–218, *218*, *219*
overcurrent protective devices (OCPDs)
    corrosion, 90
    effects of time on incident energy, 90–92, *91*, *92*
    lubrication, *89*, 89–90
    maintenance of. *See* maintenance of OCPDs
    modification, 216–218, *218*, *219*
    operating time, *17*, 17, 43
    STD (short-time delay) function, 92–94, *93*, *94*
overhead lines, uninsulated. *See* uninsulated overhead lines
ozone cutting, 233, *234*

P

panelboards, 253
panel removal from electrical enclosures, 295–296
permits for energized electrical work, *64*, 73–75, *74*, *150*, 150–152
personal protective equipment (PPE). *See* PPE (personal protective equipment)
personnel safety, 158–163
    alertness, 158
    anticipating failures, 162
    blind reaching, 158, *159*
    conductive materials, 160, *161*
    confined or enclosed work spaces, 160
    doors, 161
    flammable materials, 162
    hinged panels, 161
    housekeeping, 162
    illumination, 158, *159*
    metal-framed eyeglasses, 160, *161*
    obstructed views of work areas, 158, *159*
    reclosing circuits, 163
    routine opening and closing of circuits, 162
    working spaces, 161
philosophy sections in ESPs, 29
physical barriers, 166

PIs (Public Inputs), 3
plastic guards, *282*, 282
policies sections in ESPs, 35
portable electric equipment per NFPA
    attachment plugs, 214
    conductive work locations, 214
    conductor sizing for flexible cord sets, 213, *214*
    defective equipment, 214
    GFCI protection devices, *215*, 215–216, *216*
    grounding-type equipment, 211
    handling and storing equipment, 210
    inspection of, 211–213, *212*, *213*
    overcurrent protection modification, 216–218, *218*, *219*
    proper mating, 214
portable electric equipment per OSHA, 203, *204*
    attachment plugs, *207*, 207, 208–209, *209*
    conductive work locations, 206
    grounding-type equipment, *205*, 205–206
    handlamps, 207–208, *208*
    portable and vehicle-mounted generators, 209–210, *210*
    receptacles, cord connectors, and attachment plugs, 208–209, *209*
    visual inspection of portable electric tools and flexible cords, 204–205
portable electric tools, 204–205
portable electric tool testers, *213*, 213
portable generators, 209–210, *210*
portable GFCIs, 215
portable handlamps, 207–208, *208*
portable ladders, 165, *166*
power circuit breakers. *See* drawout-type circuit breakers
power formula, 213
power strips, 200
PPE (personal protective equipment)
    absence-of-voltage testing and, 131, *132*, 289–291
    AC drives and, 287–288
    arc flash protective clothing. *See separate entry*
    arc flash risk assessments and, *155*, 155–156, 247–248
    arc flash suits, 258
    arc-rated clothing. *See separate entry*
    arm protection, *250*, 250, 258
    body protection, 250
    care of, *260*, 260–261
    categories of, 69, 251
    chin protection, 249
    circuit troubleshooting and, 285–286
    clothing and other apparel not permitted, 259–260
    cover and panel removal from electrical enclosures and, 296
    defined, 2, 223–224, *224*
    employer responsibility per OSHA, 88–89
    equipment rated between 240 V and 600 V and, 293
    equipment rated 240 V and less and, 292
    ESP (electrical safety program) controls, 33
    eye protection, *240*, 240–241, *241*, 249
    face protection, 258
    foot protection, 251, 259

PPE (personal protective equipment) (*continued*)
  hand protection, *250*, 250, 258–259, *259*
  head protection, 226–229, *227*, *228*, *229*, 249
  hearing protection, 249
  incident energy and, 247, 248
  inspection of, 242–243, *243*, *244*
  IR scans and, 96–97
  leather protectors, 237, *238*
  light ballast replacement and, 296
  low-voltage drawout-type circuit breakers and, 273–274
  maintenance of, 226
  material, 259
  MCC buckets and, 283
  medium-voltage air-break switches and, 279–280
  medium-voltage drawout-type circuit breakers and, 277
  motor replacement and, *298*, 298, 299
  movement and visibility, *249*, 249
  neck protection, 249
  NEMA E2 (fused contactor) motor starters 2.3 kV to 7.2 kV and, 295
  per NFPA 70E. *See* NFPA 70E: PPE (personal protective equipment)
  normally operating equipment and, 149
  not needed, 44
  not permitted, 259–260
  per OSHA, 224–226
  other protective equipment, 248
  panel removal from electrical enclosures and, 296
  repairing, 245–246
  rubber insulating blankets and sleeves, 238–240, *239*
  rubber insulating gloves. *See separate entry*
  selection of, *256*, 256–258
  standards for, 251
  storage of, *245*, 245
  table method of selection. *See* table method
  troubleshooting AC drives and, 287–288, *289*
  troubleshooting circuits, 285–286
  wearing, 244–245
principles sections in ESPs, 34
probability of avoidance (Av) risk factor, 97
probability (Pr) risk factor, 97
procedures for lockout/tagout, 125, *126*, 129–130
procedures in ESPs, 29–31, *30*, *32*
proficiency defined, 39
proof test voltages, 235, *236*
properly installed, 69
properly maintained, 69–70, *70*
proposals. *See* Public Inputs (PIs)
protective clothing. *See* PPE (personal protective equipment)
protective device coordination, 92–95, *93*, *94*
protective equipment. *See also* PPE (personal protective equipment)
  alerting techniques, 166–168, *167*, *168*
  barriers, 166
  fiberglass-reinforced plastic (FRP) rods, 165

  fuse and fuseholder equipment, *165*, 165
  incident energy and, 248
  insulated tools and equipment, *163*, 163–164, *164*
  portable ladders, 165, *166*
  protective shields, 165
  ropes and handlines, 165
  rubber insulating equipment, 165
  voltage-rated plastic guard equipment, 166
protective grounding equipment, temporary, 135–138, *136*, *137*
protective shields, 165
proximity testers (noncontact voltage testers), 119, *120*, *122*, 122–123
Public Inputs (PIs), 3

Q

qualified electrical workers, 41–43, 45–47, 48
qualified persons
  electrical workers, 41–43, 45–47, 48
  Limited Approach Boundary and, *64*, 64
  per OSHA, 39–40, 44–45
  Restricted Approach Boundary and, *65*, 65

R

racking
  defined, 267
  drawout-type circuit breakers, 267–277, *268*
    low-voltage. *See* low-voltage drawout-type circuit breakers
    medium-voltage, 274–277, *275*, *276*, *277*
  MCCs, *281*, 281–283, *282*, *283*
racking screws, *275*, 275–276
reasonable care standard, 24
receptacles, 208–209, *209*
reclosing circuits, 163
recognized hazards, 84
recordable accidents, 4
reenergizing equipment, 120–121, 135
releasing stored electric energy, 114
remote earth, 175
removal or installation of CBs or switches, 252, 254
removing lockout/tagout devices, 134
repairing equipment, 151
  arc flash protective clothing, 245–246
  flexible cords, 202
  rubber insulating blankets, 239–240
Report on Proposals. *See* First Revisions (FRs)
rescue hooks, 177
resolved Public Inputs (PIs), 3
responsibility sections in ESPs, 29
Restricted Approach Boundaries, *63*, *64*, *65*, 65, 151, *311*
Restricted Space, *65*, 65, *311*
retraining for lockout/tagout procedures, 126–127
returning equipment to service after lockout/tagout, 134–135
risk assessment matrix, 105–106, *106*, *306*

risk assessments
  per ANSI/AIHA Z10—Appendix E, 107
  appropriate PPE, 87–88
  arc flash hazards, 86–87, 153–157, *155*, *156*
  assessing hazards, 87–88
  common electric tasks, 265–267, *267*
  defined, 83
  determining hazards, 94–95
  determining risk, 95–96, *96*
  forms. *See* forms for performing risk assessments
  electrical hazards, *84*, 84–86, *85*
  electrical safety programs (ESPs), 97
  energized electrical work permits (EEWPs), 73, *304*
  General Duty Clause, 88
  per NFPA 70E, 67–68, 83–84, 96–98, 153–157, *155*, *156*
  shock risk assessments, 151
  short-circuit current from transformers calculation, *102*, 102–103
  written certification, 88–89
risk-reduction strategies, 98
ropes, 165
rubber insulating blankets, 238–240, *239*
rubber insulating equipment, 165
rubber insulating gloves
  chemical damage to, *232*, 232–233, *233*
  corona, 233
  defined, 229–230, *230*
  inspecting, 230–232, *231*, *232*
  leather protectors and, 255, *258*, 258
  ratings for, *250*, 250
  storing, *234*, *236*, 236
  sunlight and, *234*, 234
  testing, *235*, 235, *236*
rubber insulating sleeves, 238–240

S

safety
  culture of, 4
  electrically safe work conditions per NFPA 70E, 125, 145–149, *146*
  electrically safe work conditions per OSHA. *See separate entry*
  personnel and. *See* personnel safety
  temporary protective grounding equipment and, 136–138, *137*
safety barricade tape, 77–78, *78*, 168
safety glasses, 240
safety per OSHA 1910.269, 176–191
  capacitors, 189–190, *190*
  current transformer secondary windings, *190*, 190–192, *191*
  existing conditions, 177–178
  job briefings, *178*, 178–179
  live-line tools, *179*, 179–181
  temporary protective grounds, *181*, 181–184, *183*, *184*
  testing and test facilities, 184–189, *185*, *186*, *188*
  training, 176–177

safety signs, 166, *167*
safety tags, 166, *167*, 168
safety teams, 29
safety training defined, 40, *41*
safe work zones, 76–78, *77*, *78*
scope sections in ESPs, 28
secondary windings, current transformers, *190*, 190–191, *191*
Second Draft meetings, 3
Second Revisions (SRs), 3
selective coordination, 217
selective tripping, 92–93
severity (Se) risk factor, 97
shift changes during lockout/tagout, 133
shock hazards, 84
shock protection, 151–152, *153*
shock risk assessments, 73, 83, 151
shocks, 8, 8–9, *9*, *10*, 11–12, *12*
short-circuit current from transformers calculation, *102*, 102–103
short-circuit currents, 247
short-time delay (STD) function, 92–95, *93*, *94*, 218, *219*
shutdowns, 130
shutters, *276*, 276
signs, safety, 166, *167*
simple lockout/tagout procedure, 125
single-line diagrams, 103, *104*, 129–130
sleeves, rubber insulating, 238–240
solenoid voltage testers, *123*, 123
specifications labels, 243, *244*
squeeze tests, 227
SRs (Second Revisions), 3
stab finger clusters (connector) assemblies, *276*, 276
Standard for Electrical Safety in the Workplace. *See* NFPA 70E
standards for ESPs, 35–36
standards of performance, 34
standards organizations, *310–311*
static grounds, 187
STD (short-time delay) function, 92–95, *93*, *94*, 218, 219
step potential, 172–173, *173*
stored energy, 114, *130*, 130
storing equipment
  arc flash protective clothing, *245*, 245
  portable electric equipment, 210
  rubber insulating gloves, *234*, *236*, 236
strain-relief devices, 203, *204*
substitution of less hazardous systems, 32
switches, 252, 254

T

table method
  arc flash warning labels and, 155–156
  importance of, 2
  incident energy and, 247
  PPE categories and, 69, 293–294
  PPE required/not required, 265–266

table method (*continued*)
 Table 130.7(C)(15)(A)(a), 251–253
 Table 130.7(C)(15)(A)(b), 253–255
 Table 130.7(C)(16), 255–256
 task categories, 266–267
tagouts
 attaching locks/tags, *115*, 115–116
 devices, *128*, 128–129
 per OSHA, *113*, 113
tags, safety, 166, *167*, 168
tank-loss-index (TLI), 187
task groups, 3
technical skills, 39–40
temporary protective grounding equipment, 135–138, *136*, *137*
temporary protective grounds, *181*, 181–184, *183*, *184*
temporary release from lockout/tagouts, 135
temporary wiring, 197–198. *See also* flexible cords
terminal guards, *284*, 284
testing
 instruments and equipment for, 149
 nominal voltage, 183
 per OSHA, 184–189, *185*, *186*, *188*
 rubber insulating blankets and sleeves, 239
 rubber insulating gloves, *235*, 235, *236*
 temporary protective grounds, 182–183, *184*
test instruments, 149
thermographic infrared (IR) scans, *96*, 96–97
TLI (tank-loss-index), 187
tools and equipment, *163*, 163–164, *164*, 204–205
tool testers, 180, *213*, 213
touch potential, 172, *173*
tracking numbers for gloves, *235*, 235
training
 different types of equipment, 40
 energized electrical work, 112
 ESPs, 34
 initial training, 185
 per NFPA 70E, 52–53
 per OSHA, 39–40, 44–45, 176–177
 types, 48–51, *49*
 unqualified persons, 51
training for lockout/tagout procedures, 126–127
trapped-key interlock systems, *288*, 288
troubleshooting
 AC drives, 286–289, *287*, *288*
 circuits, 283–286, *284*
 versus repair work, 151
tulips. *See* stab finger clusters (connector) assemblies
turns ratio, *191*, 191

U

UL (Underwriters Laboratories) labels, *212*, 212
underlayers for arc-rated clothing, 257

uninsulated overhead lines
 approach distances for unqualified persons, 169, *170*
 deenergizing and guarding, 169
 employer and employee responsibility, 169
 insulation ratings, 169
 vehicular and mechanical equipment, 170–175, *171*, *173*, *174*, *175*
unqualified persons
 approach boundaries and, 152, *153*
 approach distances and, 169, *170*
 Limited Approach Boundary and, 60, *64*, 64, 152
 training for, 51

V

vehicle-mounted generators, 209–210
vehicular equipment, 170–175, *171*
 equipment contact, 172
 equipment grounding, 172
 equipotential zone, *174*, 174–175, *175*
 step potential, 172–173, *173*
 touch potential, 172, *173*
Velcro inspection, 242, *243*
verifying equipment is deenergized, 116–117, *117*
verifying lockout/tagout, 131
viewing/inspection windows, *278*, 278
visual inspections, 204–205, 211–213, *212*
voltage creep, 237
voltage detectors, *32*
voltage drops, *174*, 174, *313*
voltage-rated plastic guard equipment, 166
voltage testers, 118–119, *119*, *122*, 122–123

W

warning labels, 32, *33*, 266, *267*
wet locations, 206
windings, current transformer, *190*, 190–192, *191*
wiring, temporary, 197–201
working distances, 16
working spaces, 161
work locations, conductive, 214
work permits, electrical, *150*, 150–152
work zone safety, 76–78, *77*, *78*
written hazard assessments, 88